AF206635

Technische Universität Dresden

A Box-Integration/WENO solver for the Boltzmann Transport Equation and its Application to High-Speed Heterojunction Bipolar Transistors

Gerald Wedel

der Fakultät Elektrotechnik und Informationstechnik der
Technischen Universität Dresden

zur Erlangung des akademischen Grades eines

Doktoringenieurs

(Dr.-Ing.)

genehmigte Dissertation

Vorsitzender: Prof. Dr.-Ing. habil. Mayr
Gutachter: Prof. Dr.-Ing. habil. Schröter   Tag der Einreichung: 02.05.2016
Gutachter: Ao. Univ.-Prof. Dr. Schürrer   Tag der Verteidigung: 07.11.2016

Bibliographic information published by the Deutsche Nationalbibliothek:
The Deutsche Nationalbibliothek lists this publication
in the Deutsche Nationalbibliografie; detailed bibliographic
data are available on the Internet at http://dnb.dnb.de .

Bibliografische Information der Deutschen Nationalbibliothek:
Die Deutsche Nationalbibliothek verzeichnet diese Publikation
in der Deutschen Nationalbibliografie; detaillierte bibliografische
Daten sind im Internet über http://dnb.dnb.de abrufbar.

© 2017 Gerald Wedel
Herstellung und Verlag:
BoD – Books on Demand, Norderstedt

ISBN: 978-3-7448-7372-7

In Memoriam
Joachim Wedel

Abstract

The ongoing trend for high-frequency (HF) applications drives the development of high-speed devices. Therefore, trustworthy device simulations are inevitable for understanding and designing future HF devices. During the last decade, the predictive capabilities of Drift-Diffusion (DD) and Hydrodynamic (HD) transport models turned out to be insufficient for state-of-the-art high-frequency transistors. Consequently, a more physics based transport model helps to counter these issues and thus, the Boltzmann transport equation (BTE) comes into focus. In this thesis, a deterministic solution method for the BTE is pursued.

First, physical fundamentals and mathematical preconsiderations for the treatment of the BTE are reviewed. This covers the calculation of band structures/dispersion relations, an overview of scattering mechanisms and a detailed description of the coordinate transformations required for analyzing prominent semiconducting materials, such as Silicon-Germanium and III-V compounds, like Indium-Phosphide.

The second part focuses on the numerical treatment of the BTE. Besides the employed normalization strategy, the discretization of the BULK BTE is described in detail. Based on the latter, the extensions for the device BTE are specified. A method for the direct calculation of stationary BTE solutions – for the BULK and device case – is introduced and an overview of the WENO method is outlined.

The third part is dedicated to the applications of the deterministic solution method and simulation results of the BTE. Recipes for calculating the most important quantities, like current/electron densities, are given. Simulation results for the BULK case and for hetero-junction bipolar transistors are presented and analyzed. Here, the focus is put on both Silicon/Silicon-Germanium and Indium-Phosphide/Indium-Gallium-Arsenide material systems. The part is concluded by a critical review on the current field of application.

A summary and an outlook on future extensions concludes the thesis. Besides pointing out the achievements of this work, the last section also gives a short motivation for adapting the method to 1D semiconductors, like carbon nanotubes.

Zusammenfassung

Die anhaltende Trend zu Hochfrequenzanwendungen treibt die Entwicklung von Hochgeschwindigkeitsbauelementen voran. Daher sind vertrauenswürdige Bauelementsimulationen für das Verstehen und Entwerfen zukünftiger Hochfrequenzbauelemente notwendig. Die Vorhersagefähigkeit von Drift-Diffusions (DD) sowie von hydrodynamischen (HD) Transportmodellen hat sich in der letzten Dekade als unzureichend für moderne Hochfrequenztransistoren herausgestellt. Um diesem Problem zu begegnen, gerät die Boltzmann Transportgleichung (BTG) als ein Transportmodell mit vergleichsweise stärkerer physikalischer Basis in den Fokus. In dieser Arbeit wird eine deterministische Lösungsmethode der BTG betrachtet.

Zunächst werden die physikalischen Grundlagen und mathematischen Vorbetrachtungen für die Behandlung der BTG beschrieben. Dies beinhaltet die Berechnung von Bandstrukturen/Dispersionsrelationen, einen Überblick über Streumechanismen und eine detaillierte Beschreibung der nötigen Koordinatentransformationen für die Untersuchung bedeutender Halbleitermaterialien, wie zum Beispiel Silizium-Germanium Heterostrukturen und III-V Materialien, wie Indiumphosphid.

Der zweite Teil ist der numerische Behandlung der BTG gewidmet. Neben der verwendeten Normierungsstrategie wird ebenfalls die Diskretisierung der BULK BTG detailliert beschrieben. Basierend darauf werden die Erweiterungen der BTG für die Bauelementsimulation näher beschrieben. Eine Methode zur direkten Berechnung der stationären Lösungen der BTG – für BULK- und Bauelementsimulationen – wird vorgestellt und ein Überblick über die WENO Methode gegeben.

Der dritte Teil widmet sich der Anwendung der deterministischen Lösungsmethode und den Simulationsergebnissen der BTG. Formeln zur Berechnung der wichtigsten Größen wie Strom- und Elektronendichten werden angegeben. Simulationsergebnisse für den BULK Fall und für Heterostruktur-Biploartransistoren werden gezeigt und analysiert. Dabei liegt das Hauptaugenmerk auf Silizium/Silizium-Germanium und Indiumphosphid/Indiumgalliumarsenid Materialsystemen. Eine kritische Bewertung des Anwendungsbereiches schließt den dritten Teil der Arbeit ab.

Eine Zusammenfassung und ein Ausblick auf zukünftige Erweiterungen schließt die Arbeit ab. Im letzten Abschnitt wird, neben dem Hervorheben der Ergebnisse dieser Arbeit, kurz die Adaption der Methode auf 1D Halbleiter, wie Kohlenstoffnanoröhren, motiviert.

Acknowledgment

First, I would like to thank my supervisor Prof. Micheal Schröter for giving me the opportunity to work on this interesting and challenging topic as well as the freedom to develop own ideas and solutions. I would also like to thank my reviewer Prof. Ferdinand Schürrer for the scientific assessment of my thesis and his subsequent support.

Special thanks go to Tommy Rosenbaum. He became not only my colleague, but also one of my friends. I really appreciate that he and Yves Zimmermann spent time for proofreading my thesis. I would also like to thank Dr. Martin Claus, who always had an open door for thought experiments and discussions. I would also like to acknowledge my colleagues Tobias Nardmann and Andreas Pawlak for the more application oriented discussions. Special thanks go to Ria Lykowski, who brilliantly managed the whole bureaucracy and whose discussions were always a welcome distraction. My thanks also go to all remaining members of the chair for providing a stimulating working atmosphere.

I am also indebted to Prof. Jungemann (RWTH Aachen) for providing access to the SHE solver. His work was an inspiration and motivation for this thesis.

Last but not least, I would like to thank my family, Vera and Tim. Their support and patience were invaluable for the completion of this thesis, especially during the final phase.

This thesis is dedicated to my father, Dipl.-Ing. Joachim Wedel, who unfortunately could not be with me to experience my PhD graduation.

Contents

CONTENTS

List of Symbols

Abbreviations

GaAs   Gallium-Arsenide
Ge   Germanium
InGaAs Indium-Gallium-Arsenide
InP   Indium-Phosphide
Si   Silicon
SiGe   Silicon-Germanium
ALY   Alloy scattering
BC   Base collector
BE   Base emitter
BJT   Bipolar junction transistor
BTE   Boltzmann transport equation
BZ   Brillouin-zone
CNT   Carbon nanotube
DD   Drift-Diffusion (transport model)
DOS   Density of states
EPM   Empirical pseudo-potential method
HBT   Heterojunction bipolar transistor
HD   Hydrodynamic (transport model)
HV   Herring-Vogt
IMP   Impurity scattering
MC   Monte-Carlo (method)
POP   Polar optical phonon scattering

PZ	Piezoelectric scattering
SHE	Spherical harmonics expansion
WENO	Weighted essentially non-oscillatory (method)

Constants

$\hbar$	Reduced Planck constant $(h/2\pi)$ $1.05 \times 10^{-34}\,\mathrm{J\,s}$, $6.58 \times 10^{-16}\,\mathrm{eV\,s}$
i	Imaginary unit $\sqrt{-1}$
ε_0	Vacuum permittivity $8.85 \times 10^{-12}\,\mathrm{F\,m^{-1}}$
h	Planck constant $6.62 \times 10^{-34}\,\mathrm{J\,s}$, $4.136 \times 10^{-15}\,\mathrm{eV\,s}$
k_b	Boltzmann constant $1.38 \times 10^{-23}\,\mathrm{J\,K^{-1}}$
m_0	Free electron mass $9.109 \times 10^{-31}\,\mathrm{kg}$
q	Elementary charge $1.602 \times 10^{-19}\,\mathrm{A\,s}$

Variables

α_ν	Non-parabolicity factor of the valley ν (dispersion relation)
dos^ν	Density of state of the valley ν
$\vec{E}^\nu_{\mathrm{eff.}}$	Effective field/driving force associated with the valley ν
ϵ^ν	Kinetic energy associated with the valley ν
μ	Cosine of the polar angle (spherical coordinates)
ν	Valley label
φ	Azimuthal angle (spherical coordinates)
$\tilde{\vec{k}}^\nu$	$\vec{k}^\nu$ after the HV transformation has been applied
$\vec{k}$	Reciprocal vector measured from the center of the 1st BZ
$\vec{k}^\nu$	Reciprocal vector measured from minimum of the valley ν
$\vec{r}$	Real space vector
a_ν	Normalized α_ν
C	Collision term associated with the BTE
f^ν	Electron distribution function of the valley ν
m^*_ν	Effective (isotropic) electron mass
n^ν	Solution variable of the discretized BTE within the valley ν
S^X	Transition rate of the scattering mechanism labeled by X
T^{HV}	Herring-Vogt transformation matrix
w^ν	Normalized ϵ^ν

CHAPTER 1

Introduction

The ongoing trend for high-speed communication, both cabled and wireless, is driving the development of integrated high-speed devices. Although layout and circuit optimization are efficient measures to improve the circuit performance, ultimately, the performance is linked to single device behavior and size. In order to relax the vertical scaling as much as possible by still meeting both the desired device performance and the production constraints, accurate and predictive device simulations are inevitable.

The Drift-Diffusion (DD) and Hydrodynamic (HD) transport models have served as work horses in industry for decades. These transport models offer simulation times suitable for the every-day engineering work. However, during the last ten years, DD and HD simulations became increasingly unreliable for heterojunction bipolar transistors (HBTs) regarding the performance prediction. Both the DD and the HD transport models are derived by moment expansions and simplifications of the Boltzmann transport equation (BTE). Compared to solutions of the BTE, the physical accuracy and the accompanied complexity is traded in for small simulation times and adjustable model equations. Thus, the basis of fast DD and HD simulations is also the reason for their failure in predicting the performance of advanced devices. In [73], a set of physical models for the HD transport model adjusted to BTE results and an adjustment strategy for advanced SiGe-devices is given. However, the adjustment cycle of the HD transport model relies on BTE simulations performed first. This leaves the following two options:

i.) Stick with existing simulation approaches. [47] argues that even complex transport mechanisms can still be accounted for by the DD transport model and that one should turn back to a profound physical understanding in combination with existing device simulation tools. This makes the electronic design automation (EDA) vendors breathe a sigh of relief. Nevertheless, also in this case more complex transport models are needed to adjust DD simulations for reasonable results.

ii.) Proceed with a more complex transport model, e.g. the BTE, in order to handle shrinking device dimensions. This is the most extensive, time consuming and also expensive option. On the other hand, the effort spent pays out, since the insight gained into the underlying physics enables a more profound understanding of the device behavior.

This work addresses the second option by presenting a solver for the BTE. For solving the BTE, two kinds of solution methods exist: stochastic solvers based on the Monte-Carlo (MC) method and deterministic solvers. The MC method is more commonly used due to the availability of detailed literature (e.g. [34, 37, 69]), low implementation effort and the manageable amount of memory required by the method. Its major drawbacks are long simulation times, since MC is an inherently transient method, and noisy results due to its stochastic nature. Due to these drawbacks, a shift towards the development of deterministic solvers for the BTE occurred more than 25 years ago. The pioneering work of Goldsmann [23], using Legendre polynomials, and of Fatemi and Odeh [17], using a finite-difference scheme, opened the era of the deterministic BTE solvers. At that time, memory was limited and those methods could only be applied using coarse simulation grids, which in turn led to rough solutions of the BTE. Thus, deterministic solvers of the BTE played a minor role. About 10 years ago, feasible deterministic and stationary simulation results of the BTE (based on [23]) were presented in [38] by using the spherical harmonic expansion (SHE), employing Legendre polynomials for expanding the solution variable of the BTE on equi-energy spheres in the reciprocal space. Limiting the number of harmonics, a trade-off between memory requirement/simulation time and accuracy was obtained.

Besides the SHE-approach, the development of direct solvers based on the work of [17] continued. In this context, the work of Banoo [2] and of Majorana et. al. [48] should be noted. Although not suitable for stationary solutions of the BTE, the approach of Majorana has been pursued further on by [19, 20].

In order to handle nearly abrupt spatial variations of the distribution function (solution variable of the BTE), the approach of [48] has been successfully extended in the work of Galler [19, 20] using the *weighted essentially non- oscillatory* (WENO) method [36]. Compared to SHE, the independence of a limited number of harmonics and the resemblance to usually employed DD/HD discretization schemes were the decisive factors for pursuing the latter approach in the present work.

The thesis is organized as follows: In chapter 2 the preliminary considerations for solving the BTE are given from both the physical and the mathematical point of view.

Chapter 3 focuses on the numerical treatment of the BTE. Besides the employed normalization strategy, a clear distinction between the BULK- and Device-BTE is made, in order to emphasize the necessary extensions for Device-BTE. Special emphasis is put on the numerical treatment of the most important scattering mechanisms for SI/SIGE and III-V materials. In addition, extensions to the scheme of [19, 20] are introduced to account for some of its original limitations. Amongst others, extensions for the stationary solutions of the BTE, non-uniform discretization of the real-space, arbitrary step-size of the energy discretization as well as position dependent band structure/dispersion relation are illustrated.

In chapter 4, simulation results for SI/SIGE and INP/INGAAS material combinations are presented. Before focusing on simulation results of HBTs, BULK simulation results of the associated materials are compared with published data.

The thesis is concluded with a short summary and an outlook on future extensions of the simulation scheme described in this work. In addition, an elaborate appendix is given, which aims to provide supplementary derivations and an entry point into the scheme used for solving the BTE.

CHAPTER 2

Fundamentals and the Boltzmann transport equation

2.1 Reciprocal lattice/space

One basic property of a semiconducting crystal lattice is, like for other crystals, its periodicity. For the sake of simplicity, these lattices are usually assumed to be perfectly arranged in a periodic manner (*Barvais* lattice). With a defined set of basis vectors (**primitive translation vectors**) the structure of the crystal, its **lattice points**, can be described by [39]

$$
\begin{aligned}
\vec{r} &= \vec{r'} + n_1\vec{a_1} + n_2\vec{a_2} + n_3\vec{a_3} \\
&= \vec{r'} + \vec{R},
\end{aligned}
\tag{2.1.0-1}
$$

with the integers n_i, the primitive translation vectors $\vec{a_i}$, which are pointing from one **basis** to an adjacent one, and two vectors in the real space $\vec{r}$, $\vec{r'}$. Note, that the **basis** of a lattice is not necessarily an atom of the lattice. The **basis** is an arrangement of atoms, which can be found at any lattice point defined by $\vec{r}$. Diamond and Zinc-blende crystals have a **face centered cubic** structure (FCC), where the basis consists of two atoms. The second atom is displaced from the first one by one quarter of the space diagonal (see figure 2.1). If both atoms are identical, a Diamond and otherwise a Zinc-blende structure is obtained. Equation (2.1.0-1) states that the crystal at $\vec{r}$ is identical to the crystal at $\vec{r'}$, due to its periodicity, and is thus invariant under translation. Figure 2.1 shows the Diamond and Zinc-blende structures, covering both silicon and III-V semiconductors, respectively. In the case of

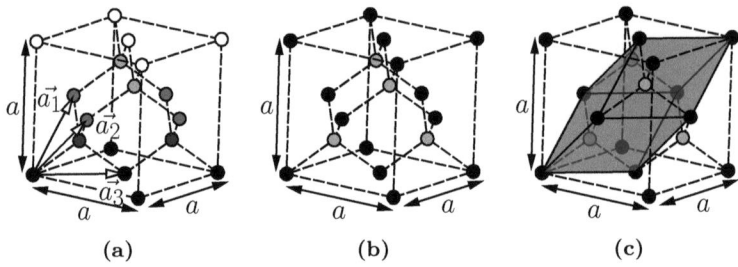

(a)   (b)   (c)

Figure 2.1: Schematic illustrations for Diamond and Zinc-blende structures: (a) position of atoms and their height (shading) within the crystallographic unit cell including the primitive translation vectors; (b) two different atoms (**basis**), e.g. Indium and Phosphide, indicated by black and gray spheres; (c) primitive unit cell for Diamond and Zinc-blende structures shown by black lines and spheres.

Diamond and Zinc-blende structures, the primitive translation vectors read

$$\vec{a_1} = \frac{a}{2} \begin{pmatrix} 0 \\ 1 \\ 1 \end{pmatrix} \ , \ \vec{a_2} = \frac{a}{2} \begin{pmatrix} 1 \\ 0 \\ 1 \end{pmatrix} \ , \ \vec{a_3} = \frac{a}{2} \begin{pmatrix} 1 \\ 1 \\ 0 \end{pmatrix} \ , \tag{2.1.0-2}$$

with the lattice constant a. Once the primitive translation vectors ($\vec{a_i}$, $i = 1, 2, 3$) are known for the crystal under investigation, the reciprocal lattice vectors can be constructed [39]:

$$
\begin{aligned}
\vec{a_1^*} &= 2\pi \frac{\vec{a_2} \times \vec{a_3}}{\vec{a_1} \cdot (\vec{a_2} \times \vec{a_3})} = 2\pi \frac{\vec{a_2} \times \vec{a_3}}{V_{\text{cell}}} \ , \\
\vec{a_2^*} &= 2\pi \frac{\vec{a_3} \times \vec{a_1}}{\vec{a_1} \cdot (\vec{a_2} \times \vec{a_3})} = 2\pi \frac{\vec{a_3} \times \vec{a_1}}{V_{\text{cell}}} \ , \\
\vec{a_3^*} &= 2\pi \frac{\vec{a_1} \times \vec{a_2}}{\vec{a_1} \cdot (\vec{a_2} \times \vec{a_3})} = 2\pi \frac{\vec{a_1} \times \vec{a_2}}{V_{\text{cell}}} \ ,
\end{aligned}
\tag{2.1.0-3}
$$

where $\vec{a}_i^*$ ($i = 1, 2, 3$) are the **basis vectors of the reciprocal lattice** and V_{cell} is the volume of the parallelepiped – the primitive unit cell – formed by the primitive translation vectors $\vec{a}_i$ (see figure 2.1(c)). Inserting (2.1.0-2) into (2.1.0-3) leads to

$$\vec{a}_1^* = \frac{2\pi}{a} \begin{pmatrix} -1 \\ 1 \\ 1 \end{pmatrix} , \ \vec{a}_2^* = \frac{2\pi}{a} \begin{pmatrix} 1 \\ -1 \\ 1 \end{pmatrix} , \ \vec{a}_3^* = \frac{2\pi}{a} \begin{pmatrix} 1 \\ 1 \\ -1 \end{pmatrix} . \quad (2.1.0\text{-}4)$$

Another way to visualize the crystal structure is to use the *Wigner-Seitz* cell for the real space and the first *Brillouin*-zone for the reciprocal lattice, respectively [39]. Both are constructed in the same manner. In the case of the *Wigner-Seitz* cell, the next nearest neighboring lattice points from the origin are obtained by all integer linear combinations of the primitive translation vectors after (2.1.0-2). These vectors serve as normal vectors of planes, which are placed at the center of the distance between the origin and the respective neighboring lattice point. Once the planes have been placed for all next neighboring lattice points, a polyhedron is obtained called the *Wigner-Seitz* cell. For the first *Brillouin*-zone the procedure of construction outlined above is identical, except that the reciprocal lattice vectors after (2.1.0-4) are used instead of the primitive translation vectors after (2.1.0-2). Figure 2.2 schematically illustrates the *Wigner-Seitz* cell and the first *Brillouin*-zone for Diamond and Zinc-blende structures.

The importance of the reciprocal lattice vectors becomes apparent when considering functions which are periodic on the lattice. First, with the definitions after (2.1.0-3) the following relation holds:

$$\vec{a}_i \cdot \vec{a}_j^* = 2\pi \delta_{ij} , \quad (2.1.0\text{-}5)$$

where δ_{ij} is the Kronecker delta. Based on the relation (2.1.0-5) and a vector $\vec{K}$ in the reciprocal space

$$\vec{K} = l_1 \vec{a}_1^* + l_2 \vec{a}_2^* + l_3 \vec{a}_3^* , \quad (2.1.0\text{-}6)$$

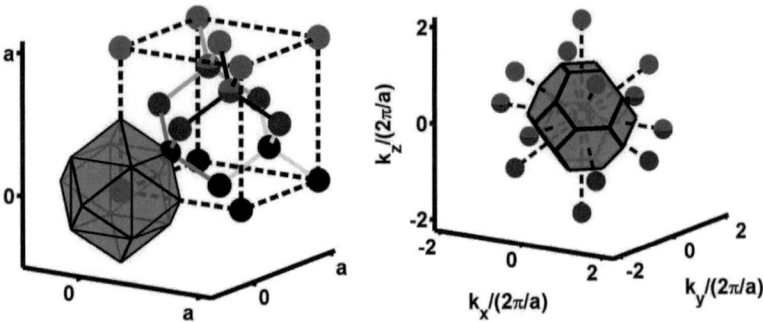

Figure 2.2: Schematic illustration of the *Wigner-Seitz* cell (left) and the first *Brillouin*-zone (right) for Diamond and Zinc-blende structures (FCC -lattice).

with integers l_i $(i = 1, 2, 3)$, the dot-product of $\vec{R}$ and $\vec{K}$ leads to

$$\vec{R} \cdot \vec{K} = 2\pi \left(n_1 l_1 + n_2 l_2 + n_3 l_3 \right) = 2\pi n \ , \tag{2.1.0-7}$$

where n is an integer. If one considers a function V being periodic with the lattice

$$V(\vec{r} + \vec{R}) = V(\vec{r}) \ , \tag{2.1.0-8}$$

one can expand V in a *Fourier*-series

$$V(\vec{r}) = \sum_{\vec{K}} V_{\vec{K}} \exp\left(i\vec{K}\vec{r} \right) \ , \tag{2.1.0-9}$$

with the *Fourier*-coefficients $V_{\vec{K}}$ and the imaginary unit i. Combining (2.1.0-8) with (2.1.0-9) leads to

$$
\begin{aligned}
V(\vec{r} + \vec{R}) &= \sum_{\vec{K}} V_{\vec{K}} \exp\left(i\vec{K} \left(\vec{R} + \vec{r} \right) \right) = \sum_{\vec{K}} V_{\vec{K}} \exp\left(i\vec{K}\vec{r} \right) \underbrace{\exp\left(i\vec{K}\vec{R} \right)}_{=1} \\
&= \sum_{\vec{K}} V_{\vec{K}} \exp\left(i\vec{K}\vec{r} \right) = V(\vec{r}) \ . \tag{2.1.0-10}
\end{aligned}
$$

In principle, the *Fourier*-series (2.1.0-9) of a lattice periodic function is performed in the same manner as the *Fourier*-series of a time dependent signal. For a time dependent signal a summation over all harmonics is performed. The same is done in (2.1.0-9), where the vectors $\vec{K}$ can be regarded as the harmonics. Due to the definition of $\vec{K}$ after (2.1.0-6) and the basis of reciprocal lattice vectors in (2.1.0-3) the *Fourier*-series (2.1.0-9) only consists of periodic functions with multiples of the lattice periodicity. The longest (shortest) period of the functions involved in (2.1.0-9) equals the largest (smallest) absolute value of the the primitive translation vectors $|\vec{a_i}|$ and is with (2.1.0-3) described by the smallest (largest) absolute value of the reciprocal basis vectors. Analogously to the *Fourier*-series of a time dependent signal, the the *Fourier*-coefficients $V_{\vec{K}}$ can be obtained by

$$V_{\vec{K}} = \frac{1}{V_{\text{cell}}} \int_{V_{\text{cell}}} V(\vec{r}) \exp\left(-\mathrm{i}\vec{K}\vec{r}\right) dV \ . \qquad (2.1.0\text{-}11)$$

Note, the unit cell contains all periods considered for the *Fourier*-series (2.1.0-9). The *Fourier*-series plays an important role for the calculation of the band structure and the *Debye*-diffraction. However, electrons are usually assumed to be free carriers being influenced by a periodic field that originates from the periodic lattice. Therefore, their reciprocal vectors are not restricted to an integer linear combination of the reciprocal lattice vectors $\vec{K}$. Instead, a continuous **reciprocal space**, containing $\vec{k}$, is needed for describing the electron transport.

2.2 Band structure and dispersion relation

From the quantum mechanical point of view, electrons are described by a wave-function $\psi_{\vec{k}}(\vec{r})$, which has to fulfill the stationary *Schrödinger* equation [4, 39, 46]:

$$-\frac{\hbar^2}{2m_0}\mathrm{div}_{\vec{r}}\left(\mathrm{grad}_{\vec{r}}\left(\psi_{\vec{k}}(\vec{r})\right)\right) + V_C(\vec{r})\psi_{\vec{k}}(\vec{r}) = E_{\vec{k}}\psi_{\vec{k}}(\vec{r}) \ , \qquad (2.2.0\text{-}12)$$

with the free electron mass m_0, the periodic crystal potential $V_C(\vec{r})$ and the energy of the electron $E_{\vec{k}}$ for a considered $\vec{k}$-vector. In [4], Bloch showed that the wave-function of an electron consists of a *plane*-wave part modulated by an amplitude $u_{\vec{k}}(\vec{r})$, which is periodic with the lattice:

$$\psi_{\vec{k}}(\vec{r}) = u_{\vec{k}}(\vec{r}) \exp\left(i\vec{k}\vec{r}\right) . \qquad (2.2.0\text{-}13)$$

Expanding the **Bloch-wave** $\psi_{\vec{k}}(\vec{r})$ and the crystal potential $V_C(\vec{r})$ in a *Fourier*-series results in

$$\psi_{\vec{k}}(\vec{r}) = \exp\left(i\vec{k}\vec{r}\right) \left[\sum_{\vec{K}} U_{\vec{K}} \exp\left(i\vec{K}\vec{r}\right) \right] , \qquad (2.2.0\text{-}14)$$

$$\mathrm{grad}_{\vec{r}}\left(\psi_{\vec{k}}(\vec{r})\right) = \exp\left(i\vec{k}\vec{r}\right) \left[i\sum_{\vec{K}} \vec{K} U_{\vec{K}} \exp\left(i\vec{K}\vec{r}\right) \right] + i\vec{k}\psi_{\vec{k}}(\vec{r}) , \qquad (2.2.0\text{-}15)$$

$$\mathrm{div}_{\vec{r}}\left(\mathrm{grad}_{\vec{r}}\left(\psi_{\vec{k}}(\vec{r})\right)\right) = -\exp\left(i\vec{k}\vec{r}\right) \left[\sum_{\vec{K}} \left|\vec{K}+\vec{k}\right|^2 U_{\vec{K}} \exp\left(i\vec{K}\vec{r}\right) \right] , \qquad (2.2.0\text{-}16)$$

$$V_C(\vec{r}) = \sum_{\vec{K}'} V_{\vec{K}'} \exp\left(i\vec{K}'\vec{r}\right) . \qquad (2.2.0\text{-}17)$$

Combining the relation above with the *Schrödinger* equation (2.2.0-12) leads to

$$\frac{\hbar^2}{2m_0} \left[\sum_{\vec{K}} \left|\vec{K}+\vec{k}\right|^2 U_{\vec{K}} \exp\left(i(\vec{K}+\vec{k})\vec{r}\right) \right]$$

$$+ \sum_{\vec{K}} \sum_{\vec{K}'} U_{\vec{K}} V_{\vec{K}'} \exp\left(i(\vec{K}'+\vec{K}+\vec{k})\vec{r}\right) = E_{\vec{k}} \left[\sum_{\vec{K}} U_{\vec{K}} \exp\left(i(\vec{K}+\vec{k})\vec{r}\right) \right] , \qquad (2.2.0\text{-}18)$$

which is a scalar equation consisting of sums over the reciprocal lattice vectors $\vec{K}$ and $\vec{K}'$. By considering (2.2.0-18) separately for each reciprocal lattice vector $\vec{K}$, the equation is split into several sub-problems.
From the mathematical point of view, the separation is performed by con-

sidering a plane-wave of the form (2.2.0-19) with a reciprocal lattice vector $\vec{K}$ [26]

$$\psi^* = \exp\left(-\mathrm{i}(\vec{k} + \vec{K})\vec{r}\right) , \qquad (2.2.0\text{-}19)$$

and the orthogonality relation

$$\frac{1}{V} \int_V \exp\left(\mathrm{i}\vec{q}\vec{r}\right) \exp\left(-\mathrm{i}(\vec{k} + \vec{K})\vec{r}\right) dV = \frac{1}{V} \int_V \exp\left(\mathrm{i}(\vec{q} - \vec{k} - \vec{K})\vec{r}\right) dV$$
$$= \delta_{\vec{q},\vec{k}+\vec{K}} . \qquad (2.2.0\text{-}20)$$

The integrals in equation (2.2.0-20) evaluate to one if $\vec{q} = \vec{k} + \vec{K}$ (and the arguments of the exponential function become zero) and otherwise to zero, which is described by the Kronecker δ-function. The usage of the δ-function becomes more clear if one considers the reciprocal vector $\vec{\beta} = \vec{q} - \vec{k} - \vec{K} \neq \vec{0}$. In this case, the integral involved in (2.2.0-20) reads

$$\frac{1}{V} \int_V \exp\left(\mathrm{i}\vec{\beta}\vec{r}\right) dV = \frac{1}{V} \int_V \left(\cos\left(\vec{\beta}\vec{r}\right) + \mathrm{i}\sin\left(\vec{\beta}\vec{r}\right)\right) dV = 0 . \quad (2.2.0\text{-}21)$$

Assuming a sufficiently large crystal volume V, the contribution from both the sine- and cosine-terms cancel out by themselves due to their periodicity. Therefore, multiplying the *Schrödinger* equation (2.2.0-18) by the plane-wave (2.2.0-19) and employing the orthogonality relation (2.2.0-20) leads to

$$\frac{\hbar^2}{2m_0}\left[\sum_{\vec{K}} \left|\vec{K} + \vec{k}\right|^2 U_{\vec{K}} \delta_{\vec{K},\vec{k}}\right] + \sum_{\vec{K}}\sum_{\vec{K}'} U_{\vec{K}} V_{\vec{K}'} \delta_{\vec{K}',\vec{K}-\vec{k}}$$
$$= E_{\vec{k}}\left[\sum_{\vec{K}} U_{\vec{K}} \delta_{\vec{K},\vec{k}}\right] . \quad (2.2.0\text{-}22)$$

Finally, after evaluating the δ-functions [66], (2.2.0-23) is obtained:

$$\left(\frac{\hbar^2}{2m_0}\left|\vec{K} + \vec{k}\right|^2\right) U_{\vec{K}} + \sum_{\vec{K}} U_{\vec{K}} V_{\vec{K}-\vec{K}} = E_{\vec{k}} U_{\vec{K}} . \qquad (2.2.0\text{-}23)$$

The scheme outlined previously corresponds to a transformation of the real space formulation of the *Schrödinger* equation (2.2.0-18) to its formulation in the reciprocal space via the *Fourier*-transformation

$$\mathcal{F}(\vec{k}') = \frac{1}{V} \int_V F(\vec{r}) \exp\left(-i\vec{k}'\vec{r}\right) dV , \qquad (2.2.0\text{-}24)$$

with a reciprocal vector $\vec{k}'$. (2.2.0-24) is used for each reciprocal vector $\vec{K}+\vec{k}$ in (2.2.0-18). Thus, instead of dealing with a scalar equation consisting of several sums (see (2.2.0-18)), a set of linear equations is obtained (see (2.2.0-23)). For each reciprocal lattice vector $\vec{K}$ equation (2.2.0-23) is considered separately and via the sum over the reciprocal lattice vectors $\vec{K}$, a coupling to all other linear equations is present. The form of the system described by (2.2.0-23) is schematically shown in (2.2.0-25).

$$\begin{pmatrix} a_{-N,-N} & a_{-N,-N+1} & \cdots & & \cdots & a_{-N,N} \\ \ddots & & \ddots & & \ddots & \ddots \\ \cdots & & a_{0,-1} & a_{0,0} & a_{0,1} & \cdots \\ \cdots & & & \ddots & \ddots & \ddots & \cdots \\ a_{N,-N} & & \cdots & & \ddots & a_{N,N-1} & a_{N,N} \end{pmatrix} \begin{pmatrix} U_{-N} \\ \vdots \\ U_0 \\ \vdots \\ U_N \end{pmatrix} = E_{\vec{k}} \begin{pmatrix} U_{-N} \\ \vdots \\ U_0 \\ \vdots \\ U_N \end{pmatrix}$$

$$(2.2.0\text{-}25)$$

The system (2.2.0-25) is a *special eigenvalue problem* [8] and its eigenvalues are the energies $E_{\vec{k}}$. The dimension of the system is $2N + 1$, since one needs to consider N reciprocal lattice vectors $\vec{K}$ in one and its opposite direction including $\vec{K} = \vec{0}$, where the $+1$ is originating from. Each element in a row of (2.2.0-25) consists of a term $U_{\vec{K}} V_{\vec{K}-\vec{K}}$ (see (2.2.0-23)), where in the case of the main diagonal the term $\frac{\hbar^2}{2m_0} \left|\vec{K} + \vec{k}\right|^2$ is added. Theoretically, an infinite set of reciprocal lattice vectors $\vec{K}$ and $\vec{K}$ is considered.

Solving the system (2.2.0-25) for its eigenvalues results in up to $2N+1$ energies $E_{\vec{k}}$. In order to obtain the energy **dispersion relation** $E(\vec{k})$ for the crystal under investigation, the system has to be solved separately for each considered vector $\vec{k}$. Based on the energies $E_{\vec{k}}$ for each vector $\vec{k}$ the eigenvectors

(*Fourier*-coefficients $U_{\vec{K}}$ in (2.2.0-23) / U_i in (2.2.0-25)) of the Bloch-waves for electrons can be obtained.

However, the difficult part in calculating an accurate dispersion relation $E(\vec{k})$ is a proper description of the periodic crystal potential $V_C(\vec{r})$ and its *Fourier*-coefficients $V_{\vec{K}'}$. In the **empty lattice approximation** [26, 39] $V_C(\vec{r})$ is assumed to be negligible and thus set to zero. In this special case, the *Schrödinger* equation (2.2.0-23) simplifies for each of the $2N + 1$ reciprocal lattice vectors $\vec{K}$ to

$$E_i(\vec{k}) = \frac{\hbar^2}{2m_0} \left| \vec{K}_i + \vec{k} \right|^2 , \tag{2.2.0-26}$$

with $i = -N, \ldots, 0, \ldots, N$. For a given $\vec{k}$, $2N + 1$ energies given by the parabolas (2.2.0-26), with vertices at $S(-\vec{K}_i, 0)$, are obtained. The vector $\vec{k}$ itself may be composed of a reciprocal lattice vector $\vec{K}$ and the vector $\vec{k}_{BZ1}$ located within the first *Brillouin*-zone

$$\begin{aligned}
\vec{k} &= (\hat{l}_1 \vec{a}_1^* + \hat{l}_2 \vec{a}_2^* + \hat{l}_3 \vec{a}_3^*) + \vec{k}_{BZ1} \\
&= \vec{K} + \vec{k}_{BZ1} , \tag{2.2.0-27} \\
\rightarrow \vec{K}_i + \vec{k} &= \left((\tilde{l}_1 + \hat{l}_1)\vec{a}_1^* + (\tilde{l}_2 + \hat{l}_2)\vec{a}_2^* + (\tilde{l}_3 + \hat{l}_3)\vec{a}_3^* \right) + \vec{k}_{BZ1} \\
&= \left(\bar{l}_1 \vec{a}_1^* + \bar{l}_2 \vec{a}_2^* + \bar{l}_3 \vec{a}_3^* \right) + \vec{k}_{BZ1} \\
&= \vec{K}_j + \vec{k}_{BZ1} , \tag{2.2.0-28} \\
\rightarrow \vec{K}_j &= \vec{K} + \vec{K}_i , \tag{2.2.0-29}
\end{aligned}$$

with the integers $\tilde{l}_{i'}$, $\hat{l}_{i'}$ and $\bar{l}_{i'}$ ($i' = 1, 2, 3$). Therefore, the result (2.2.0-26) can be rewritten in terms of a reciprocal lattice vector and a vector $\vec{k}_{BZ1}$ located within the first *Brillouin*-zone:

$$E_j(\vec{k}_{BZ1}) = \frac{\hbar^2}{2m_0} \left| \vec{K}_j + \vec{k}_{BZ1} \right|^2 . \tag{2.2.0-30}$$

Equation (2.2.0-26) describes the periodicity of the dispersion relation in the reciprocal space. Considering an infinite number of reciprocal lattice vectors ($N \rightarrow \infty$), a reduction/enlargement of $\vec{k}$ in (2.2.0-26) by any reciprocal lattice vector results in $2N + 1$ energies, identical to the ones obtained initially. The reduction/enlargement of $\vec{k}$ by any reciprocal lattice vector corresponds to a shift of the vertices $S(-\vec{\tilde{K}}_i, 0)$ of (2.2.0-26). Therefore, $\vec{k}$ is decomposed into a reciprocal lattice vector $\hat{K}$ and the reciprocal vector $\vec{k}_{\mathrm{BZ1}}$ by equation (2.2.0-27). $\hat{K}$ corresponds to the aforementioned reduction/enlargement and the shift of the parabola vertices is mathematically described by the equations (2.2.0-28) and results in $\vec{\tilde{K}}_j$ after (2.2.0-29).

Figure 2.3 schematically illustrates the result obtained by (2.2.0-26) and (2.2.0-30) for the one dimensional case. Since only one spatial component, for example in x-direction, is considered the arrows for indicating a vector are omitted. In figure 2.3, $k = a^* + k_{\mathrm{BZ1}}$ ($\hat{K} = a^*$) is assumed to consist of the component of the reciprocal lattice vector (labeled a^*) and a component of the first *Brillouin*-zone k_{BZ1}. On the left-hand side of figure 2.3, the energies after (2.2.0-26) obtained for $\tilde{K}_i = -3a^*$, $-2a^* - a^*$ and 0 are shown. By using (2.2.0-29) and (2.2.0-30) identical energies are obtained, but the parabolas associated to the energies initially are shifted by a^* in negative k-direction, as shown on the right-hand side of figure 2.3. Therefore, it is sufficient to

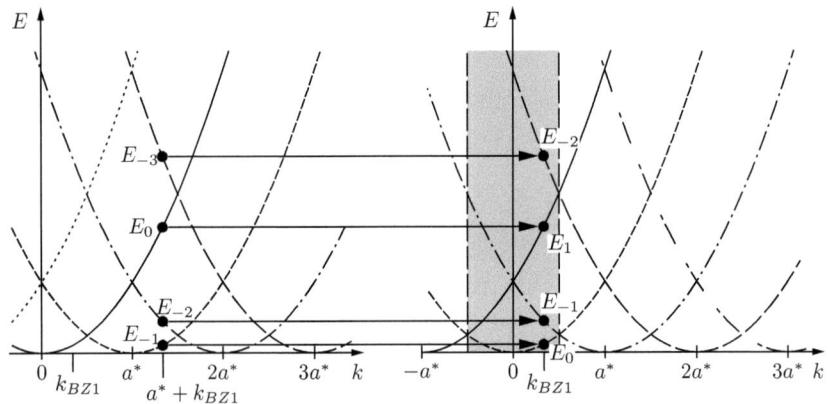

Figure 2.3: Schematic illustration of the periodicity of (2.2.0-26) (left) and the reduction to the first Brillouin zone after (2.2.0-30) (right) in the one dimensional case.

consider equation (2.2.0-30) within the first *Brillouin*-zone (see figure 2.3, grey shaded region on the right-hand side).

With the **empty lattice approximation** some basic properties of the dispersion relation, like its periodicity or the influence of different reciprocal lattice vectors on the energy diagram, can be explained. However, it is a simplified method and is therefore not capable of reproducing the dispersion relation in actual crystals accurately. More accurate results can be obtained by, for example, the **k·p** method [26] or by the empirical pseudo-potential method (**EPM**) [14, 26, 66, 70]. The latter can be viewed as the more practical approach for calculating the dispersion relation in a crystal. The basic idea behind the EPM is to directly use the *Fourier*-expansion of the crystal potential after (2.2.0-17), where the *Fourier*-coefficients $V_{\vec{K}'}$ are adjusted to measurements [14], like reflectivity measurements of the material under investigation [9]. EPM has been mainly developed and studied for semiconductors with Diamond and Zinc-blende structures, which captures both silicon and III-V semiconductors, respectively.

As described in section 2.1, the basis of Diamond and Zinc-blende lattices consists of two atoms (displaced by one quarter of the space diagonal), which are located at each lattice point of the FCC lattice. For both atoms a crystal potential is assumed, which is spherical and periodic with the lattice. Consequently, the resulting crystal potential of the basis is a superposition of the two crystal potentials [14, 26]

$$V_{C,\textbf{basis}}(\vec{r}) = V_{C,\alpha}(\vec{r} - \vec{r}_\alpha) + V_{C,\beta}(\vec{r} - \vec{r}_\beta) \,, \qquad (2.2.0\text{-}31)$$

with the crystal potentials $V_{C,\alpha}$, $V_{C,\beta}$ associated to the atom α and β within the primitive unit cell and their respective displacement vectors $\vec{r}_\alpha$, $\vec{r}_\beta$. The *Fourier*-coefficients of the resulting crystal potential become with (2.1.0-11)

$$
\begin{aligned}
V_{C,\textbf{basis},\vec{K}} &= \frac{1}{V_{\text{cell}}} \int_{V_{\text{cell}}} (V_{C,\alpha}(\vec{r} - \vec{r}_\alpha) + V_{C,\beta}(\vec{r} - \vec{r}_\beta)) \exp\left(-\mathrm{i}\vec{K}\vec{r}\right) dV \\
&= \frac{1}{V_{\text{cell}}} \left[\int_{V_{\text{cell}} - \vec{r}_\alpha} V_{C,\alpha}(\vec{\tilde{r}}_\alpha) \exp\left(-\mathrm{i}\vec{K}(\vec{\tilde{r}}_\alpha + \vec{r}_\alpha)\right) dV \right. \\
&\quad \left. + \int_{V_{\text{cell}} - \vec{r}_\beta} V_{C,\beta}(\vec{\tilde{r}}_\beta) \exp\left(-\mathrm{i}\vec{K}(\vec{\tilde{r}}_\beta + \vec{r}_\beta)\right) dV \right] \,, \qquad (2.2.0\text{-}32)
\end{aligned}
$$

with the substitutions $\vec{\tilde{r}}_\alpha = \vec{r} - \vec{r}_\alpha$ and $\vec{\tilde{r}}_\beta = \vec{r} - \vec{r}_\beta$. Since lattice periodic crystal potentials $V_{\mathrm{C},\alpha}$ and $V_{\mathrm{C},\beta}$ are assumed, the integration limits in (2.2.0-32) can be shifted to any position, as long as a full period is covered (primitive unit cell). Therefore, equation (2.2.0-32) is rewritten as

$$V_{C,\text{basis},\vec{K}} = \frac{1}{V_{\text{cell}}} \int_{V_{\text{cell}}} \left(V_{\mathrm{C},\alpha}(\vec{r}) \exp\left(-\mathrm{i}\vec{K}\vec{\tilde{r}}_\alpha\right) \right. $$
$$\left. + V_{\mathrm{C},\beta}(\vec{r}) \exp\left(-\mathrm{i}\vec{K}\vec{\tilde{r}}_\beta\right) \right) \exp\left(-\mathrm{i}\vec{K}\vec{r}\right) dV . \qquad (2.2.0\text{-}33)$$

For further calculations, it is convenient to place the origin in the middle between the two atoms α and β, which is schematically shown in figure 2.4. Thus, for Diamond and Zinc-blende structures, the positions of the atoms

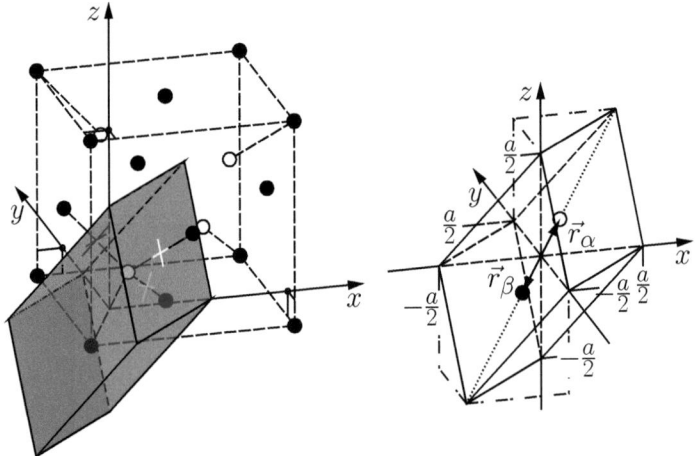

Figure 2.4: Schematic illustration of the primitive unit cell centered halfway between the two atoms (basis) of Diamond and Zinc-blende structures (FCC -lattice).

w.r.t. the shifted origin are

$$\vec{r}_\alpha = -\vec{r}_\beta = \vec{\tau} = \frac{a}{8} \begin{pmatrix} 1 \\ 1 \\ 1 \end{pmatrix} . \qquad (2.2.0\text{-}34)$$

Equation (2.2.0-33) becomes

$$V_{\text{C,basis},\vec{K}} = \frac{1}{V_{\text{cell}}} \int_{V_{\text{cell}}} \bigg((V_{\text{C},\alpha}(\vec{r}) + V_{\text{C},\beta}(\vec{r})) \cos(\vec{K}\vec{r})$$

$$+ \, \text{i} \, (V_{\text{C},\beta}(\vec{r}) - V_{\text{C},\alpha}(\vec{r})) \sin(\vec{K}\vec{r}) \bigg) \exp\left(-\text{i}\vec{K}\vec{r}\right) dV$$

$$= V_{\text{S},\vec{K}} S_{\text{S}}(\vec{K}) + \text{i}V_{\text{A},\vec{K}} S_{\text{A}}(\vec{K}) \, , \qquad (2.2.0\text{-}35)$$

with the symmetric and asymmetric pseudo-potential form factors $V_{\text{S},\vec{K}}$, $V_{\text{A},\vec{K}}$ defined by

$$V_{\text{S},\vec{K}} \quad = \quad \frac{1}{V_{\text{cell}}} \int_{V_{\text{cell}}} (V_{\text{C},\alpha}(\vec{r}) + V_{\text{C},\beta}(\vec{r})) \exp\left(-\text{i}\vec{K}\vec{r}\right) dV \, , \quad (2.2.0\text{-}36)$$

$$V_{\text{A},\vec{K}} \quad = \quad \frac{1}{V_{\text{cell}}} \int_{V_{\text{cell}}} (V_{\text{C},\beta}(\vec{r}) - V_{\text{C},\alpha}(\vec{r})) \exp\left(-\text{i}\vec{K}\vec{r}\right) dV \, , \quad (2.2.0\text{-}37)$$

and the associated structure factors $S_{\text{S}}(\vec{K})$, $S_{\text{A}}(\vec{K})$

$$S_{\text{S}}(\vec{K}) \quad = \quad \cos(\vec{K}\vec{r}) \, , \qquad (2.2.0\text{-}38)$$

$$S_{\text{A}}(\vec{K}) \quad = \quad \sin(\vec{K}\vec{r}) \, . \qquad (2.2.0\text{-}39)$$

The symmetric and asymmetric form factors after (2.2.0-36) and (2.2.0-37), respectively, are adjusted to experimental data. Therefore, no assumptions regarding the crystal potentials $V_{\text{C},\alpha}$ and $V_{\text{C},\beta}$ have to be made, which leads to a high degree of flexibility and accuracy compared to other methods. For the *Schrödinger* equation (2.2.0-23), the *Fourier*-coefficients after (2.2.0-35) have to be known for the distances of the reciprocal lattice vectors $\vec{K} - \vec{K}$. For the latter, only a few coefficients for the smallest possible absolute values of $|\vec{K} - \vec{K}|$ are considered [14]: $\frac{2\pi}{a}\sqrt{3}$, $\frac{2\pi}{a}2$, $\frac{2\pi}{a}\sqrt{8}$ and $\frac{2\pi}{a}\sqrt{11}$. Note, $|\vec{K} - \vec{K}| = 0$ is excluded and set to zero, since its contribution only results in a shift of the pseudo-potential (2.2.0-31). The possible resulting reciprocal lattice vectors are listed in table 2.1. Note, due to the considered materials (lattice and basis), the number of pseudo-potential form factors $V_{\text{S},\vec{K}-\vec{K}}$, $V_{\text{A},\vec{K}-\vec{K}}$ is reduced due to the structure factors after (2.2.0-38) and (2.2.0-39).

$\lvert \vec{\tilde{K}} - \vec{K} \rvert / \frac{2\pi}{a}$	permutations of reciprocal lattice vectors					
$\sqrt{3}$	(-1,-1,-1)	(-1,-1,1)	(-1,1,-1)	(1,-1,-1)	(-1,1,1)	(1,-1,1)
	(1,1,-1)	(1,1,1)				
2	(-2,0,0)	(0,-2,0)	(0,0,-2)	(0,0,2)	(0,2,0)	(2,0,0)
$\sqrt{8}$	(-2,-2,0)	(-2,0,-2)	(0,-2,-2)	(-2,0,2)	(0,-2,2)	(-2,2,0)
	(2,-2,0)	(0,2,-2)	(2,0,-2)	(0,2,2)	(2,0,2)	(2,2,0)
$\sqrt{11}$	(-3,-1,-1)	(-1,-3,-1)	(-3,-1,1)	(-1,-3,1)	(-1,-1,-3)	(-3,1,-1)
	(-3,1,1)	(1,-3,-1)	(1,-3,1)	(-1,-1,3)	(-1,1,-3)	(1,-1,-3)
	(-1,1,3)	(1,-3,3)	(1,1,-3)	(-1,3,-1)	(-1,3,1)	(3,-1,-1)
	(3,-1,1)	(1,1,3)	(1,3,-1)	(3,1,-1)	(1,3,1)	(3,1,1)

Table 2.1: Considered reciprocal lattice vectors $\vec{\tilde{K}} - \vec{K}$ (normalized to $\frac{2\pi}{a}$) for the EPM.

For example, for $\lvert \vec{\tilde{K}} - \vec{K} \rvert = \frac{2\pi}{a} 2$ the symmetric structure factor (2.2.0-38) becomes zero, since $\cos\left(\vec{\tau}(\vec{\tilde{K}} - \vec{K})\right) = \cos\left(\pm\frac{\pi}{2}\right) = 0$. In the same manner, the asymmetric pseudo-potential form factors vanish for $\lvert \vec{\tilde{K}} - \vec{K} \rvert = \frac{2\pi}{a}\sqrt{8}$, due to $\sin\left(\vec{\tau}(\vec{\tilde{K}} - \vec{K})\right) = \sin\left(\pm\pi\right) = 0$. Table 2.2 summarizes the required pseudo-potential form factors as functions of the reciprocal lattice vectors $\vec{\tilde{K}} - \vec{K}$. Additionally, the asymmetric pseudo-potential form factors $V_{A,\vec{\tilde{K}}-\vec{K}}$ are zero for all reciprocal lattice vectors, if the two atoms within the primitive unit cell are identical (c.g. (2.2.0-37) for $V_{C,\alpha}(\vec{r}) = V_{C,\beta}(\vec{r})$). Thus, for pure Silicon or Germanium only three symmetric pseudo-potential form factors have to be adjusted, whereas for compound semiconductors like Indium-Phosphide or Gallium-Arsenid also the three asymmetric pseudo-potential form factors have to be adjusted. Figure 2.5 shows the calculated band structure for Sil-

$\lvert \vec{\tilde{K}} - \vec{K} \rvert$	$\frac{2\pi}{a}\sqrt{3}$	$\frac{2\pi}{a}2$	$\frac{2\pi}{a}\sqrt{8}$	$\frac{2\pi}{a}\sqrt{11}$	comment
$V_{S,\vec{\tilde{K}}-\vec{K}}$	x	-	x	x	-
$V_{A,\vec{\tilde{K}}-\vec{K}}$	x	x	-	x	$V_{C,\alpha}(\vec{r}) \neq V_{C,\beta}(\vec{r})$

Table 2.2: Pseudo-potential form factor pattern as function of $\lvert \vec{\tilde{K}} - \vec{K} \rvert$.

icon and Indium-Phosphide obtained by the EPM. The traced path within

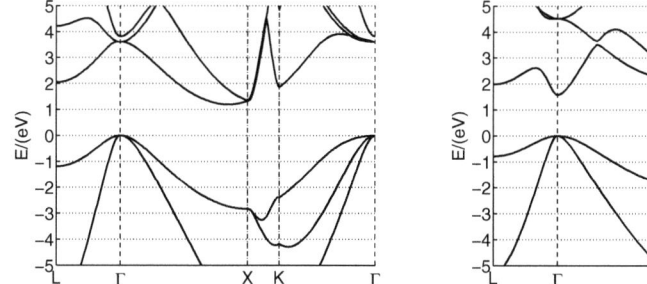

Figure 2.5: Band structures obtained by the EPM for Silicon (left) and Indium-Phosphide (right). The pseudo-potential form factors for Silicon and Indium-Phosphide are taken from [70] and [14], respectively.

the first *Brillouin*-zone, for which the band structures in figure 2.5 are illustrated, is drawn in figure 2.6 by a solid black line. Due to the symmetry, it

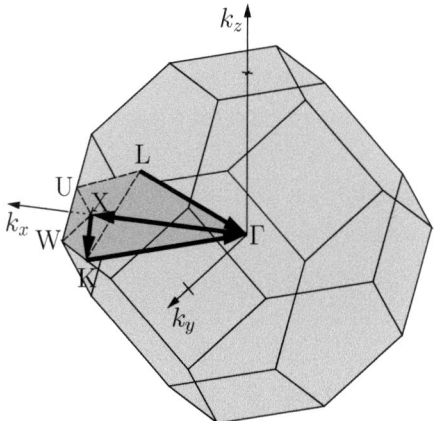

Figure 2.6: First irreducible wedge and its points of the first *Brillouin*-zone including the traced path used for the band structure diagram in figure 2.5.

is sufficient to consider the first irreducible wedge of the first *Brillouin*-zone for the band structure [37]. The missing edges of the first irreducible wedge, which were not traced for the band structures in figure 2.5, are sketched in figure 2.6 by dashed lines.

The marked reciprocal lattice points in figure 2.5 and 2.6 read [37]

$$\Gamma = \frac{2\pi}{a} \begin{pmatrix} 0 \\ 0 \\ 0 \end{pmatrix}, \; X = \frac{2\pi}{a} \begin{pmatrix} 1 \\ 0 \\ 0 \end{pmatrix}, \; L = \frac{2\pi}{a} \begin{pmatrix} \frac{1}{2} \\ \frac{1}{2} \\ \frac{1}{2} \end{pmatrix}$$

$$K = \frac{2\pi}{a} \begin{pmatrix} \frac{3}{4} \\ \frac{3}{4} \\ 0 \end{pmatrix}, \; W = \frac{2\pi}{a} \begin{pmatrix} 1 \\ \frac{1}{2} \\ 0 \end{pmatrix}, \; U = \frac{2\pi}{a} \begin{pmatrix} 1 \\ \frac{1}{4} \\ \frac{1}{4} \end{pmatrix} . \tag{2.2.0-40}$$

The band structures in figure 2.5 are showing the energies (often also called states) that an electron can have for a given vector $\vec{k}$ pointing to any point of a traced path within the first *Brillouin*-zone. The continuous curves are associated with bands. It can also be taken from figure 2.5 that for both considered materials an energetic region exists, which is not passed by any curve. Therefore, the latter region is called the **band-gap**. The bands below and above the band-gap are called **valence** and **conduction** bands, respectively. For Silicon the valence band maximum and the conduction band minimum are not located at the same reciprocal point, in contrast to Indium-Phosphide, where these can be found at the Γ-point. Consequently, Silicon (and likewise Germanium) is called an **indirect** and Indium-Phosphide (and several other III-V compound semiconductors) a **direct** semiconductor, respectively.

Depending on the transport model used for the description of the electron transport, simplifications regarding the band structure are performed. For example, in the Drift-Diffusion (DD) and Hydrodynamic (HD) transport model, which are momentum-based transport models derived from the Boltzmann transport equation (BTE) [37, 46], only **band-edges** are considered. For these transport models, the conduction and valence band-edge coincide with the corresponding band minimum and maximum, respectively. Nevertheless, more sophisticated transport models, like the Boltzmann transport equation, demand more information about the band structure, since the electron transport is described in both real and reciprocal space. Due to the high dimensionality of the BTE and the limited computational resources, the BTE is mainly solved by the Monte-Carlo method (MC), which on one hand offers a high degree of flexibility and low implementation effort, but suffers on the other hand from long simulation times and noisy results [34, 37, 69]. Nevertheless, MC offers the possibility to directly incorporate the valence and conduction bands obtained for example by EPM [37], but at the expense of the originally low implementation effort and simulation time. Therefore, simplified band structures are often used, which are based on the first conduction band (for electron transport). Furthermore, the first conduction band is separated into **valleys**, which represent the corresponding minima [34, 69]. These valleys are described by a dispersion relation $E^\nu(\vec{k}_\nu)$, where ν labels the currently considered valley. In the simplest case, the results of the **empty lattice approximation** are adopted:

$$
\begin{aligned}
E^\nu(\vec{k}) &= E_0^\nu + \frac{\hbar^2}{2m_0 m_\nu^*}|\vec{k} - \vec{k}^{\nu,0}|^2 \\
&= E_0^\nu + \underbrace{\frac{\hbar^2}{2m_0 m_\nu^*}|\vec{k}^\nu|^2}_{\text{kinetic carrier energy}} \quad ,
\end{aligned}
\tag{2.2.0-41}
$$

with the potential energy E_0^ν (energetic minimum) of the considered valley, its position in the reciprocal space $\vec{k}^{\nu,0}$ and the resulting reciprocal vector $\vec{k}^\nu$, measured from the location of the valley minimum. The parameter m_ν^* models the curvature of the parabola for the currently considered valley and is called *effective mass*, where m_0 is again the free electron mass. Therefore, the dispersion relation after (2.2.0-41) is called **parabolic dispersion relation**. m_ν^* is adjusted, for example, to the first conduction band obtained by the EPM for the material under investigation. Figure 2.7 illustrates the first conduction band of Silicon and Indium-Phosphide including the valley approximations with (2.2.0-41). Another modeling approach for the valleys, which gives a higher degree of flexibility regarding the adjustment, is the **non-parabolic dispersion relation**

$$
E^\nu(\vec{k}^\nu) = E_0^\nu - \frac{1}{2\alpha_\nu} + \sqrt{\frac{1}{4\alpha_\nu^2} + \frac{1}{\alpha_\nu}\frac{\hbar^2}{2m_0 m_\nu^*}|\vec{k}^\nu|^2} \quad ,
\tag{2.2.0-42}
$$

with the non-parabolicity factor α_ν. The difference between the two ap-

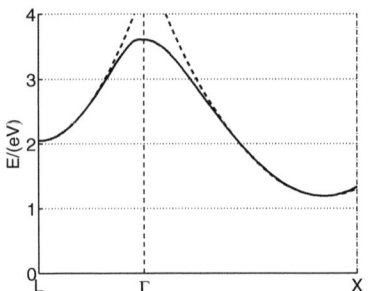 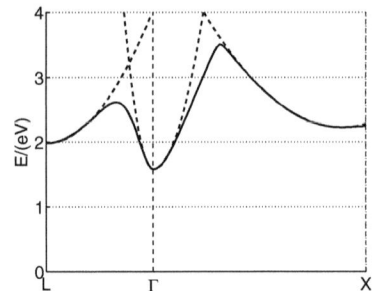

Figure 2.7: First conduction band (solid line, obtained by EPM) and parabolic approximation of the valleys (dashed lines) for Silicon (left) and Indium-Phosphide (right).

proaches is the modeling of the kinetic carrier energy. For (2.2.0-41) a simple parabola is assumed, whereas for (2.2.0-42) the implicit form of kinetic energy reads

$$\epsilon^\nu \left(1 + \alpha_\nu \epsilon^\nu\right) = \frac{\hbar^2}{2m_0 m_\nu^*} |\vec{k}^\nu|^2 \; , \tag{2.2.0-43}$$

with the kinetic energy ϵ^ν. Thus, in addition to the effective mass m_ν^* another adjustment parameter, the non-parabolicity factor α_ν, can be employed to improve the agreement between the first conduction band and the valley approximations.

However, both (2.2.0-41) and (2.2.0-42) assume a spherical shape of the kinetic energy as function of $\vec{k}_\nu$ due to the square of its absolute value. This isotropic approximation is sufficient for some materials and kinetic energies in the vicinity of the valley minimum. However, with increasing kinetic energies major deviations occur due to the anisotropy of the first conduction band in the reciprocal space. Figure 2.8 shows the equi-energy lines of the kinetic energy for Silicon at the X- and L-valley, respectively. In the case of Silicon, both the X- and L-valley are anisotropic starting from low energies. In contrast to Silicon, the Γ-valley of Indium-Phosphide is isotropic, which is illustrated in figure 2.9. Nevertheless, for the L-valley a high anisotropy is also apparent at low energies, like in the case of Silicon. In order to take

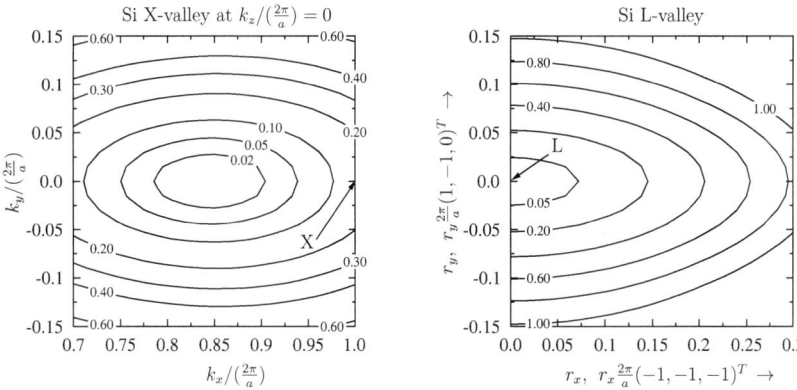

Figure 2.8: Equi-energy lines of the X- (left) and L-valley (right) of Silicon. The labels at the equi-energy lines mark the kinetic energy in eV.

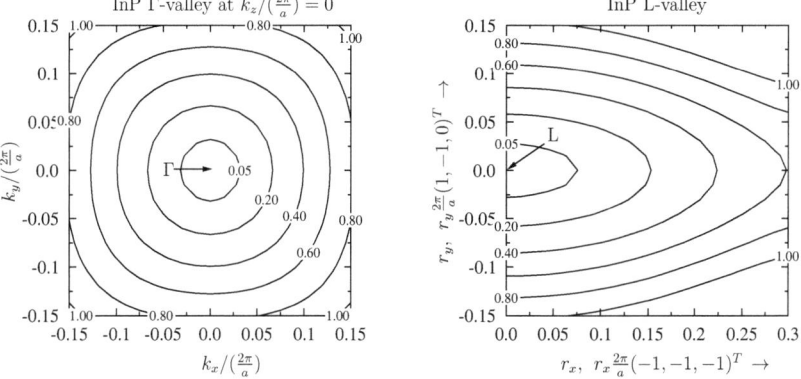

Figure 2.9: Equi-energy lines of the Γ- (left) and L-valley (right) of Indium-Phosphide. The labels at the equi-energy lines mark the kinetic energy in eV.

the anisotropy into account and due to the shape of the equi-energy lines, ellipsoidal non-parabolic surfaces are considered for the kinetic energy:

$$\epsilon^\nu \left(1 + \alpha_\nu \epsilon^\nu\right) = \frac{\hbar^2}{2m_0} \left(\frac{k_l^{\nu\,2}}{m_l^\nu} + \frac{k_t^{\nu\,2}}{m_t^\nu}\right) , \qquad (2.2.0\text{-}44)$$

23

with the longitudinal and transverse effective masses m_l^ν and m_t^ν and the respective components of the vector k_l^ν and k_t^ν [34, 46, 69]. The ellipsoids of the X-valleys in Silicon and L-valleys in Silicon and Indium-Phosphide are schematically shown in figure 2.10(a) and 2.10(b), respectively. In the case of silicon (see figure 2.10(a)), 6 equivalent X-valleys are present. In contrast to the X-valleys, the minima of L-valleys for both Silicon and Indium-Phosphide are directly located at the lateral face of the first Brillouin-zone (c.g. figure 2.6). Therefore, only one half of the ellipsoids is contributing to the currently considered first *Brillouin*-zone, where the rest contributes to adjacent first *Brillouin*-zones. Thus, the number of equivalent L-valleys is reduced from 8 to 4. Compared to the a spherical dispersion relation ((2.2.0-41) or

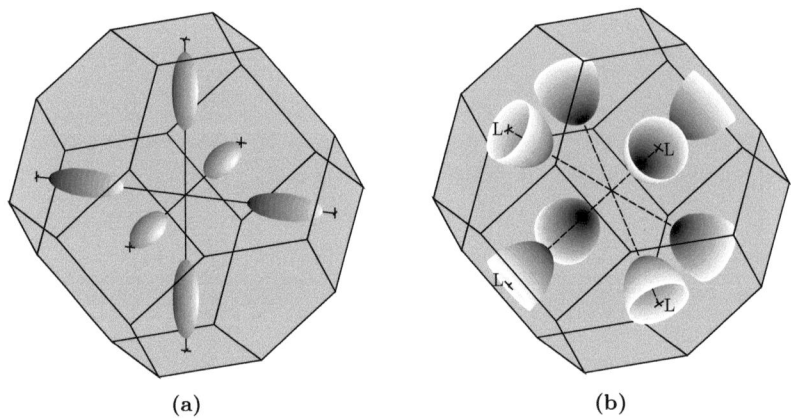

(a)   (b)

Figure 2.10: Schematic illustration of the equi-energy surfaces of the X-valleys in Silicon (left) and the L-valleys in Silicon and Indium-Phosphide (right).

(2.2.0-42)), the anisotropic dispersion relation (2.2.0-44) makes further calculations in the reciprocal space more complicated. Therefore, a transformation is often employed, which maps the ellipsoidal equi-energy surfaces to spheres. Latter is called **Herring-Vogt transformation** [30, 34]. The basic idea behind the Herring-Vogt transformation is to find a transformation matrix T^{HV}, which scales the reciprocal vectors involved in (2.2.0-44) in such a way, that the anisotropic dispersion relation (2.2.0-44) reduces to the isotropic description after (2.2.0-43). Equating a more general form of the right-hand side of

(2.2.0-44) with the right-hand side of (2.2.0-43) gives

$$
\frac{|\vec{\tilde{k}}^{\nu}|^2}{m_\nu^*} = \left(\frac{k_x^{\nu 2}}{m_x^\nu} + \frac{k_y^{\nu 2}}{m_y^\nu} + \frac{k_z^{\nu 2}}{m_z^\nu} \right) \text{ or}
$$

$$
(\tilde{k}_x^\nu)^2 + (\tilde{k}_y^\nu)^2 + (\tilde{k}_y^\nu)^2 = \frac{m_\nu^*}{m_x^\nu} k_x^{\nu 2} + \frac{m_\nu^*}{m_y^\nu} k_y^{\nu 2} + \frac{m_\nu^*}{m_z^\nu} k_z^{\nu 2} \text{ , (2.2.0-45)}
$$

where $\vec{\tilde{k}}^{\nu}$ is the transformed reciprocal vector. Considering the spatial components of $\vec{\tilde{k}}^{\nu}$ and $\vec{k}^{\nu}$ in (2.2.0-45) separately implies the transformation matrix

$$
\vec{\tilde{k}}^{\nu} = \begin{pmatrix} \sqrt{\frac{m_\nu^*}{m_x^\nu}} & 0 & 0 \\ 0 & \sqrt{\frac{m_\nu^*}{m_y^\nu}} & 0 \\ 0 & 0 & \sqrt{\frac{m_\nu^*}{m_z^\nu}} \end{pmatrix} \vec{k}^{\nu} \text{ , (2.2.0-46)}
$$

$$
\vec{\tilde{k}}^{\nu} = T^{\mathrm{HV},\nu} \vec{k}^{\nu} \text{ , (2.2.0-47)}
$$

with the transformation matrix $T^{\mathrm{HV},\nu}$ for the valley ν. The isotropic effective mass m_ν^* involved in the transformation can be obtained by demanding that the volume of the reciprocal space has to be invariant under the transformation. This demand results in

$$
d\tilde{k}_x^\nu d\tilde{k}_y^\nu d\tilde{k}_z^\nu = \underbrace{\sqrt{\frac{m_\nu^{*3}}{m_x^\nu m_y^\nu m_z^\nu}}}_{=1} dk_x^\nu dk_y^\nu dk_z^\nu \text{ ,}
$$

$$
m_\nu^* = \sqrt[3]{m_x^\nu m_y^\nu m_z^\nu} \text{ . (2.2.0-48)}
$$

The differential operator in the reciprocal space has to be modified, for consistency. For example, the divergence of a vector field $\vec{F}_v(\vec{k}^{\nu})$ reads

$$
\mathrm{div}_{\vec{k}^{\nu}} \left(\vec{F}_v(\vec{k}^{\nu}) \right) = T_{1,1}^{\mathrm{HV},\nu} \frac{\partial F_{v,k_x}(\vec{\tilde{k}}^{\nu})}{\partial \tilde{k}_x^\nu} + T_{2,2}^{\mathrm{HV},\nu} \frac{\partial F_{v,k_y}(\vec{\tilde{k}}^{\nu})}{\partial \tilde{k}_y^\nu} + T_{3,3}^{\mathrm{HV},\nu} \frac{\partial F_{v,k_z}(\vec{\tilde{k}}^{\nu})}{\partial \tilde{k}_z^\nu}
$$

$$
= \mathrm{div}_{\vec{\tilde{k}}^{\nu}} \left(T^{\mathrm{HV},\nu} \vec{F}_v(\vec{k}^{\nu}) \right) \text{ . (2.2.0-49)}
$$

Thus, the components of a vector field $\vec{F}_v$ ($F_{v,k_{\{x,y,z\}}}$) in the original k^{ν}-space need to be expressed in terms of $\vec{\tilde{k}}^{\nu}$ instead of $\vec{k}^{\nu}$ for the right-hand side of (2.2.0-49). Likewise, the gradient of a scalar function $F_{\mathrm{s}}(\vec{\tilde{k}}^{\nu})$ in the reciprocal space becomes

$$\mathrm{grad}_{\vec{\tilde{k}}^{\nu}}\left(F_{\mathrm{s}}(\vec{\tilde{k}}^{\nu})\right) = \begin{pmatrix} T_{1,1}^{\mathrm{HV},\nu}\frac{\partial F_{\mathrm{s}}(\vec{\tilde{k}}^{\nu})}{\partial \tilde{k}_x^{\nu}} \\ T_{2,2}^{\mathrm{HV},\nu}\frac{\partial F_{\mathrm{s}}(\vec{\tilde{k}}^{\nu})}{\partial \tilde{k}_y^{\nu}} \\ T_{3,3}^{\mathrm{HV},\nu}\frac{\partial F_{\mathrm{s}}(\vec{\tilde{k}}^{\nu})}{\partial \tilde{k}_z^{\nu}} \end{pmatrix} = T^{\mathrm{HV},\nu}\mathrm{grad}_{\vec{\tilde{k}}^{\nu}}\left(F_{\mathrm{s}}(\vec{\tilde{k}}^{\nu})\right) \ .$$

$$(2.2.0\text{-}50)$$

A more detailed view on the Herring-Vogt transformation is given in the appendix A.1, which explains the results for the transformed differential operators above in detail. By obeying relation (2.2.0-49) and (2.2.0-50), a spherical dispersion relation can be considered, which simplifies further calculations.

In this work, the non-parabolic dispersion relation after (2.2.0-43) is employed in combination with the Herring-Vogt transformation (if an anisotropic band structure is considered). As will be shown in the subsequent sections, an integration over the reciprocal space has to be performed in order to obtain relevant quantities, for example the electron density at a certain real space point or the *momentum relaxation time* of a scattering mechanism. Instead of integrating over the $\vec{\tilde{k}}^{\nu}$-vector directly, it is useful to employ the isotropy of the band structure and the dispersion relation (2.2.0-43). This allows to express the transformed $\vec{\tilde{k}}^{\nu}$-vectors after (2.2.0-46) within a valley in spherical coordinates:

$$|\vec{\tilde{k}}^{\nu}| = \frac{\sqrt{2m_0 m_{\nu}^*}}{\hbar}\sqrt{\epsilon^{\nu}\left(1 + \alpha_{\nu}\epsilon^{\nu}\right)} \ , \qquad (2.2.0\text{-}51)$$

$$\vec{\tilde{k}}^{\nu} = |\vec{\tilde{k}}^{\nu}|\begin{pmatrix} \cos(\varphi)\sin(\theta) \\ \sin(\varphi)\sin(\theta) \\ \cos(\theta) \end{pmatrix} \ . \qquad (2.2.0\text{-}52)$$

By calculating the functional determinant of the Jacobian

$$
J = \begin{pmatrix} \frac{\partial \tilde{k}_x^\nu}{\partial \epsilon^\nu} & \frac{\partial \tilde{k}_x^\nu}{\partial \varphi} & \frac{\partial \tilde{k}_x^\nu}{\partial \theta} \\ \frac{\partial \tilde{k}_y^\nu}{\partial \epsilon^\nu} & \frac{\partial \tilde{k}_y^\nu}{\partial \varphi} & \frac{\partial \tilde{k}_y^\nu}{\partial \theta} \\ \frac{\partial \tilde{k}_z^\nu}{\partial \epsilon^\nu} & \frac{\partial \tilde{k}_z^\nu}{\partial \varphi} & \frac{\partial \tilde{k}_z^\nu}{\partial \theta} \end{pmatrix}
\tag{2.2.0-53}
$$

$$
= \begin{pmatrix} \frac{\sqrt{2m_0 m_\nu^*}}{\hbar} \frac{(1+2\alpha_\nu \epsilon^\nu)\cos(\varphi)\sin(\theta)}{2\sqrt{\epsilon^\nu(1+\alpha_\nu\epsilon^\nu)}} & -|\vec{k}^\nu|\sin(\varphi)\sin(\theta) & |\vec{k}^\nu|\cos(\varphi)\cos(\theta) \\ \frac{\sqrt{2m_0 m_\nu^*}}{\hbar} \frac{(1+2\alpha_\nu \epsilon^\nu)\sin(\varphi)\sin(\theta)}{2\sqrt{\epsilon^\nu(1+\alpha_\nu\epsilon^\nu)}} & |\vec{k}^\nu|\cos(\varphi)\sin(\theta) & |\vec{k}^\nu|\sin(\varphi)\cos(\theta) \\ \frac{\sqrt{2m_0 m_\nu^*}}{\hbar} \frac{(1+2\alpha_\nu \epsilon^\nu)\cos(\theta)}{2\sqrt{\epsilon^\nu(1+\alpha_\nu\epsilon^\nu)}} & 0 & -|\vec{k}^\nu|\sin(\theta) \end{pmatrix} ,
$$

$$
\tag{2.2.0-54}
$$

one obtains

$$
d\tilde{k}_x^\nu d\tilde{k}_y^\nu d\tilde{k}_z^\nu = \frac{1}{2}\left(\frac{\sqrt{2m_0 m_\nu^*}}{\hbar}\right)^3 \sqrt{\epsilon^\nu(1+\alpha_\nu\epsilon^\nu)}(1+2\alpha_\nu\epsilon^\nu)\sin(\theta)d\epsilon^\nu d\theta d\varphi ,
$$

$$
\tag{2.2.0-55}
$$

with the **density of states** (DOS) for the valley ν

$$
\begin{aligned}
\mathrm{dos}^\nu(\epsilon^\nu) &= \frac{1}{2}\left(\frac{\sqrt{2m_0 m_\nu^*}}{\hbar}\right)^3 \sqrt{\epsilon^\nu(1+\alpha_\nu\epsilon^\nu)}(1+2\alpha_\nu\epsilon^\nu) , \tag{2.2.0-56} \\
&= \frac{1}{2}\left(\frac{\sqrt{2m_0}}{\hbar}\right)^3 \sqrt{m_x^\nu m_y^\nu m_z^\nu}\sqrt{\epsilon^\nu(1+\alpha_\nu\epsilon^\nu)}(1+2\alpha_\nu\epsilon^\nu) .
\end{aligned}
$$

The DOS after (2.2.0-56) indicates the number of available states per (kinetic) unit energy ϵ^ν and angles θ (polar), φ (azimuthal) within the valley ν at the ϵ^ν. Thus, the total number of available states within one valley ν can be obtained by integrating the right-hand side of (2.2.0-55) within the limits $\epsilon^\nu \in [0,\infty]$, $\varphi \in [0,2\pi]$ and $\theta \in [0,\pi]$. From the mathematical point of view, the DOS is a remnant of the coordinate transformation after (2.2.0-51)-(2.2.0-52). Note, the DOS after (2.2.0-56) is valid for **parabolic** ($\alpha_\nu = 0$) and **non-parabolic** ($\alpha_\nu > 0$) band structures.

Up to now, only pure materials were considered. However, also practically more relevant compound semiconductors, like Silicon-Germanium hetero-structures (SiGe), need to be examined. A prominent candidate is the SiGe heterojunction-bipolar transistor (HBT), where a SiGe layer (base) is located in between a relaxed Si emitter and collector. As the lattice constant of the SiGe layer is larger than the one of pure Si, biaxial-compressive strain occurs (cf. figure 2.11). The consequence is, that 4 of the 6 X-valleys are shifted to lower energies, whereas the 2 remaining X-valleys(in growth direction, x-direction in figure 2.11) are shifted to higher energies [10, 11, 31]. The energetic downward shift ($\Delta E_{C_{6 \to 4}}$) is modeled as Ge induced change of the electron affinity [37], whereas the upward shift is approximated by [31]

$$\Delta E_{C_{4 \to 2}} \approx 0.63 \, \text{eV} \, x_{\text{Ge}} \, , \tag{2.2.0-57}$$

where x_{Ge} is the mole fraction of Germanium. Following [11] and [52], the non-parabolicity factor α_ν and the anisotropic masses are assumed to remain unaffected by Ge, respectively.

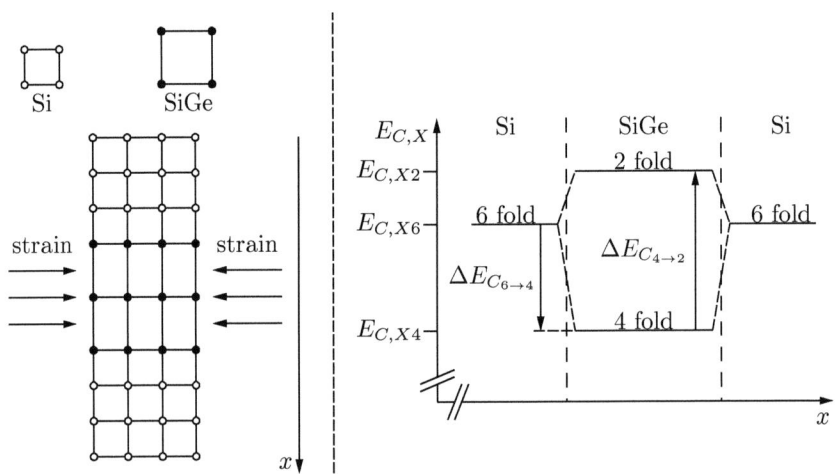

Figure 2.11: Schematic visualization of the biaxial-compressive strain in SiGe HBTs (left) and the resulting conduction band splitting (right).

2.3   Electron-lattice interaction

The interaction between electrons and the lattice is quantum-mechanically described by the first-order perturbation theory [51]. The origin of these interactions is from the displacement of atoms in a lattice around their equilibrium position, lattice defects or from impurities. For the Boltzmann transport equation, it is important to know the relation of a scattering mechanism when an electron with a given k-vector $\vec{k}$ in the valley ν is scattered to $\vec{k}'$ in a valley ν'. Note, that $\vec{k}$ and $\vec{k}'$ are defined from the center of the first *Brillouin-zone* (Γ point) and not from the respective position of the valley minimum in the reciprocal space. The transition rate is obtained by *Fermi's Golden Rule* [46, 51, 69]

$$S_{\mathrm{X}}^{\nu \to \nu'}(\vec{k}, \vec{k}') = \frac{2\pi}{\hbar} |M_{\mathrm{X}}(\vec{k}, \vec{k}')|^2 \delta \left(E_{\mathrm{i}}^{\nu}(\vec{k}) - E_{\mathrm{f}}^{\nu'}(\vec{k}') \right) , \qquad (2.3.0\text{-}58)$$

with the *matrix element* $M_{\mathrm{X}}(\vec{k}, \vec{k}')$ of a scattering type labeled by "X" and the initial/final energy $E_{\mathrm{i}}/E_{\mathrm{f}}$, respectively. For more information about the calculation of the *matrix elements* of different scattering mechanisms the reader is referred to [46, 51, 69]. Here, only the properties for the most important scattering mechanisms and the transition rates are listed. The general form of the transition rates is covered by (2.3.0-58). Based on (2.3.0-58), the basic properties of the transition rates are discussed first.

Depending on the *matrix element* – the square of its absolute value $|M_{\mathrm{X}}(\vec{k}, \vec{k}')|^2$ – a scattering mechanism can be characterized as **isotropic** or **anisotropic**. The scattering mechanism is **anisotropic**, if $|M_{\mathrm{X}}(\vec{k}, \vec{k}')|^2$ depends on $(\vec{k}, \vec{k}')$. Otherwise, the scattering mechanism is **isotropic**. In the latter case, no direction in the reciprocal space is preferred to be scattered to during a scattering event, contrary to an **anisotropic** scattering event.

By the δ-function in (2.3.0-58), the initial energy E_{i} and the final energy E_{f} are correlated with each other. If those energies are identical, the energy of the electron is preserved. Such kind of scattering events are called **elastic**. Otherwise, if an electron absorbs or emits energy due to scattering, the event is called **inelastic**.

Additionally, depending on the considered scattering mechanism, an electron can be scattered within the initial valley or to another valley. Thus, scattering mechanisms can be categorized as an **intra-** or **intervalley** scattering. Depending on the material under investigation and the considered valleys, intervalley scattering can occur between two **non-equivalent** valleys (e.g. Γ- and L-valley for INP & GAAS) or between **equivalent** valleys (e.g. X-valleys in SI or L-valleys in INP & GAAS). In the case of SI it needs to be distinguished between **f-type** and **g-type** intervalley scattering: For **g-type** scattering, the electron is scattered to the opposite valley within the first *Brillouin-zone*, whereas **f-type** scattering implies the electron to be scattered to all other valleys except the opposite one. Figure 2.12 summarizes the different possible properties of a scattering mechanism.

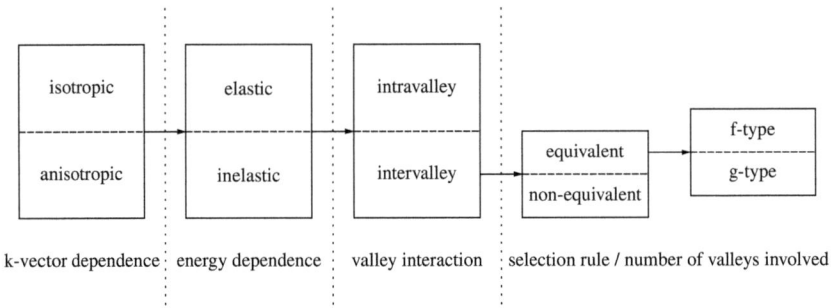

Figure 2.12: Transition rate properties for scattering mechanisms.

2.3.1 Phonon scattering

In the case of scattering events due to lattice vibrations (**phonons**), an energy exchange usually takes place. The amount of energy exchanged due to **phonon-scattering** depends on the scattering vector $\vec{\beta}$:

$$\vec{\beta} = \vec{k}' - \vec{k} \tag{2.3.1-59}$$

and on the associated **phonon dispersion relation**. Note again, $\vec{k}$ and $\vec{k}'$ in (2.3.1-59) are defined from the center of the first *Brillouin-zone* (Γ point) and not from the position of the valley minima in the reciprocal space. Similar to the E-$\vec{k}$ relation, a dispersion relation within the first *Brillouin-zone* has to be considered for phonons, too. Several approaches are existing for the calculation of the **phonon dispersion relation**, as outlined in [26, 39, 51, 74]. Here, only the basic ideas and properties are given. Since a three dimensional semiconductor crystal is considered here, three phonon modes are existing: one *longitudinal* and two *transverse* modes. For the *longitudinal* mode, the direction of the energy transfer $(\vec{\beta})$ coincides with the direction of the atom displacement, whereas for the *transverse* modes the atom displacement is perpendicular to the energy transfer direction [26]. In order to obtain the corresponding dispersion relations, a spring mass model is used (see e.g. [39]), with one atom being connected to its neighbors by springs. Depending on the number of atoms n_a within the primitive unit cell, 3 acoustic and $3n_a - 3$ optical branches in the phonon dispersion relation occur. Therefore, 3 acoustic and 3 optical phonon branches are existing for most semiconductors, as $n_a = 2$ applies. In the case of acoustic phonons, n_a atoms (and their neighbors) are displaced in the same direction and in phase, but the magnitude of displacement varies from atom to atom. On the other hand, the displacement of the n_a atoms within the primitive unit cell is out of phase for optical phonons. Figure 2.13 shows the phonon dispersion relation of acoustic and optical phonons in SI and GE, with data from [74]. If an electron with re-

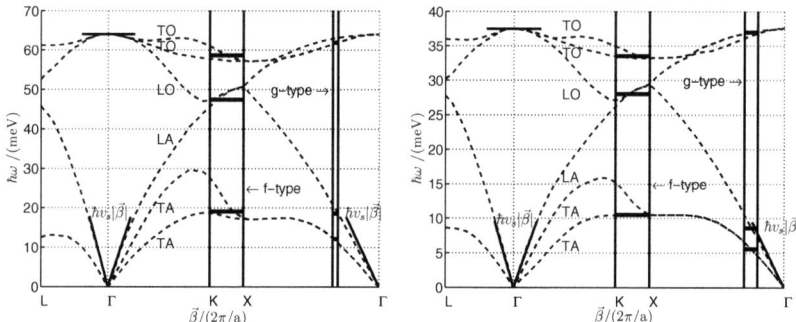

Figure 2.13: Phonon dispersion relations for SI (left) and GE (right) of acoustic and optical phonons [74].

ciprocal vector $\vec{k}$ is scattered by a phonon scattering event, both the energy and the momentum have to be conserved (see for example [46]). The first demand originates from the δ-function in (2.3.0-58) and the second one from the definition of the scattering vector $\vec{\beta}$ after (2.3.1-59). Thus, the following relation has to be fulfilled:

$$E_{\mathrm{f}}^{\nu'}(\vec{k'}) = E_{\mathrm{i}}^{\nu}(\vec{k}) \pm \hbar\omega(\vec{\beta} = \vec{k'} - \vec{k}) \,, \qquad (2.3.1\text{-}60)$$

where the $+/-$ sign refers to an energy absorption/emission process, respectively. Equation (2.3.1-60) relates the phonon dispersion relation to the band structure of the material under investigation. In order to accommodate the phonon dispersion relation to the E-$\vec{k}$ relation, some simplifications of the phonon dispersion relation can be made based on the scattering type as outlined in the following subsections [37, 46, 69].

2.3.1.1 Acoustic deformation potential scattering

In the case of **acoustic phonon scattering**, the phonon dispersion relation can be linearized in the vicinity of the Γ-point by

$$\hbar\omega(\vec{\beta}) \approx \hbar v_{\mathrm{s}}|\vec{\beta}| \,, \qquad (2.3.1\text{-}61)$$

with the material sound velocity v_{s} (Γ-point, see figure 2.13). Additionally, at room temperature the phonon energy is smaller than the thermal energy $k_{\mathrm{b}}T_{\mathrm{L}}$ ($\approx 26\,\mathrm{meV}$). Based on these considerations, acoustic phonon scattering can be regarded as **elastic** in the vicinity of the Γ-point [69]. Since the scattering vectors $\vec{\beta}$ are small in this case, electrons cannot be transferred to adjacent valleys and thus, the considered scattering process is an **intravalley** scattering process. The transition rate reads [34, 37, 46, 69]

$$S_{\mathrm{ADP}}^{\nu \to \nu'}(\vec{k}, \vec{k'}) = \frac{2\pi D_{\mathrm{A}}^2 k_{\mathrm{b}} T_{\mathrm{L}}}{\Omega \hbar \rho v_{\mathrm{s}}^2} \delta\left(E_{\mathrm{i}}^{\nu}(\vec{k}) - E_{\mathrm{f}}^{\nu'}(\vec{k'})\right) \delta_{\nu, \nu'} \,, \qquad (2.3.1\text{-}62)$$

with the acoustic deformation potential D_{A}, the mass density of the considered material ρ and the *sample* volume Ω. The Kronecker delta $\delta_{\nu, \nu'}$ implies that an intravalley process is considered. Acoustic phonon scattering is an isotropic scattering process, since the transition rate after (2.3.1-62) is independent of the scattering vector $\vec{\beta}$.

2.3.1.2 Intervalley phonon scattering

Considering the **acoustic** and **optical** phonon branches near the boundary of the first *Brillouin-zone* (K- and X-points in figure 2.13), one can approximate the corresponding phonon energies $\hbar\omega$ by constant values (see black horizontal bars in figure 2.13). In addition, the scattering vector $\vec{\beta}$ is large enough enabling intervalley transitions of electrons. For example, in the case of silicon *f-type* scattering of an electron from the current X-valley to any other X-valley occurs, except for the one lying in opposite direction (see figure 2.14).

Due to the periodicity of the first *Brillouin-zone* in the reciprocal space, *g-type* scattering occurs by short scattering vectors (in X-direction) from adjacent *Brillouin-zones*. The corresponding phonon energies $\hbar\omega$ are also indicated by black horizontal bars in figure 2.13 and the scattering process itself is also schematically shown in figure 2.14. Since only X-valleys are involved, *f-type* and *g-type* scattering in SI are classified as **equivalent intervalley scattering**.

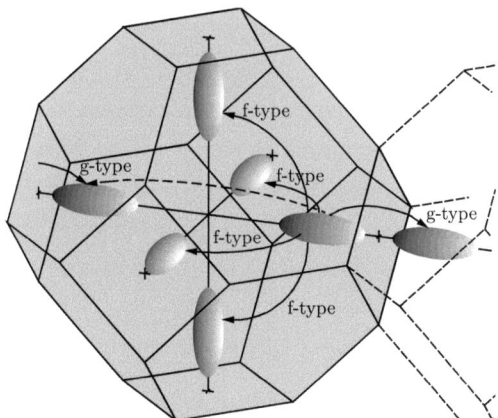

Figure 2.14: Schematic illustration of f- and g-type intervalley scattering in SI.

In the case of polar semiconductors, like INP or GAAS, intervalley scattering processes occur due to scattering vectors $\vec{\beta}$ in the vicinity of the L-points. In contrast to SI, where scattering is taking place between equivalent X-valleys, a transition of electrons between the Γ- and L-valleys is observed for polar semiconductors, instead. Thus, the process is a **non-equivalent intervalley scattering** mechanism.

In literature, the above described intervalley scattering mechanisms are usually labeled as **non-polar optical phonon scattering**, although in some cases acoustic branches of the phonon dispersion relation are involved [31, 51, 69]. The corresponding transition rate reads [34, 37, 46, 69]

$$S_{\mathrm{NOP}}^{\nu \to \nu'}(\vec{k}, \vec{k}') = \frac{\pi Z^{\nu'} D_{ij}^2}{\Omega \rho \omega_{ij}} \left(n_{\mathrm{ph}}(\hbar \omega_{ij}) + \frac{1}{2} \pm \frac{1}{2} \right) \delta \left(E_{\mathrm{f}}^{\nu'}(\vec{k}') - E_{\mathrm{i}}^{\nu}(\vec{k}) \pm \hbar \omega_{ij} \right) ,$$

$$(2.3.1\text{-}63)$$

with the number of final valleys $Z^{\nu'}$, the coupling constant D_{ij} (in literature often labeled as $D_t K$), the phonon frequency ω_{ij}, the mass density ρ of the material under investigation and the *sample* volume Ω. The phonon occupation number $n_{\mathrm{ph}}(\hbar \omega_{ij})$ is approximated by a *Bose-Einstein* distribution [34, 37, 46, 69]:

$$n_{\mathrm{ph}}(\hbar \omega_{ij}) = \frac{1}{\exp \left(\frac{\hbar \omega_{ij}}{k_{\mathrm{b}} T_{\mathrm{L}}} \right) - 1} . \qquad (2.3.1\text{-}64)$$

The upper sign in (2.3.1-63) corresponds to an emission process (the electron emits a phonon of $\hbar \omega_{ij}$ and transfers to a lower energetic state), whereas the lower sign stands for an absorption process, where the final electron energy is $\hbar \omega_{ij}$ higher than the initial one. Non-polar optical phonon scattering is an isotropic scattering process, since (2.3.1-63) does not depend on the scattering vector $\vec{\beta}$.

2.3.1.3 Intravalley optical phonon scattering

According to the phonon dispersion relation shown in figure 2.13, also intravalley scattering by optical phonons (for scattering vectors $\vec{\beta}$ near the center of the *Brillouin-zone* (Γ-point)) is possible. However, such an intravalley scattering mechanism is only allowed for L-valleys and is prohibited for Γ- or X-valleys [69]. Therefore, this scattering mechanism plays an important role for III-V semiconductors, where L-valleys have an influence on the electron transport. The transition rate is modeled in the same way as for non-polar optical phonon scattering (2.3.1-63)

$$S_{\mathrm{OP}}^{\nu \to \nu'}(\vec{k}, \vec{k}') = \frac{\pi D_{\mathrm{O}}^2}{\Omega \rho \omega_{\mathrm{O}}} \left(n_{\mathrm{ph}}(\hbar \omega_{\mathrm{O}}) + \frac{1}{2} \pm \frac{1}{2} \right) \delta \left(E_{\mathrm{f}}^{\nu'}(\vec{k}') - E_{\mathrm{i}}^{\nu}(\vec{k}) \pm \hbar \omega_{\mathrm{O}} \right) \delta_{\nu,\nu'} \ ,$$

$$(2.3.1\text{-}65)$$

with the deformation potential D_{O}, the phonon energy $\hbar \omega_{\mathrm{O}}$ and the phonon occupation number $n_{\mathrm{ph}}(\hbar \omega_{\mathrm{O}})$ after (2.3.1-64). Similar to non-polar optical phonon scattering, also intravalley optical phonon scattering is an isotropic and inelastic scattering process.

2.3.1.4 Polar optical phonon scattering (POP)

In the case of polar semiconductors, which are mainly III-V compound semiconductors, polar optical phonon scattering is the dominating scattering process near room temperature [46, 51, 69]. The transition rate reads [69]

$$S_{\mathrm{POP}}^{\nu \to \nu'}(\vec{k}, \vec{k}') = \frac{C_{\mathrm{POP}}}{|\vec{k} - \vec{k}'|^2} \left(n_{\mathrm{ph}}(\hbar \omega_{\mathrm{p}}) + \frac{1}{2} \pm \frac{1}{2} \right) \delta \left(E_{\mathrm{f}}^{\nu'}(\vec{k}') - E_{\mathrm{i}}^{\nu}(\vec{k}) \pm \hbar \omega_{\mathrm{p}} \right) \delta_{\nu,\nu'} \ ,$$

$$(2.3.1\text{-}66)$$

$$C_{\mathrm{POP}} = \frac{\pi q^2 \omega_{\mathrm{p}}}{\Omega \epsilon_0} \left(\frac{1}{\epsilon_{\mathrm{hf}}} - \frac{1}{\epsilon_{\mathrm{st}}} \right) \ , \qquad (2.3.1\text{-}67)$$

with the elementary charge q, the polar optical phonon energy $\hbar \omega_{\mathrm{p}}$, the phonon occupation number $n_{\mathrm{ph}}(\hbar \omega_{\mathrm{p}})$ after (2.3.1-64), the vacuum permittivity ϵ_0 and the relative permittivity at low/high frequencies of the material under investigation $\epsilon_{\mathrm{st}}/\epsilon_{\mathrm{hf}}$, respectively. In contrast to the previously considered phonon scattering processes, polar optical phonon scattering is an anisotropic scattering process, due to the term $|\vec{k} - \vec{k'}|^2$ in the denominator of (2.3.1-66). Thus, the transition rate has a maximum, if $\vec{k'}$ (both for emission and absorption) points into the same direction as $\vec{k}$. Note, the term $|\vec{k} - \vec{k'}|^2$ cannot become zero, due to the inelastic nature of this scattering process (2.3.1-66) and the strictly increasing dispersion relation as function of $|\vec{k}|$. Additionally, POP is an intravalley scattering process implying

$$\vec{k'}(E_{\mathrm{f}}^{\nu'}) = \vec{k}(E_{\mathrm{i}}^{\nu} \mp \hbar\omega_{\mathrm{p}}) \neq \vec{k}(E_{\mathrm{i}}^{\nu}) \, , \qquad (2.3.1\text{-}68)$$

with the inverse dispersion relation $\vec{k}(E)$. The latter fact is schematically sketched in figure 2.15, with the scattering vector $\vec{\beta} = \vec{k'} - \vec{k}$. $\vec{\beta}_{\mathrm{EM}}$ and $\vec{\beta}_{\mathrm{AB}}$ are the scattering vectors associated with the emission and absorption of the phonon energy $\hbar\omega_{\mathrm{p}}$, respectively. Their minimum absolute values are labeled by $\vec{\beta}_{\mathrm{EM,min}}$ and $\vec{\beta}_{\mathrm{AB,min}}$, where the transition rate (2.3.1-66) peaks. $\vec{k'}_{\mathrm{EM}}$ and $\vec{k'}_{\mathrm{AB}}$, measured from the valley minimum E_0^{ν}, are the reciprocal vectors after the scattering event and are thus pointing to the equi-energy circles $E_{\mathrm{f,EM}}^{\nu} = E_{\mathrm{i}}^{\nu} - \hbar\omega_{\mathrm{p}}$ and $E_{\mathrm{f,AB}}^{\nu} = E_{\mathrm{i}}^{\nu} + \hbar\omega_{\mathrm{p}}$, where E_{i}^{ν} is the initial energy.

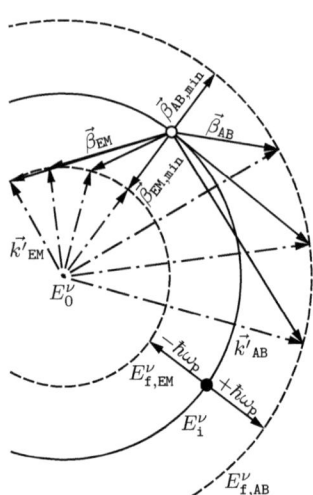

Figure 2.15: Schematic illustration of the scattering vectors $\vec{\beta}$ for POP-scattering.

2.3.2 Impurity scattering (IMP)

This scattering mechanism describes scattering of carriers by charged impurities in doped material. Impurity scattering is modeled as an intravalley scattering process and has the same form for polar and non-polar semiconductors. The transition rate after Brooks and Herring [50] is given by

$$S_{\text{IMP}}^{\nu \to \nu'}(\vec{k}, \vec{k}') = \frac{C_{\text{IMP}}}{\left(|\vec{k} - \vec{k}'|^2 + q_{\text{d}}^2\right)^2} \delta(E_{\text{f}}^{\nu'}(\vec{k}') - E_{\text{i}}^{\nu}(\vec{k})) \delta_{\nu,\nu'} , \quad (2.3.2\text{-}69)$$

$$C_{\text{IMP}} = \frac{2\pi N_{\text{I}}}{\Omega \hbar} \left(\frac{q^2}{\varepsilon_0 \varepsilon_{\text{r}}}\right)^2 , \text{ and} \quad (2.3.2\text{-}70)$$

$$q_{\text{d}} = \sqrt{\frac{q^2 N_{\text{I}}}{\varepsilon_0 \varepsilon_{\text{r}} k_{\text{b}} T}}. \quad (2.3.2\text{-}71)$$

Here, N_{I} is the total doping concentration ($N_{\text{A}} + |N_{\text{D}}|$), q is the elementary charge, ε_0 is the permittivity and ε_{r} is the relative permittivity of the semiconductor material under investigation. All remaining symbols have their usual meaning. Due to the difference between the initial and the final k-vector ($|\vec{k} - \vec{k}'|$ in (2.3.2-69)), the *strength* of the scattering mechanisms is modulated, similar to polar optical phonon scattering (2.3.1-66) as discussed in the previous section. Therefore, depending on the locations of the final k-vector w.r.t. the initial one, different transition rates are obtained although the initial and final energies are identical ($\delta(E_{\text{f}}^{\nu'}(\vec{k}') - E_{\text{i}}^{\nu}(\vec{k}))$ in (2.3.2-69)). Thus, impurity scattering is an anisotropic but elastic scattering process. Note, $|\vec{k} - \vec{k}'|$ can become zero, as IMP is an elastic scattering process. However, due to the non-zero *Debye-length* q_{d} the denominator of (2.3.2-69) cannot become zero.

2.3.3 Alloy scattering (ALY)

Alloy scattering is used to describe scattering of a carrier due to material composition, here GE in SI (SIGE heterostructures), and is assumed to be isotropic and elastic. In addition, alloy scattering contributes to both intra- and intervalley scattering. The transition rate reads [10, 31, 37]

$$S_{\text{ALY}}^{\nu \rightarrow \nu'}(\vec{k}, \vec{k}') = \frac{\pi a_\perp a_\parallel^2 U_{\text{ALY}}^2}{4\Omega\hbar} \left(1 - x_{\text{Ge}}\right) x_{\text{Ge}} \delta(E_f^{\nu'}(\vec{k}') - E_i^{\nu}(\vec{k})) , \quad (2.3.3\text{-}72)$$

with the alloy scattering potential U_{ALY}, the material composition content x_{Ge} and the parallel/perpendicular lattice constants $a_\parallel / a_\perp$, respectively. Latter are defined w.r.t. the interface layer of the SI/ SIGE transition (see figure 2.16). The corresponding lattice constants can be calculated by [10]

$$a_\parallel = a_{0,\text{Si}} , \quad (2.3.3\text{-}73)$$

$$a_\perp = a_{0,\text{SiGe}}(x_{\text{Ge}}) \left[1 - 2\frac{C_{12}(x_{\text{Ge}})}{C_{11}(x_{\text{Ge}})} \frac{a_\parallel - a_{0,\text{SiGe}}(x_{\text{Ge}})}{a_{0,\text{SiGe}}(x_{\text{Ge}})}\right] , \quad (2.3.3\text{-}74)$$

$$a_{0,\text{SiGe}}(x_{\text{Ge}}) = a_{0,\text{Si}} + m_{\text{SiGe}} x_{\text{Ge}}(1 - x_{\text{Ge}}) + (a_{0,\text{Ge}} - a_{0,\text{Si}})x_{\text{Ge}}^2 , \quad (2.3.3\text{-}75)$$

$$C_{ij}(x_{\text{Ge}}) = (1 - x_{\text{Ge}})C_{ij,\text{Si}} + x_{\text{Ge}}C_{ij,\text{Ge}} . \quad (2.3.3\text{-}76)$$

The parameters used for the equations (2.3.3-73)-(2.3.3-76) are summarized

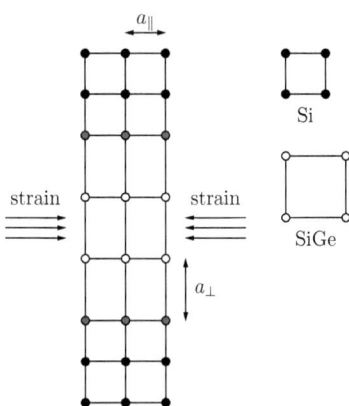

Figure 2.16: Schematic illustration of the lattice constants $a_\parallel / a_\perp$ and of the biaxial-compressive strain in SI/ SIGE/ SI heterostructures.

in table 2.3.

$a_{0,\text{Si}}$	$a_{0,\text{Ge}}$	m_{SiGe}	
5.43 Å	5.66 Å	0.200 326 Å	
$C_{11,\text{Si}}$	$C_{12,\text{Si}}$	$C_{11,\text{Ge}}$	$C_{12,\text{Ge}}$
165.77 GPa	63.93 GPa	128.53 GPa	48.26 GPa

Table 2.3: Parameters used for the calculation of the lattice constants $a_\parallel / a_\perp$ in Si/SiGe/Si heterostructures [10].

2.3.4 Piezoelectric scattering (PZ)

Piezoelectric scattering is an anisotropic intravalley scattering process occurring in polar semiconductors, which is dominant at low temperatures. Therefore, it is often neglected for investigations close to room temperature. However, for the sake of completeness, this scattering process is also considered here. The transition rate is modeled by [51]

$$S_{\text{PZ}}^{\nu \to \nu'}(\vec{k}, \vec{k}') = \frac{C_{\text{PZ}}}{2} \frac{|\vec{k} - \vec{k}'|^2}{\left(|\vec{k} - \vec{k}'|^2 + q_{\text{d}}^2 \right)^2} \delta(E_{\text{f}}^{\nu'}(\vec{k}') - E_{\text{i}}^{\nu}(\vec{k}) \pm \hbar \omega_{\text{PZ}}) \delta_{\nu, \nu'},$$

$$\text{(2.3.4-77)}$$

$$C_{\text{PZ}} = \frac{2\pi}{\hbar} \frac{k_{\text{b}} T}{\rho v_{\text{s,av}}^2 \Omega} \left(\frac{q \varepsilon_{\text{PZ}}}{\varepsilon_0 \varepsilon_{\text{r}}} \right)^2, \qquad \text{(2.3.4-78)}$$

with the mass density ρ, the piezoelectric coupling constant ε_{PZ}, an average sound velocity $v_{\text{s,av}}$ and the *Debye-length* q_{d} after (2.3.2-71). The sound velocity $v_{\text{s,av}}$ is used to ease the complicated directional dependence. Similar to acoustic deformation potential scattering, the phonon energy near room temperature can be neglected. Thus, the emission and absorption processes are overlapping and the transition rate is doubled. Therefore, one obtains

$$S_{\text{PZ}}^{\nu \to \nu'}(\vec{k}, \vec{k}') = C_{\text{PZ}} \frac{|\vec{k} - \vec{k}'|^2}{\left(|\vec{k} - \vec{k}'|^2 + q_{\text{d}}^2 \right)^2} \delta(E_{\text{f}}^{\nu'}(\vec{k}') - E_{\text{i}}^{\nu}(\vec{k})) \delta_{\nu, \nu'}, \qquad \text{(2.3.4-79)}$$

which is an elastic scattering process, due to the assumption of $\hbar\omega_{PZ} \approx 0\,\mathrm{eV}$. Note, in contrast to the previously considered anisotropic scattering mechanisms (polar optical phonon and impurity scattering), the transition rate after (2.3.4-79) has a maximum for $|\vec{k} - \vec{k'}| = q_d$ and is zero for $\vec{k} = \vec{k'}$.

2.3.5 Isotropic approximations of anisotropic scattering events

As mentioned before, anisotropic scattering mechanisms depend on the absolute value of the scattering vector $|\vec{\beta}| = |\vec{k} - \vec{k'}|$. Due to $|\vec{\beta}|$, the preparation of anisotropic scattering mechanisms for the implementation in a simulator becomes a cumbersome task and even not manageable in the case of anisotropic band structures. However, isotropic approximations can be used instead. The main idea behind the isotropic approximation is to find an isotropic scattering rate, which has the same *momentum relaxation time* [46]

$$
\tau_m = \left(\frac{\Omega}{8\pi^3} \int S_X^{\nu \to \nu'}(\vec{k}, \vec{k'}) \frac{\Delta k}{|\vec{k}|} d\vec{k'} \right)^{-1}, \qquad (2.3.5\text{-}80)
$$

as the originally considered anisotropic scattering mechanism. Using the momentum relaxation time is motivated by obtaining an identical carrier mobility for identical distribution functions in the k-space [41]. To interpret (2.3.5-80), the scattering rate

$$
\tau = \left(\frac{\Omega}{8\pi^3} \int S_X^{\nu \to \nu'}(\vec{k}, \vec{k'}) d\vec{k'} \right)^{-1}, \qquad (2.3.5\text{-}81)
$$

is considered first. The scattering rate τ measures the average time needed to scatter a carrier ensemble with $\vec{k}$-vectors parallel to the direction of transport to all possible $\vec{k'}$-directions (including backscattering). However, anisotropic mechanisms tend to scatter the carrier ensemble in a distinct direction (for the maximum transition rate). Consequently, after the time τ (2.3.5-81) has elapsed the carrier ensemble is scattered to all directions, but most of them in the direction of the maximum transition rate. In order to find the average time after which the carrier ensemble is completely randomized (momentum relaxation time), the transition rate is weighted with the fractional change of k-vector in transport direction ($\frac{\Delta k}{|\vec{k}|}$ in (2.3.5-80)). Δk represents the signed magnitude of the change of the k-vector component aligned with the transport direction (see figure 2.17). For the sake of simplicity (and without any loss of generality), it is assumed that the initial vector $\vec{k}$ only has a z-component and $\vec{k'}$ is expressed in spherical coordinates after (2.2.0-52). For the sake of readability and since all anisotropic scattering mechanisms listed before are intravalley scattering mechanisms, the labels for the valleys (ν, ν') are omitted. In addition, an isotropic dispersion relation is assumed. Thus, Δk

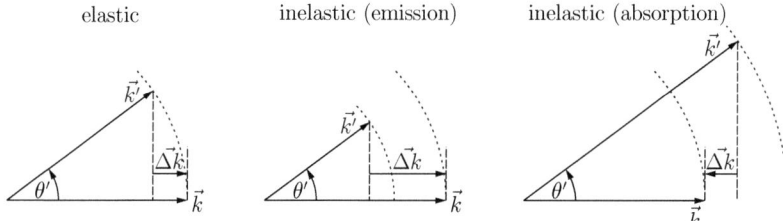

Figure 2.17: Schematic illustration of the vectors $\vec{k}$ and $\vec{k'}$ for momentum relaxation time calculation. Three different cases are shown: elastic scattering process (left), inelastic scattering (emission, middle) and an inelastic for the case of absorbtion (right). The dotted arcs are indicating the equi-energy surfaces before and after a scattering event.

and the weighting term $\frac{\Delta k}{|\vec{k}|}$, with $\Delta k = |\vec{\Delta k}|$, can be expressed by

$$\vec{\Delta k} = \Delta k \frac{\vec{k}}{|\vec{k}|} \, , \tag{2.3.5-82}$$

$$\Delta k = |\vec{k}| - |\vec{k'}| \cos(\theta') \text{ and} \tag{2.3.5-83}$$

$$\frac{\Delta k}{|\vec{k}|} = \left(1 - \frac{|\vec{k'}|}{|\vec{k}|} \cos(\theta')\right) \, . \tag{2.3.5-84}$$

The weighting term after (2.3.5-84) is schematically illustrated in figure 2.18 for the three cases considered in figure 2.17. With the weighting term (2.3.5-84) and the above mentioned assumptions, the inverse momentum relaxation time

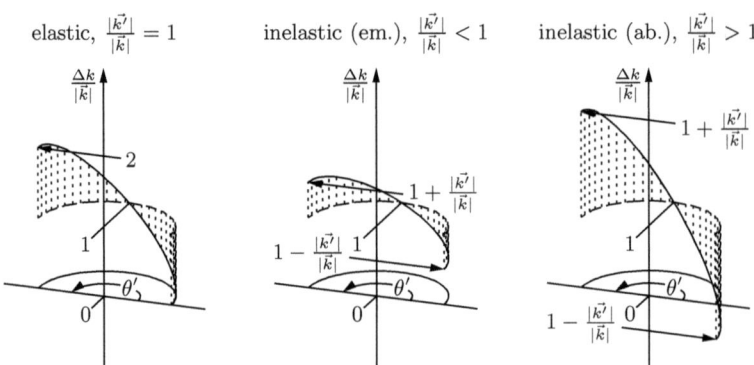

Figure 2.18: Schematic illustration of the weighting term $\frac{\Delta k}{|\vec{k}|}$ after (2.3.5-84) (solid curve). Note, since $\vec{k'}$ is expressed in spherical coordinates after (2.2.0-52), $0 \leq \theta' \leq \pi$ holds and is identical to θ' in figure 2.17. In addition, the dotted area for $0 \leq \theta' \leq \frac{\pi}{2}$ is identical to the one for $\frac{\pi}{2} \leq \theta' \leq \pi$ for all of the three cases.

(see equation (2.3.5-80)) becomes

$$
\begin{aligned}
\frac{1}{\tau_{\mathrm{m}}} &= \frac{\Omega}{8\pi^3} \int_0^{2\pi} \int_0^{\pi} \int_0^{\epsilon_{\max}} S_X(\vec{k}, \vec{k'}) \left(1 - \frac{|\vec{k'}|}{|\vec{k}|}\cos(\theta')\right) \\
&\qquad \underbrace{\frac{1}{2}\left(\frac{\sqrt{2m_0 m^*}}{\hbar}\right)^3 \sqrt{\epsilon'\,(1+\alpha\epsilon')}(1+2\alpha\epsilon')\sin(\theta')d\epsilon'd\theta'd\varphi'}_{d\vec{k'}} , \\
&= \frac{\Omega}{8\pi^3} \int_0^{2\pi} \int_0^{\pi} \int_0^{\epsilon_{\max}} S_X(\vec{k}, \vec{k'}) \left(1 - \frac{|\vec{k'}|}{|\vec{k}|}\cos(\theta')\right) \\
&\qquad \mathrm{dos}(\epsilon')\sin(\theta')d\epsilon'd\theta'd\varphi' , \qquad\qquad (2.3.5\text{-}85)
\end{aligned}
$$

where (2.2.0-55) has been employed for $d\vec{k'}$ and the density-of-states after (2.2.0-56). As shown in figure 2.18 (indicated by the dotted areas), the average value of the weighting term (2.3.5-84) equals 1. The consequence is that the momentum relaxation time is identical to the transition time τ after (2.3.5-81) for isotropic scattering mechanisms. This is consistent, as isotropic scattering mechanisms scatter in all directions equally and thus the carrier ensemble is already completely randomized after τ has elapsed. Nevertheless, carriers are scattered in a distinct direction for anisotropic scattering mechanisms, since the transition rate $S_X(\vec{k}, \vec{k}')$ is a function of $|\vec{k} - \vec{k}'|$ evaluating to

$$|\vec{k} - \vec{k}'| = \sqrt{|\vec{k}'|^2 - 2|\vec{k}'||\vec{k}|\cos(\theta') + |\vec{k}|^2} \ . \qquad (2.3.5\text{-}86)$$

Note, for (2.3.5-86) it is assumed that $\vec{k}$ only has a z-component. Thus, $S_X(\vec{k}, \vec{k}')$ is θ'-dependent for anisotropic scattering. Via (2.3.5-84), the contribution of $S_X(\vec{k}, \vec{k}')$ to τ_m for low θ' ($\theta' < \frac{\pi}{2}$: $\vec{k}'$ has a remnant of the transport direction) is weighted by $\frac{\Delta k}{|\vec{k}|} < 1$ in contrast to $\theta' > \frac{\pi}{2}$ ($\vec{k}'$ points in opposite direction of the transport), where $\frac{\Delta k}{|\vec{k}|} > 1$ holds (see figure 2.18). The signs and magnitudes of the θ'-dependent weights depend on the considered scattering mechanism via the ratio $\frac{|\vec{k}'|}{|\vec{k}|}$.

Negative weights only occur for inelastic scattering by absorption. In this case and for low θ', carriers are accelerated in transport direction and thus have a higher kinetic energy at $\vec{k}'$ than before. From the mathematical point of view, a negative momentum relaxation can be obtained due to the negative weights for a strong anisotropic scattering mechanism by absorption. However, this is completely nonphysical, since such a process would violate the first law of thermodynamics (the device under investigation would generate energy).

If the projection of $\vec{k}'$ and $\vec{k}$ in transport direction are identical, the weighting term becomes zero. This can occur for both elastic and inelastic (absorption) scattering events. In this case, the carrier ensemble moves on in transport direction as before.

The weighting term only gives positive weights if the projection of $\vec{k}'$ on $\vec{k}$ has a smaller magnitude. In this case, the movement of the carrier ensemble becomes less aligned with the undisturbed transport direction. For $\theta' = \frac{\pi}{2}$, all possible $\vec{k}'$ are lying within a layer perpendicular to the transport direction and the weighting term becomes one. Increasing θ' ($\theta' > \frac{\pi}{2}$) leads to back-scattering and weights larger than one are obtained.

In order to find an isotropic approximation for an anisotropic scattering mechanism, the transition rate

$$S_{\text{iso., ela.}}(\vec{k}, \vec{k}') = A\delta\left(\epsilon' - \epsilon\right) \qquad (2.3.5\text{-}87)$$

for an elastic and

$$S_{\text{iso., inela.}}(\vec{k}, \vec{k}') = A\delta\left(\epsilon' - \epsilon \pm \Delta\epsilon\right) \qquad (2.3.5\text{-}88)$$

for an inelastic scattering process is assumed, with the initial/final energy ϵ/ϵ' and the energy exchange $\Delta\epsilon$. The upper/lower sign in (2.3.5-88) represents an emission/absorption process, respectively. Thus, with (2.3.5-85) the respective inverse momentum relaxation times are

$$\frac{1}{\tau_{\text{m,iso., ela.,inela.}}} = \frac{A\Omega}{2\pi^2}\text{dos}(\epsilon') , \qquad (2.3.5\text{-}89)$$

with ϵ' as the final energy (also for an emission/absorption process) and the parameter A in (2.3.5-87)-(2.3.5-89) (*strength* of the isotropic approximated transition rate). A is determined by comparing (2.3.5-89) with the momentum relaxation times for the anisotropic transition rates. In the following sections, isotropic approximations are given for all anisotropic scattering mechanisms listed previously.

2.3.5.1 Polar optical phonon scattering

Employing the transition rate for polar optical phonon (POP) scattering after (2.3.1-66) in section 2.3.1.4 and the inverse momentum relaxation time after (2.3.5-85) one obtains:

$$\frac{1}{\tau_\text{m}} = \frac{\Omega}{8\pi^3}\int_0^{2\pi}\int_0^\pi\int_0^{\epsilon_\text{max}} \frac{C_\text{POP}}{|\vec{k}'|^2 - 2|\vec{k}'||\vec{k}|\cos(\theta') + |\vec{k}|^2}$$

$$\left(n_\text{ph}(\hbar\omega_\text{p}) + \frac{1}{2} \pm \frac{1}{2}\right)\delta\left(\epsilon' - \epsilon \pm \hbar\omega_\text{p}\right)\left(1 - \frac{|\vec{k}'|}{|\vec{k}|}\cos(\theta')\right)$$

$$\text{dos}(\epsilon')\sin(\theta')d\epsilon'd\theta'd\varphi' , \qquad (2.3.5\text{-}90)$$

where the relation (2.3.5-86) has been used and the dispersion relation (2.2.0-43) is assumed. The integration w.r.t. φ' results in a multiplication of 2π and the integration w.r.t. ϵ' leads to $\epsilon' = \epsilon \mp \hbar\omega_\text{p}$ due to the Dirac-function. Equation

(2.3.5-90) evaluates to

$$
\frac{1}{\tau_\mathrm{m}} = \frac{\Omega}{4\pi^2} \frac{C_\mathrm{POP}}{|\vec{k}|^2} \left(n_\mathrm{ph}(\hbar\omega_\mathrm{p}) + \frac{1}{2} \pm \frac{1}{2} \right) \mathrm{dos}(\epsilon')
$$

$$
\int_0^\pi \frac{1 - \tilde{a}\cos(\theta')}{1 - 2\tilde{a}\cos(\theta') + \tilde{a}^2} \sin(\theta')d\theta' , \qquad (2.3.5\text{-}91)
$$

$$
\tilde{a} = \frac{|\vec{k'}|}{|\vec{k}|} = \sqrt{\frac{\epsilon'(1 + \alpha\epsilon')}{\epsilon(1 + \alpha\epsilon)}} \geq 0 . \qquad (2.3.5\text{-}92)
$$

By using the substitution $z = -\cos(\theta')$ and adjusting the integration limits, one obtains

$$
\frac{1}{\tau_\mathrm{m}} = \frac{\Omega}{4\pi^2} \frac{C_\mathrm{POP}}{|\vec{k}|^2} \left(n_\mathrm{ph}(\hbar\omega_\mathrm{p}) + \frac{1}{2} \pm \frac{1}{2} \right) \mathrm{dos}(\epsilon')
$$

$$
\left[1 + \frac{(1 - \tilde{a})(1 + \tilde{a})}{2\tilde{a}} \log\left(\left| \frac{1 + \tilde{a}}{1 - \tilde{a}} \right| \right) \right] . \qquad (2.3.5\text{-}93)
$$

Comparing the inverse momentum relaxation time (2.3.5-93) with (2.3.5-89) results in

$$
A = \frac{C_\mathrm{POP}}{2|\vec{k}|^2} \left(n_\mathrm{ph}(\hbar\omega_\mathrm{p}) + \frac{1}{2} \pm \frac{1}{2} \right) \left[1 + \frac{(1 - \tilde{a})(1 + \tilde{a})}{2\tilde{a}} \log\left(\left| \frac{1 + \tilde{a}}{1 - \tilde{a}} \right| \right) \right] ,
$$

$$
(2.3.5\text{-}94)
$$

and leads thus to the isotropic approximated transition rate (with (2.3.5-88))

$$
S_\mathrm{POP}^{\mathrm{ISO},\nu \to \nu'}(\vec{k}, \vec{k'}) = \frac{C_\mathrm{POP}}{2|\vec{k}|^2} \left[1 + \frac{(1 - \tilde{a})(1 + \tilde{a})}{2\tilde{a}} \log\left(\left| \frac{1 + \tilde{a}}{1 - \tilde{a}} \right| \right) \right]
$$

$$
\left(n_\mathrm{ph}(\hbar\omega_\mathrm{p}) + \frac{1}{2} \pm \frac{1}{2} \right) \delta\left(E_\mathrm{f}^{\nu'}(\vec{k'}) - E_\mathrm{i}^{\nu}(\vec{k}) \pm \hbar\omega_\mathrm{p} \right) \delta_{\nu,\nu'} ,
$$

$$
(2.3.5\text{-}95)
$$

with

$$
\tilde{a} = \sqrt{\frac{E_\mathrm{f}^{\nu'}\left(1 + \alpha_{\nu'} E_\mathrm{f}^{\nu'}\right)}{E_\mathrm{i}^{\nu}\left(1 + \alpha_\nu E_\mathrm{i}^{\nu}\right)}} . \qquad (2.3.5\text{-}96)
$$

2.3.5.2 Impurity scattering

The inverse momentum relaxation time for impurity scattering (with the transition rate after (2.3.2-69) from section 2.3.2) evaluates to

$$\frac{1}{\tau_m} = \frac{\Omega}{4\pi^2} \frac{C_{\text{IMP}}}{4|\vec{k}|^4} \text{dos}(\epsilon) \int_0^\pi \frac{1 - \cos(\theta')}{\left(1 + \frac{1}{2}\left(\frac{q_d}{|\vec{k}|}\right)^2 - \cos(\theta')\right)^2} \sin(\theta')d\theta' \,,$$

$$(2.3.5\text{-}97)$$

where the integrations w.r.t. ϵ' and φ' originating from (2.3.5-85) have already been performed. Note, due to the integration over the Dirac-function $\delta(\epsilon' - \epsilon)$ associated to the transition rate (2.3.2-69), $\epsilon' = \epsilon$ is obtained. Using the substitution $z = -\cos(\theta')$ (for θ'), (2.3.5-97) becomes

$$\frac{1}{\tau_m} = \frac{\Omega}{4\pi^2} \frac{C_{\text{IMP}}}{4|\vec{k}|^4} \text{dos}(\epsilon) \left[\log\left(\frac{\tilde{b} + 2}{\tilde{b}}\right) - \frac{2}{\tilde{b} + 2}\right] \,, \text{ with} \quad (2.3.5\text{-}98)$$

$$\tilde{b} = \frac{1}{2}\left(\frac{q_d}{|\vec{k}|}\right)^2 \,. \qquad (2.3.5\text{-}99)$$

Thus, the coefficient A in (2.3.5-89) becomes

$$A = \frac{C_{\text{IMP}}}{8|\vec{k}|^4} \left[\log\left(\frac{\tilde{b} + 2}{\tilde{b}}\right) - \frac{2}{\tilde{b} + 2}\right] \,. \qquad (2.3.5\text{-}100)$$

With (2.3.5-87) the iostropic approximation of the transition rate reads

$$S_{\text{IMP}}^{\text{ISO},\nu \to \nu'}(\vec{k}, \vec{k}') = \frac{C_{\text{IMP}}}{8|\vec{k}|^4} \left[\log\left(\frac{\tilde{b} + 2}{\tilde{b}}\right) - \frac{2}{\tilde{b} + 2}\right] \delta(E_f^{\nu'}(\vec{k}') - E_i^{\nu}(\vec{k}))\delta_{\nu,\nu'} \,.$$

$$(2.3.5\text{-}101)$$

2.3.5.3 Piezoelectric scattering

Performing the integration w.r.t. ϵ', φ' and using the elastic nature of PZ-scattering (see section 2.3.4, equation (2.3.4-79)) the inverse momentum relaxation time after (2.3.5-85) reads

$$\frac{1}{\tau_{\mathrm{m}}} = \frac{\Omega}{4\pi^2} \frac{C_{\mathrm{PZ}}}{2|\vec{k}|^2} \mathrm{dos}(\epsilon) \int_0^\pi \frac{(1 - \cos(\theta'))^2}{\left(1 + \frac{1}{2}\left(\frac{q_{\mathrm{d}}}{|\vec{k}|}\right)^2 - \cos(\theta')\right)^2} \sin(\theta') d\theta' \ .$$

$$(2.3.5\text{-}102)$$

Substituting $z = -\cos(\theta')$ leads to

$$\frac{1}{\tau_{\mathrm{m}}} = \frac{\Omega}{4\pi^2} \frac{C_{\mathrm{PZ}}}{|\vec{k}|^2} \mathrm{dos}(\epsilon) \left[\tilde{b} \log\left(\frac{\tilde{b}}{\tilde{b}+2}\right) + 2\frac{\tilde{b}+1}{\tilde{b}+2}\right] \quad (2.3.5\text{-}103)$$

with

$$\tilde{b} = \frac{1}{2}\left(\frac{q_{\mathrm{d}}}{|\vec{k}|}\right)^2 \ . \quad (2.3.5\text{-}104)$$

Comparing the result (2.3.5-103) with (2.3.5-89) implies

$$A = \frac{C_{\mathrm{PZ}}}{2|\vec{k}|^2} \left[\tilde{b} \log\left(\frac{\tilde{b}}{\tilde{b}+2}\right) + 2\frac{\tilde{b}+1}{\tilde{b}+2}\right] \ . \quad (2.3.5\text{-}105)$$

With (2.3.5-87) the isotropic approximation of PZ-scattering results in

$$S_{\mathrm{PZ}}^{\mathrm{ISO},\nu\to\nu'}(\vec{k}, \vec{k'}) = \frac{C_{\mathrm{PZ}}}{2|\vec{k}|^2} \left[\tilde{b} \log\left(\frac{\tilde{b}}{\tilde{b}+2}\right) + 2\frac{\tilde{b}+1}{\tilde{b}+2}\right] \delta(E_{\mathrm{f}}^{\nu'}(\vec{k'}) - E_{\mathrm{i}}^{\nu}(\vec{k}))\delta_{\nu,\nu'} \ .$$

$$(2.3.5\text{-}106)$$

2.4 The Boltzmann transport equation (BTE)

The Boltzmann transport equation (BTE) is a seven dimensional integro-differential equation. It is defined over three spatial dimensions in real space $\vec{r}$, three dimensions in the reciprocal space $\vec{k}$ and in time t. For electrons, the BTE reads [19, 31, 34, 37, 46, 69]

$$\underbrace{\frac{\partial f^\nu(\vec{r}, \vec{k}^\nu, t)}{\partial t}}_{\text{transient term}} + \underbrace{\vec{v}_g^\nu(\vec{r}, \vec{k}^\nu) \text{grad}_{\vec{r}} \left(f^\nu(\vec{r}, \vec{k}^\nu, t) \right)}_{\text{diffusion term}}$$

$$\underbrace{- \frac{q}{\hbar} \vec{E}_{\text{eff.}}^\nu(\vec{r}, \vec{k}^\nu, t) \text{grad}_{\vec{k}^\nu} \left(f^\nu(\vec{r}, \vec{k}^\nu, t) \right)}_{\text{drift term}} = \underbrace{C(\vec{r}, f^\nu(\vec{r}, \vec{k}^\nu, t), f^{\nu'}(\vec{r}, \vec{k}^{\nu'}, t))}_{\text{collision term}}$$

$$(2.4.0\text{-}107)$$

where f^ν is the carrier distribution function of the valley ν, that describes the occupation probability of available states (in the reciprocal space) for a given $\vec{r}$ and $\vec{k}$ at the time t. v_g^ν is the group velocity of electrons in the valley ν and $\vec{E}_{\text{eff.}}^\nu$ is the driving force for electrons in the valley ν:

$$\vec{E}_{\text{eff.}}^\nu(\vec{r}, \vec{k}^\nu, t) = \text{grad}_{\vec{r}} \left(-\psi(\vec{r}, t) + \frac{E_{C,0}^\nu(\vec{r})}{q} + \frac{\varepsilon^\nu(\vec{r}, \vec{k}^\nu)}{q} \right) \ , \quad (2.4.0\text{-}108)$$

which usually varies with $\vec{r}$ and t. ψ in (2.4.0-108) is the electrostatic potential (obtained by the *Poisson* equation), $E_{C,0}^\nu$ is the bias independent (but material dependent) part of the conduction-band edge and ε^ν is the kinetic energy given by the dispersion relation for the valley ν. C in equation (2.4.0-107) is the collision term that describes the interaction between the currently observed state at $\vec{r}$, $\vec{k}$ and t, occupied with f^ν, and all other available states in the reciprocal space (even valleys ν' different than ν) for the currently observed $\vec{r}$ and t. All other symbols have their usual meaning.
The group velocity v_g^ν in (2.4.0-107) is defined as follows

$$\vec{v}_g^\nu(\vec{r}, \vec{k}^\nu) = \frac{1}{\hbar} \text{grad}_{\vec{k}^\nu} \left(\varepsilon^\nu(\vec{r}, \vec{k}^\nu) \right) \ . \quad (2.4.0\text{-}109)$$

For the numerical treatment of the BTE, it is advantageous to use the conservative form of (2.4.0-107). With the definition of the driving force $\vec{E}_{\text{eff.}}^{\nu}$ after (2.4.0-108) and the group velocity $\vec{v}_{\text{g}}^{\nu}$ after (2.4.0-109) the BTE (2.4.0-107) can be rewritten as

$$\frac{\partial f^{\nu}(\vec{r}, \vec{k}^{\nu}, t)}{\partial t} + \text{div}_{\vec{r}}\left(\vec{v}_{\text{g}}^{\nu}(\vec{r}, \vec{k}^{\nu}) f^{\nu}(\vec{r}, \vec{k}^{\nu}, t)\right) - \frac{q}{\hbar}\text{div}_{\vec{k}^{\nu}}\left(\vec{E}_{\text{eff.}}^{\nu}(\vec{r}, \vec{k}^{\nu}, t) f^{\nu}(\vec{r}, \vec{k}^{\nu}, t)\right)$$
$$= C(\vec{r}, f^{\nu}(\vec{r}, \vec{k}^{\nu}, t), f^{\nu'}(\vec{r}, \vec{k}^{\nu'}, t)) . \tag{2.4.0-110}$$

A more detailed derivation for the conservative form of the BTE is presented in appendix A.2. No assumptions have been made regarding the dispersion relation in (2.4.0-110). However, in this work the dispersion relation is restricted to the spherical and non-parabolic formulation after (2.2.0-43). Additionally, it is assumed that a possible anisotropy of the considered valleys is representable by Herring-Vogt transformation (2.2.0-46)[1]. Using the transformed divergence after (2.2.0-49) and the transformed gradient after (2.2.0-50) for the group velocity $\vec{v}_{\text{g}}^{\nu}$ (2.4.0-109) the BTE becomes

$$\frac{\partial f^{\nu}(\vec{r}, \vec{\tilde{k}}^{\nu}, t)}{\partial t} + \text{div}_{\vec{r}}\left(T^{\text{HV},\nu}(\vec{r})\vec{\tilde{v}}_{\text{g}}^{\nu}(\vec{r}, \vec{\tilde{k}}^{\nu}) f^{\nu}(\vec{r}, \vec{\tilde{k}}^{\nu}, t)\right)$$
$$-\frac{q}{\hbar}\text{div}_{\vec{\tilde{k}}^{\nu}}\left(T^{\text{HV},\nu}(\vec{r})\vec{E}_{\text{eff.}}^{\nu}(\vec{r}, \vec{\tilde{k}}^{\nu}, t) f^{\nu}(\vec{r}, \vec{\tilde{k}}^{\nu}, t)\right) = C(\vec{r}, f^{\nu}(\vec{r}, \vec{\tilde{k}}^{\nu}, t), f^{\nu'}(\vec{r}, \vec{\tilde{k}}^{\nu'}, t)) ,$$
$$\tag{2.4.0-111}$$

with

$$\vec{v}_{\text{g}}^{\nu}(\vec{r}, \vec{\tilde{k}}^{\nu}) = T^{\text{HV},\nu}(\vec{r})\vec{\tilde{v}}_{\text{g}}^{\nu}(\vec{r}, \vec{\tilde{k}}^{\nu}) . \tag{2.4.0-112}$$

The BTE after (2.4.0-111) also captures a possible position dependent aniostropy of the dispersion relation due the position dependent transformation matrix $T^{\text{HV},\nu}(\vec{r})$. The formulation (2.4.0-111) will be used in this work, although it is not as generic as the one of (2.4.0-110).

---

[1]By using the Herring-Vogt transformation (2.2.0-46) it is assumed that the principal axes of the ellipsoid (kinetic equi-energy surfaces) are aligned to the axes of the reciprocal space.

2.4.1 Collision term

The collision term C in the BTE models the interaction of carriers with lattice vibrations, impurities or with other carriers [37, 46, 69]. Mathematically, the collision term is described by [46]

$$
C(\vec{r}, f^{\nu}(\vec{r}, \vec{k}^{\nu}, t), f^{\nu'}(\vec{r}, \vec{k}^{\nu'}, t)) =
$$

$$
\sum_{X} \sum_{\vec{k}^{\nu'}, \uparrow} \left(\underbrace{S^{X}(\vec{k}^{\nu'}, \vec{k}^{\nu}) f^{\nu'}(\vec{r}, \vec{k}^{\nu'}, t) \left(1 - f^{\nu}(\vec{r}, \vec{k}^{\nu}, t)\right)}_{\text{in-scattering}} \right.
$$

$$
\left. \underbrace{- S^{X}(\vec{k}^{\nu}, \vec{k}^{\nu'}) f^{\nu}(\vec{r}, \vec{k}^{\nu}, t) \left(1 - f^{\nu'}(\vec{r}, \vec{k}^{\nu'}, t)\right)}_{\text{out-scattering}} \right) , \qquad (2.4.1\text{-}113)
$$

where S^{X} is the transition rate of the currently considered scattering mechanism X (see section 2.3). The first and second term describes the in- and outscattering, respectively. Thus, the interaction of the currently observed state $\vec{k}^{\nu}$ with any state $\vec{k}^{\nu'}$ is specified. The product of the transition rate S^{X} with the distribution function $f^{\nu/\nu'}$ states the probability of the scattering event and the term $(1 - f^{\nu/\nu'})$ indicates the probability of an unoccupied final state. The arrow underneath the sum for the k-vectors in (2.4.1-113) illustrates that the spin degeneracy is not taken into account, since scattering does not change the carrier spin [46]. The summation over $\vec{k}^{\nu}$ is performed over the whole first *Brillouin*-zone, but the contribution to the collision term is restricted by the transition rates involved in the in- and out-scattering term, respectively. Since a continuous dispersion relation is used (see section 2.2) the summation is converted into an integral by

$$
\sum_{\vec{k}^{\nu'}, \uparrow} \Rightarrow \frac{\Omega}{8\pi^3} \int_{\text{BZ}} d\vec{k}^{\nu'} , \qquad (2.4.1\text{-}114)
$$

with Ω as the sample volume.

Thus, (2.4.1-113) becomes

$$C(\vec{r}, f^{\nu}(\vec{r}, \vec{k}^{\nu}, t), f^{\nu'}(\vec{r}, \vec{k}^{\nu'}, t)) =$$

$$\sum_X \frac{\Omega}{8\pi^3} \int_{BZ} \Bigg(\underbrace{S^X(\vec{k}^{\nu'}, \vec{k}^{\nu}) f^{\nu'}(\vec{r}, \vec{k}^{\nu'}, t) \left(1 - f^{\nu}(\vec{r}, \vec{k}^{\nu}, t)\right)}_{\text{in-scattering}}$$

$$\underbrace{- S^X(\vec{k}^{\nu}, \vec{k}^{\nu'}) f^{\nu}(\vec{r}, \vec{k}^{\nu}, t) \left(1 - f^{\nu'}(\vec{r}, \vec{k}^{\nu'}, t)\right)}_{\text{out-scattering}} \Bigg) d\vec{k}^{\nu'} , \quad (2.4.1\text{-}115)$$

where the integration is performed over the first *Brillouin*-zone. In the non-degenerate case, where the Fermi-level is energetically far enough away from the band-edge, the term $(1 - f^{\nu/\nu'})$ simplifies to one, since $f^{\nu/\nu'} << 1$ and hence (2.4.1-116) is obtained.

$$C(\vec{r}, f^{\nu}(\vec{r}, \vec{k}^{\nu}, t), f^{\nu'}(\vec{r}, \vec{k}^{\nu'}, t)) =$$

$$\sum_X \frac{\Omega}{8\pi^3} \int_{BZ} \Bigg(S^X(\vec{k}^{\nu'}, \vec{k}^{\nu}) f^{\nu'}(\vec{r}, \vec{k}^{\nu'}, t) - S^X(\vec{k}^{\nu}, \vec{k}^{\nu'}) f^{\nu}(\vec{r}, \vec{k}^{\nu}, t) \Bigg) d\vec{k}^{\nu'} .$$

$$(2.4.1\text{-}116)$$

In literature, considering the term $(1 - f^{\nu/\nu'})$ is referred as "inclusion of the **Pauli principle**" (see for example [31]), whereas its neglect is referred as "exclusion of the **Pauli principle**". The simplified collision term (2.4.1-116) is suitable for semiconductors with low doping concentrations, since the approximation of $f^{\nu/\nu'} << 1$ holds for energies close to the conduction band edge. However, for high doping concentrations, the approximation is not valid anymore. Therefore, the general formulation (2.4.1-115) should be used for such cases. From the numerical point of view, (2.4.1-116) is more attractive, since the collision term becomes a linear function of $f^{\nu/\nu'}$ (solution variable of the BTE), whereas (2.4.1-115) makes the collision term non-linear w.r.t. $f^{\nu/\nu'}$, which requires iterative solution techniques such as the *Newton* algorithm for finding a numerical solution. For elastic scattering processes, the formulation with the Pauli principle considered according to (2.4.1-115) automatically turns into the formulation (2.4.1-116). In this case, the transition rates associated with the in- and out-scattering terms are identical

$$S^X(\vec{k}^{\nu'}, \vec{k}^{\nu}) = S^X(\vec{k}^{\nu}, \vec{k}^{\nu'}) , \qquad (2.4.1\text{-}117)$$

since no energy exchange has to be considered in the δ-function describing the energy conservation. Rewriting the integrand of (2.4.1-115) with (2.4.1-117) leads to the integrand of (2.4.1-116)

$$S^X f^{\nu'}(1 - f^{\nu}) - S^X f^{\nu}\left(1 - f^{\nu'}\right) = S^X f^{\nu'} - S^X f^{\nu} \\ \underbrace{- S^X f^{\nu'} f^{\nu} + S^X f^{\nu} f^{\nu'}}_{=0} . \quad (2.4.1\text{-}118)$$

Basically, the decrease of in-scattering (due to the occupancy f^{ν} of the currently observed state) equals the decrease of out-scattering (due to the occupancy $f^{\nu'}$ of the final state) and both compensate each other. Therefore the terms associated with the Pauli principle are not effective. This holds for all elastic scattering mechanisms, regardless of their other properties.

For inelastic (isotropic and anisotropic) scattering mechanisms an energy exchange by a certain phonon energy is involved (emission/absorption of a phonon). The arguments of the δ-functions associated with the in- and out-scattering terms are different and therefore

$$S^X(\vec{k}^{\nu'}, \vec{k}^{\nu}) \neq S^X(\vec{k}^{\nu}, \vec{k}^{\nu'}) . \qquad (2.4.1\text{-}119)$$

In this case, (2.4.1-118) is not valid and the results obtained by the collision terms after (2.4.1-115) and (2.4.1-116) are different. In this work, both formulations of the collision term (2.4.1-115) and (2.4.1-116) are considered. Due to its importance the numerical treatment of the collision term in (2.4.1-115) is described in the main part (section 3.2.3). For inelastic scattering, the discretized collision term in (2.4.1-116) can be found in appendix B.4.

2.4.2 Coordinate transformation

The goal of the coordinate transformation is to perform the following mapping of the original (stationary) 6-dimensional space $(x, y, z, k_x^\nu, k_y^\nu, k_z^\nu)$:

$$(x, y, z, k_x^\nu, k_y^\nu, k_z^\nu) \to (x_T, y_T, z_T, \epsilon^\nu, \mu, \varphi) \,, \qquad (2.4.2\text{-}120)$$

using the subscript T for the transformed space in order to avoid confusion with the original space. The variables ϵ^ν, μ and φ stem from two subsequent transformations of the k-space:

i.) Herring-Vogt transformation after (2.2.0-46) for mapping ellipsoidal equi-energy surfaces to spheres $(k_x^\nu, k_y^\nu, k_z^\nu) \to (\tilde{k}_x^\nu, \tilde{k}_y^\nu, \tilde{k}_z^\nu)$ and

ii.) expressing the resulting spheres in terms of the kinetic energy ϵ^ν and angular dependencies μ and φ.

These two points are addressed by the inverse of the Herring-Vogt transformation matrix and $\vec{\tilde{k}}^\nu$ expressed in spherical coordinates [20, 31, 48]:

$$\vec{k}^\nu = \{T^{HV,\nu}\}^{-1} \vec{\tilde{k}}^\nu \qquad (2.4.2\text{-}121)$$

$$\vec{\tilde{k}}^\nu = |\vec{\tilde{k}}^\nu|(\epsilon^\nu, \vec{r}) \begin{pmatrix} \mu \\ \sqrt{1-\mu^2}\cos(\varphi) \\ \sqrt{1-\mu^2}\sin(\varphi) \end{pmatrix} \,, \qquad (2.4.2\text{-}122)$$

with μ as the cosine of the polar angle, φ as the azimuthal angle and the kinetic energy ϵ^ν specified by the inverse dispersion relation $|\vec{\tilde{k}}^\nu|(\epsilon^\nu, \vec{r})$. Note, in contrast to the conventional spherical coordinate system, which is also used in [48], the one employed after (2.4.2-122) is slightly modified. Here, the spherical coordinate system is rotated to measure the polar angle μ w.r.t. the $\tilde{k}_x^\nu$-axis instead of the $\tilde{k}_z^\nu$-axis. This rotation is advantageous for 1D simulations (in x-direction of the real space), since the azimuthal angle φ can be omitted due to the symmetry of the distribution function.

Figure 2.19 provides a graphical interpretation of the coordinate transformation for the k-space. Note, both the Herring-Vogt transformation matrix and the dispersion relation are assumed to be position dependent.

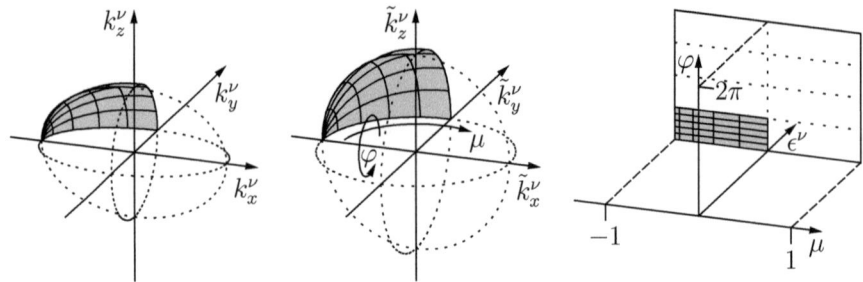

Figure 2.19: Schematic illustration of the coordinate transformation in the reciprocal space. On the left-hand side, the original k-space with an ellipsoid of constant kinetic energy is shown. In the middle, the resulting sphere after the Herring-Vogt transformation is sketched. On the right-hand side, the sphere is transformed to a plane with a constant ϵ^ν via (2.4.2-122). For consistency, the corresponding part of the grey shaded ellipsoid is also shaded for both the sphere and the transformed space.

2.4.2.1 Real space

Since the real space coordinates are not affected by the new variables of the reciprocal space, the following identity holds

$$\vec{r} = \begin{pmatrix} x \\ y \\ z \end{pmatrix} = \begin{pmatrix} x_T \\ y_T \\ z_T \end{pmatrix} = \vec{r}_T \ . \tag{2.4.2-123}$$

However, due to a position dependent band structure, the real-space nabla-operator has to be adjusted when the transformation (2.4.2-120) is used. For example, a scalar function $F(\vec{r}, \vec{k}^\nu)$ transforms in the new-space into $F(\vec{r}_T, \epsilon^\nu(\vec{r}), \mu(\vec{r}), \varphi(\vec{r}))$. The real-space gradient of the original coordinate system turns with the chain rule of differentiation into

$$\begin{aligned} \mathrm{grad}_{\vec{r}}(F) &= \mathrm{grad}_{\vec{r}_T}(F) + \frac{\partial F}{\partial \epsilon^\nu} \mathrm{grad}_{\vec{r}}(\epsilon^\nu) \\ &+ \frac{\partial F}{\partial \mu} \mathrm{grad}_{\vec{r}}(\mu) + \frac{\partial F}{\partial \varphi} \mathrm{grad}_{\vec{r}}(\varphi) \ , \end{aligned} \tag{2.4.2-124}$$

since the new coordinates ϵ^ν, μ and φ are, contrary to $\vec{k}^\nu$, position dependent.

A graphical interpretation of (2.4.2-124) is given in figure 2.20. For the sake of simplicity, a 1D dispersion relation is assumed ($\epsilon^\nu(x, k_x)$, spatially dependent non-parabolicity factor). In the original space, the derivative $\frac{\partial F}{\partial x}$ is evaluated at k_{cur}. Due to the transformation, $\frac{\partial F}{\partial x_T}$ is evaluated at a constant kinetic energy $\epsilon^\nu_{\text{cur}}$ in the transformed space. Thus, the corrective term $\frac{\partial F}{\partial \epsilon^\nu} \frac{\partial \epsilon^\nu}{\partial x}$ is added in order to preserve the derivative of the original space. Analogously, the

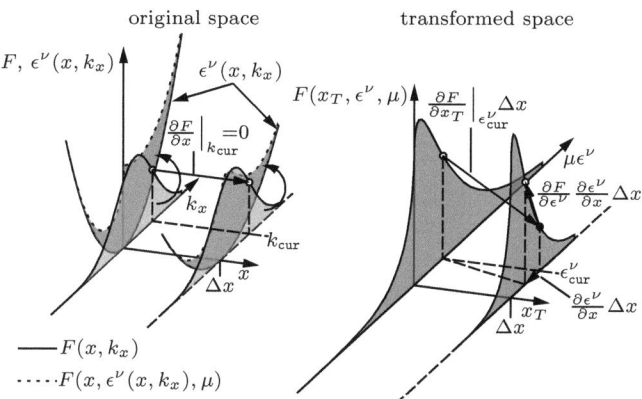

Figure 2.20: Schematic illustration of the transformed real-space gradient (2.4.2-124) for a 1D dispersion relation. Δx is an infinitesimally small change in x-direction and μ is the sign of k_x. The spatially dependent 1D dispersion relation and the mapping $F(x, k_x) \mapsto F(x_T, \epsilon^\nu, \mu)$ (indicated by rounded arrows) is also sketched in the original space.

divergence of a vector function $\vec{F}(\vec{r}, \vec{k}^\nu) \mapsto \vec{F}\big(\vec{r}_T, \epsilon^\nu(\vec{r}), \mu(\vec{r}), \varphi(\vec{r})\big)$ becomes

$$
\begin{aligned}
\text{div}_{\vec{r}}\left(\vec{F}\right) &= \left(\frac{\partial F_x}{\partial x_T} + \frac{\partial F_x}{\partial \epsilon^\nu}\frac{\partial \epsilon^\nu}{\partial x} + \frac{\partial F_x}{\partial \mu}\frac{\partial \mu}{\partial x} + \frac{\partial F_x}{\partial \varphi}\frac{\partial \varphi}{\partial x}\right) \\
&+ \left(\frac{\partial F_y}{\partial y_T} + \frac{\partial F_y}{\partial \epsilon^\nu}\frac{\partial \epsilon^\nu}{\partial y} + \frac{\partial F_y}{\partial \mu}\frac{\partial \mu}{\partial y} + \frac{\partial F_y}{\partial \varphi}\frac{\partial \varphi}{\partial y}\right) \\
&+ \left(\frac{\partial F_z}{\partial z_T} + \frac{\partial F_z}{\partial \epsilon^\nu}\frac{\partial \epsilon^\nu}{\partial z} + \frac{\partial F_z}{\partial \mu}\frac{\partial \mu}{\partial z} + \frac{\partial F_z}{\partial \varphi}\frac{\partial \varphi}{\partial z}\right) \\
&= \text{div}_{\vec{r}_T}\left(\vec{F}\right) + \frac{\partial \vec{F}}{\partial \epsilon^\nu}\cdot\text{grad}_{\vec{r}}(\epsilon^\nu) + \frac{\partial \vec{F}}{\partial \mu}\cdot\text{grad}_{\vec{r}}(\mu) + \frac{\partial \vec{F}}{\partial \varphi}\cdot\text{grad}_{\vec{r}}(\varphi)\ , \quad (2.4.2\text{-}125)
\end{aligned}
$$

with the spatial components $F_{x,y,z}$ of $\vec{F}$.

2.4.2.2 Jacobian of the transformation and its inverse

The Jacobian of the transformation is needed in order to obtain the missing derivatives in (2.4.2-124) and (2.4.2-125) associated with the spatially dependent band structure. With the transformation after (2.4.2-120) the inverse Jacobian in its general form reads

$$
J^{-1} = \begin{pmatrix}
\frac{\partial x}{\partial x_T} & 0 & 0 & 0 & 0 & 0 \\
0 & \frac{\partial y}{\partial y_T} & 0 & 0 & 0 & 0 \\
0 & 0 & \frac{\partial z}{\partial z_T} & 0 & 0 & 0 \\
\frac{\partial k_x^\nu}{\partial x_T} & \frac{\partial k_x^\nu}{\partial y_T} & \frac{\partial k_x^\nu}{\partial z_T} & \frac{\partial k_x^\nu}{\partial \epsilon^\nu} & \frac{\partial k_x^\nu}{\partial \mu} & \frac{\partial k_x^\nu}{\partial \varphi} \\
\frac{\partial k_y^\nu}{\partial x_T} & \frac{\partial k_y^\nu}{\partial y_T} & \frac{\partial k_y^\nu}{\partial z_T} & \frac{\partial k_y^\nu}{\partial \epsilon^\nu} & \frac{\partial k_y^\nu}{\partial \mu} & \frac{\partial k_y^\nu}{\partial \varphi} \\
\frac{\partial k_z^\nu}{\partial x_T} & \frac{\partial k_z^\nu}{\partial y_T} & \frac{\partial k_z^\nu}{\partial z_T} & \frac{\partial k_z^\nu}{\partial \epsilon^\nu} & \frac{\partial k_z^\nu}{\partial \mu} & \frac{\partial k_z^\nu}{\partial \varphi}
\end{pmatrix} . \qquad (2.4.2\text{-}126)
$$

Combining (2.4.2-126) with the Herring-Vogt transformation matrix (2.2.0-46) leads to

$$
J^{-1} = \begin{pmatrix}
\begin{pmatrix} 1 & 0 & 0 \\ 0 & 1 & 0 \\ 0 & 0 & 1 \end{pmatrix} & \mathbf{0} \\
C & \{T^{\mathrm{HV},\nu}\}^{-1} \cdot D
\end{pmatrix} , \qquad (2.4.2\text{-}127)
$$

with

$$
\begin{aligned}
\{T^{\mathrm{HV},\nu}\}^{-1} &= \begin{pmatrix}
\frac{1}{T_{1,1}^{\mathrm{HV},\nu}} & 0 & 0 \\
0 & \frac{1}{T_{2,2}^{\mathrm{HV},\nu}} & 0 \\
0 & 0 & \frac{1}{T_{3,3}^{\mathrm{HV},\nu}}
\end{pmatrix} \\[2mm]
&= \begin{pmatrix}
\sqrt{\frac{m_x^\nu}{m_\nu^*}} & 0 & 0 \\
0 & \sqrt{\frac{m_y^\nu}{m_\nu^*}} & 0 \\
0 & 0 & \sqrt{\frac{m_z^\nu}{m_\nu^*}}
\end{pmatrix} . \qquad (2.4.2\text{-}128)
\end{aligned}
$$

The block matrices C and D are given by

$$
C = \begin{pmatrix}
\frac{\frac{\partial |\vec{k}^\nu|}{\partial x_T}\mu}{T_{1,1}^{HV,\nu}} - \frac{|\vec{k}^\nu|\mu \frac{\partial T_{1,1}^{HV,\nu}}{\partial x_T}}{\left\{T_{1,1}^{HV,\nu}\right\}^2} \\
\frac{\frac{\partial |\vec{k}^\nu|}{\partial x_T}\sqrt{1-\mu^2}\cos(\varphi)}{T_{2,2}^{HV,\nu}} - \frac{|\vec{k}^\nu|\sqrt{1-\mu^2}\cos(\varphi)\frac{\partial T_{2,2}^{HV,\nu}}{\partial x_T}}{\left\{T_{2,2}^{HV,\nu}\right\}^2} \\
\frac{\frac{\partial |\vec{k}^\nu|}{\partial x_T}\sqrt{1-\mu^2}\sin(\varphi)}{T_{3,3}^{HV,\nu}} - \frac{|\vec{k}^\nu|\sqrt{1-\mu^2}\sin(\varphi)\frac{\partial T_{3,3}^{HV,\nu}}{\partial x_T}}{\left\{T_{3,3}^{HV,\nu}\right\}^2}
\end{pmatrix}
$$

$$
\begin{pmatrix}
\frac{\frac{\partial |\vec{k}^\nu|}{\partial y_T}\mu}{T_{1,1}^{HV,\nu}} - \frac{|\vec{k}^\nu|\mu\frac{\partial T_{1,1}^{HV,\nu}}{\partial y_T}}{\left\{T_{1,1}^{HV,\nu}\right\}^2} & \frac{\frac{\partial |\vec{k}^\nu|}{\partial z_T}\mu}{T_{1,1}^{HV,\nu}} - \frac{|\vec{k}^\nu|\mu\frac{\partial T_{1,1}^{HV,\nu}}{\partial z_T}}{\left\{T_{1,1}^{HV,\nu}\right\}^2} \\
\frac{\frac{\partial |\vec{k}^\nu|}{\partial y_T}\sqrt{1-\mu^2}\cos(\varphi)}{T_{2,2}^{HV,\nu}} - \frac{|\vec{k}^\nu|\sqrt{1-\mu^2}\cos(\varphi)\frac{\partial T_{2,2}^{HV,\nu}}{\partial y_T}}{\left\{T_{2,2}^{HV,\nu}\right\}^2} & \frac{\frac{\partial |\vec{k}^\nu|}{\partial z_T}\sqrt{1-\mu^2}\cos(\varphi)}{T_{2,2}^{HV,\nu}} - \frac{|\vec{k}^\nu|\sqrt{1-\mu^2}\cos(\varphi)\frac{\partial T_{2,2}^{HV,\nu}}{\partial z_T}}{\left\{T_{2,2}^{HV,\nu}\right\}^2} \\
\frac{\frac{\partial |\vec{k}^\nu|}{\partial y_T}\sqrt{1-\mu^2}\sin(\varphi)}{T_{3,3}^{HV,\nu}} - \frac{|\vec{k}^\nu|\sqrt{1-\mu^2}\sin(\varphi)\frac{\partial T_{3,3}^{HV,\nu}}{\partial y_T}}{\left\{T_{3,3}^{HV,\nu}\right\}^2} & \frac{\frac{\partial |\vec{k}^\nu|}{\partial z_T}\sqrt{1-\mu^2}\sin(\varphi)}{T_{3,3}^{HV,\nu}} - \frac{|\vec{k}^\nu|\sqrt{1-\mu^2}\sin(\varphi)\frac{\partial T_{3,3}^{HV,\nu}}{\partial z_T}}{\left\{T_{3,3}^{HV,\nu}\right\}^2}
\end{pmatrix},
$$

$$(2.4.2\text{-}129)$$

$$
\begin{aligned}
D &= \begin{pmatrix}
\frac{\partial \tilde{k}_x^\nu}{\partial \epsilon^\nu} & \frac{\partial \tilde{k}_x^\nu}{\partial \mu} & \frac{\partial \tilde{k}_x^\nu}{\partial \varphi} \\
\frac{\partial \tilde{k}_y^\nu}{\partial \epsilon^\nu} & \frac{\partial \tilde{k}_y^\nu}{\partial \mu} & \frac{\partial \tilde{k}_y^\nu}{\partial \varphi} \\
\frac{\partial \tilde{k}_z^\nu}{\partial \epsilon^\nu} & \frac{\partial \tilde{k}_z^\nu}{\partial \mu} & \frac{\partial \tilde{k}_z^\nu}{\partial \varphi}
\end{pmatrix} \\
&= \begin{pmatrix}
\frac{\partial |\vec{k}^\nu|}{\partial \epsilon^\nu}\mu & |\vec{k}^\nu| & 0 \\
\frac{\partial |\vec{k}^\nu|}{\partial \epsilon^\nu}\sqrt{1-\mu^2}\cos(\varphi) & \frac{-\mu}{\sqrt{1-\mu^2}}|\vec{k}^\nu|\cos(\varphi) & -|\vec{k}^\nu|\sqrt{1-\mu^2}\sin(\varphi) \\
\frac{\partial |\vec{k}^\nu|}{\partial \epsilon^\nu}\sqrt{1-\mu^2}\sin(\varphi) & \frac{-\mu}{\sqrt{1-\mu^2}}|\vec{k}^\nu|\sin(\varphi) & |\vec{k}^\nu|\sqrt{1-\mu^2}\cos(\varphi)
\end{pmatrix}.
\end{aligned}
$$

$$(2.4.2\text{-}130)$$

The sub-matrix C is obtained by calculating the partial derivatives of (2.4.2-121), in conjunction with (2.4.2-122), w.r.t. the real space coordinates and thus, originates from a position dependent band structure. For a position independent band structure, C becomes a zero matrix. The sub-matrix D contains the partial derivatives of $\tilde{k}^\nu$-space (2.4.2-122) w.r.t. the new variables ϵ^ν, μ and φ. The absolute value of the determinant of the inverse Jacobian (used to transform the infinitesimal volume elements of the original space $(x, y, z, k_x^\nu, k_y^\nu, k_z^\nu)$ to the new space $(x_T, y_T, z_T, \epsilon^\nu, \mu, \varphi)$) reads

$$
||J^{-1}|| = ||\{T^{HV,\nu}\}^{-1}\cdot D|| = ||\{T^{HV,\nu}\}^{-1}||\,||D|| = ||D||\ . \quad (2.4.2\text{-}131)
$$

The determinant of the Herring-Vogt transformation matrix $T^{HV,\nu}$ and its inverse – according to the definition of m_ν^* (2.2.0-48) – always equals to one (see appendix A.1). This is also valid for a position dependent Herring-Vogt transformation matrix $T^{HV,\nu}(\vec{r})$, since m_ν^* is a function of $\vec{r}$ in this case. Thus

$$dk_x^\nu \, dk_y^\nu \, dk_z^\nu \, dx \, dy \, dz \;\; = \;\; (||D(\vec{r}_T)|| \, d\epsilon^\nu \, d\mu \, d\varphi) \, dx_T \, dy_T \, dz_T \; , \qquad (2.4.2\text{-}132)$$

$$\text{with} \quad dk_x^\nu \, dk_y^\nu \, dk_z^\nu \;\; = \;\; (||D(\vec{r}_T)|| \, d\epsilon^\nu \, d\mu \, d\varphi) \; . \qquad (2.4.2\text{-}133)$$

Relation (2.4.2-132) is beneficial for **Finite-Volume/Box-Integration** methods that require an integration over the total space. A position dependence of $T^{HV,\nu}$ does not result in additional terms for the infinitesimal volume element due to the definition of m_ν^*. In fact, equation (2.4.2-132) is also valid for isotropic and position independent band structures. However, one should keep in mind that the determinant of the sub-matrix D becomes a function of $\vec{r}_T$ (due to $m_\nu^*(\vec{r}_T)$), if a position dependent band structure is considered. Based on (2.4.2-127), the Jacobian is obtained by blockwise inversion:

$$J = \begin{pmatrix} \begin{pmatrix} 1 & 0 & 0 \\ 0 & 1 & 0 \\ 0 & 0 & 1 \end{pmatrix} & \mathbf{0} \\ -D^{-1} \cdot \{T^{HV,\nu}\} \cdot C & D^{-1} \cdot \{T^{HV,\nu}\} \end{pmatrix} \; , \qquad (2.4.2\text{-}134)$$

where its entries can be identified as follows:

$$J = \begin{pmatrix}
\dfrac{\partial x_T}{\partial x} & 0 & 0 & & & \\
0 & \dfrac{\partial y_T}{\partial y} & 0 & & \mathbf{0} & \\
0 & 0 & \dfrac{\partial z_T}{\partial z} & & & \\[2mm]
\dfrac{\partial \epsilon^\nu}{\partial x} & \dfrac{\partial \epsilon^\nu}{\partial y} & \dfrac{\partial \epsilon^\nu}{\partial z} & \dfrac{\partial \epsilon^\nu}{\partial k_x^\nu} & \dfrac{\partial \epsilon^\nu}{\partial k_y^\nu} & \dfrac{\partial \epsilon^\nu}{\partial k_z^\nu} \\[2mm]
\dfrac{\partial \mu}{\partial x} & \dfrac{\partial \mu}{\partial y} & \dfrac{\partial \mu}{\partial z} & \dfrac{\partial \mu}{\partial k_x^\nu} & \dfrac{\partial \mu}{\partial k_y^\nu} & \dfrac{\partial \mu}{\partial k_z^\nu} \\[2mm]
\dfrac{\partial \varphi}{\partial x} & \dfrac{\partial \varphi}{\partial y} & \dfrac{\partial \varphi}{\partial z} & \dfrac{\partial \varphi}{\partial k_x^\nu} & \dfrac{\partial \varphi}{\partial k_y^\nu} & \dfrac{\partial \varphi}{\partial k_z^\nu}
\end{pmatrix} \; . \qquad (2.4.2\text{-}135)$$

The inverse of the sub-matrix D in (2.4.2-134) reads

$$
D^{-1} = \begin{pmatrix}
\frac{\partial \epsilon^\nu}{\partial k_x^\nu} & \frac{\partial \epsilon^\nu}{\partial k_y^\nu} & \frac{\partial \epsilon^\nu}{\partial k_z^\nu} \\[6pt]
\frac{\partial \mu}{\partial k_x^\nu} & \frac{\partial \mu}{\partial k_y^\nu} & \frac{\partial \mu}{\partial k_z^\nu} \\[6pt]
\frac{\partial \varphi}{\partial k_x^\nu} & \frac{\partial \varphi}{\partial k_y^\nu} & \frac{\partial \varphi}{\partial k_z^\nu}
\end{pmatrix} \qquad (2.4.2\text{-}136)
$$

$$
= \begin{pmatrix}
\boxed{\dfrac{\mu}{\frac{\partial |\vec{k}^\nu|}{\partial \epsilon^\nu}} \quad \dfrac{\sqrt{1-\mu^2}\cos(\varphi)}{\frac{\partial |\vec{k}^\nu|}{\partial \epsilon^\nu}} \quad \dfrac{\sqrt{1-\mu^2}\sin(\varphi)}{\frac{\partial |\vec{k}^\nu|}{\partial \epsilon^\nu}}} \\[14pt]
\dfrac{1-\mu^2}{|\vec{k}^\nu|} \quad \dfrac{-\mu\sqrt{1-\mu^2}\cos(\varphi)}{|\vec{k}^\nu|} \quad \dfrac{-\mu\sqrt{1-\mu^2}\sin(\varphi)}{|\vec{k}^\nu|} \\[14pt]
0 \qquad -\dfrac{\sin(\varphi)}{|\vec{k}^\nu|\sqrt{1-\mu^2}} \qquad \dfrac{\cos(\varphi)}{|\vec{k}^\nu|\sqrt{1-\mu^2}}
\end{pmatrix} \cdot (2.4.2\text{-}137)
$$

Note, the framed row of (2.4.2-137) is the transposed carrier group velocity vector in the $\tilde{k}^\nu$-space multiplied by $\hbar$: $\hbar \vec{\tilde{v}}_g^\nu{}^{\mathsf{T}}$. With the Herring-Vogt transformation matrix (in accordance with (2.4.2-134)), the carrier group velocity vector $\vec{v}_g^\nu$ in the original k-space (entries framed by solid lines in (2.4.2-135)) is obtained

$$
\begin{aligned}
\vec{v}_g^\nu &= \frac{1}{\hbar} \left(\left(\{D^{-1}\}_{1,1} \quad \{D^{-1}\}_{1,2} \quad \{D^{-1}\}_{1,3} \right) \cdot \{T^{\mathrm{HV},\nu}\} \right)^{\mathsf{T}} \\[6pt]
&= \frac{1}{\hbar} \{T^{\mathrm{HV},\nu}\} \cdot \begin{pmatrix} \{D^{-1}\}_{1,1} \\ \{D^{-1}\}_{1,2} \\ \{D^{-1}\}_{1,3} \end{pmatrix} = \{T^{\mathrm{HV},\nu}\} \cdot \vec{\tilde{v}}_g^\nu , \qquad (2.4.2\text{-}138)
\end{aligned}
$$

$$
\vec{\tilde{v}}_g^\nu = \frac{1}{\hbar} \frac{1}{\frac{\partial |\vec{k}^\nu|}{\partial \epsilon^\nu}} \begin{pmatrix} \mu \\ \sqrt{1-\mu^2}\cos(\varphi) \\ \sqrt{1-\mu^2}\sin(\varphi) \end{pmatrix} . \qquad (2.4.2\text{-}139)
$$

This result is consistent with the relations (2.2.0-50) and (2.4.0-112). The dashed box of (2.4.2-135) contains the missing derivatives to transform the real-space gradient and divergence (2.4.2-124) and (2.4.2-125), respectively.

The derivatives of the kinetic energy w.r.t. the real-space dimensions read

$$\frac{\partial \epsilon^{\nu}}{\partial d} = \frac{1}{\frac{\partial |\vec{k}^{\nu}|}{\partial \epsilon^{\nu}}} \left[|\vec{k}^{\nu}|(1-\mu^2)h_d + \left(|\vec{k}^{\nu}|\mu^2 \frac{\partial \ln\left(T_{1,1}^{\text{HV},\nu}\right)}{\partial d} - \frac{\partial |\vec{k}^{\nu}|}{\partial d} \right) \right], \quad (2.4.2\text{-}140)$$

where d is a placeholder for the real space dimensions ($d = \{x, y, z\}$) and with the term h_d being given by

$$h_d = \cos(\varphi)^2 \frac{\partial \ln\left(T_{2,2}^{\text{HV},\nu}\right)}{\partial d} + \sin(\varphi)^2 \frac{\partial \ln\left(T_{3,3}^{\text{HV},\nu}\right)}{\partial d}. \quad (2.4.2\text{-}141)$$

For μ and φ the following formulas are obtained:

$$\frac{\partial \mu}{\partial d} = \mu\left(1-\mu^2\right)\left(\frac{\partial \ln\left(T_{1,1}^{\text{HV},\nu}\right)}{\partial d} - h_d\right), \quad (2.4.2\text{-}142)$$

$$\frac{\partial \varphi}{\partial d} = \frac{1}{2}\sin(2\varphi)\frac{\partial}{\partial d}\left(\ln\left(\frac{T_{3,3}^{\text{HV},\nu}}{T_{2,2}^{\text{HV},\nu}}\right)\right), \quad (2.4.2\text{-}143)$$

where for (2.4.2-141), (2.4.2-142) and (2.4.2-143) d is again used as placeholder ($d = \{x, y, z\}$) like for (2.4.2-140). Note, the absolute value of the k-vector $|\vec{k}^{\nu}|$ (or inverse of the dispersion relation) is intentionally kept general. The reason is to emphasize the possibility to adjust $|\vec{k}^{\nu}|$ to measurements of the density of states.

2.4.2.3 Reciprocal space

The Herring-Vogt transformation maps both the initial $\vec{k}^\nu$-space and its differential operators to the isotropic $\tilde{\vec{k}}^\nu$-space after (2.4.2-122) (see appendix A.1). Here, the missing transformation $(\tilde{k}_x^\nu, \tilde{k}_y^\nu, \tilde{k}_z^\nu) \rightarrow (\epsilon^\nu, \mu, \varphi)$ is described. With (2.4.2-122), the unit-vectors of the $(\epsilon^\nu, \mu, \varphi)$-space read

$$\vec{e}_{\epsilon^\nu} \;=\; \frac{\frac{\partial \tilde{\vec{k}}^\nu}{\partial \epsilon^\nu}}{h_{\epsilon^\nu}} = \begin{pmatrix} \mu \\ \sqrt{1-\mu^2}\cos(\varphi) \\ \sqrt{1-\mu^2}\sin(\varphi) \end{pmatrix} \;, \qquad (2.4.2\text{-}144)$$

$$\vec{e}_{\mu} \;=\; \frac{\frac{\partial \tilde{\vec{k}}^\nu}{\partial \mu}}{h_{\mu}} = \begin{pmatrix} \sqrt{1-\mu^2} \\ -\mu\cos(\varphi) \\ -\mu\sin(\varphi) \end{pmatrix} \;, \qquad (2.4.2\text{-}145)$$

$$\vec{e}_{\varphi} \;=\; \frac{\frac{\partial \tilde{\vec{k}}^\nu}{\partial \varphi}}{h_{\varphi}} = \begin{pmatrix} 0 \\ -\sin(\varphi) \\ \cos(\varphi) \end{pmatrix} \;, \qquad (2.4.2\text{-}146)$$

with

$$h_{\epsilon^\nu} \;=\; \left| \frac{\partial \tilde{\vec{k}}^\nu}{\partial \epsilon^\nu} \right| = \frac{\partial |\tilde{\vec{k}}^\nu|}{\partial \epsilon^\nu} \;, \qquad (2.4.2\text{-}147)$$

$$h_{\mu} \;=\; \left| \frac{\partial \tilde{\vec{k}}^\nu}{\partial \mu} \right| = \frac{|\tilde{\vec{k}}^\nu|}{\sqrt{1-\mu^2}} \;, \qquad (2.4.2\text{-}148)$$

$$h_{\varphi} \;=\; \left| \frac{\partial \tilde{\vec{k}}^\nu}{\partial \varphi} \right| = |\tilde{\vec{k}}^\nu|\sqrt{1-\mu^2} \;. \qquad (2.4.2\text{-}149)$$

With the new set of basis vectors, a vector function $\vec{F}(\tilde{\vec{k}}^\nu)$ in $\tilde{\vec{k}}^\nu$-space transforms into the new space $(\vec{F}(\epsilon^\nu, \mu, \varphi))$ by

$$\begin{pmatrix} F_{\epsilon^\nu} \\ F_{\mu} \\ F_{\varphi} \end{pmatrix} = \begin{pmatrix} \mu & \sqrt{1-\mu^2}\cos(\varphi) & \sqrt{1-\mu^2}\sin(\varphi) \\ \sqrt{1-\mu^2} & -\mu\cos(\varphi) & -\mu\sin(\varphi) \\ 0 & -\sin(\varphi) & \cos(\varphi) \end{pmatrix} \cdot \begin{pmatrix} F_{\tilde{k}_x^\nu} \\ F_{\tilde{k}_y^\nu} \\ F_{\tilde{k}_z^\nu} \end{pmatrix} \;,$$

$$(2.4.2\text{-}150)$$

or vice versa

$$
\begin{pmatrix} F_{\tilde{k}_x^\nu} \\ F_{\tilde{k}_y^\nu} \\ F_{\tilde{k}_z^\nu} \end{pmatrix} = \begin{pmatrix} \mu & \sqrt{1-\mu^2} & 0 \\ \sqrt{1-\mu^2}\cos(\varphi) & -\mu\cos(\varphi) & -\sin(\varphi) \\ \sqrt{1-\mu^2}\sin(\varphi) & -\mu\sin(\varphi) & \cos(\varphi) \end{pmatrix} \cdot \begin{pmatrix} F_{\epsilon^\nu} \\ F_\mu \\ F_\varphi \end{pmatrix} . \quad (2.4.2\text{-}151)
$$

The divergence in $\vec{k}^\nu$-space is transformed in the same manner as for the one in real-space after (2.4.2-125). Calculating the total differential, one obtains

$$
\begin{aligned}
\operatorname{div}_{\vec{k}^\nu}\left(\vec{F}\right) &= \operatorname{grad}_{\vec{k}^\nu}(\epsilon^\nu) \cdot \frac{\partial \vec{F}}{\partial \epsilon^\nu} + \operatorname{grad}_{\vec{k}^\nu}(\mu) \cdot \frac{\partial \vec{F}}{\partial \mu} \\
&\quad + \operatorname{grad}_{\vec{k}^\nu}(\varphi) \cdot \frac{\partial \vec{F}}{\partial \varphi} ,
\end{aligned} \quad (2.4.2\text{-}152)
$$

where $\vec{F}$ on the right-hand side of (2.4.2-152) is already transformed by (2.4.2-150). Note, the gradients $(\operatorname{grad}_{\vec{k}^\nu}(\{\epsilon^\nu, \mu, \varphi\}))$ are be obtained by the Jacobian of the coordinate transformation (see section 2.4.2.2, entries of the matrices (2.4.2-136) and (2.4.2-137)). A detailed derivation of the transformation of $\operatorname{div}_{\vec{k}^\nu}$ is given in appendix A.3. However, since the new coordinate system $(\epsilon^\nu, \mu, \varphi)$ is orthogonal

$$
\vec{e}_{\epsilon^\nu} \cdot \vec{e}_\mu = \vec{e}_{\epsilon^\nu} \cdot \vec{e}_\varphi = \vec{e}_\mu \cdot \vec{e}_\varphi = 0 , \quad (2.4.2\text{-}153)
$$

the divergence in $\vec{k}^\nu$-space can also be transformed by [8]

$$
\operatorname{div}_{\vec{k}^\nu}\left(\vec{F}\right) = \frac{1}{h_{\epsilon^\nu} h_\mu h_\varphi} \left[\frac{\partial}{\partial \epsilon^\nu}(h_\mu h_\varphi F_{\epsilon^\nu}) + \frac{\partial}{\partial \mu}(h_{\epsilon^\nu} h_\varphi F_\mu) + \frac{\partial}{\partial \varphi}(h_{\epsilon^\nu} h_\mu F_\varphi) \right] ,
$$
$$
(2.4.2\text{-}154)
$$

where F_{ϵ^ν}, F_μ, F_φ are the components of $\vec{F}$ expressed in terms of the new basis vectors $\vec{e}_{\epsilon^\nu}$, $\vec{e}_\mu$ and $\vec{e}_\varphi$ (see equation (2.4.2-150)). Another advantage of the orthogonality of the coordinate system $(\epsilon^\nu, \mu, \varphi)$ is that the infinitesimal volume $d\vec{k}^\nu$ is directly transformed by [8]

$$
d\vec{k}^\nu = h_{\epsilon^\nu} h_\mu h_\varphi \, d\epsilon^\nu d\mu d\varphi = \frac{\partial |\vec{k}^\nu|}{\partial \epsilon^\nu} |\vec{k}^\nu|^2 \, d\epsilon^\nu d\mu d\varphi , \quad (2.4.2\text{-}155)
$$

where again the density of states

$$\text{dos}^\nu(\vec{r}, \epsilon^\nu) = \frac{\partial |\vec{\bar{k}}^\nu|}{\partial \epsilon^\nu} |\vec{\bar{k}}^\nu|^2 \tag{2.4.2-156}$$

is obtained. Note, the result (2.4.2-155) is identical to (2.4.2-133) given in the previous section, since the matrix D after (2.4.2-130) consists of the un-normalized versions of the unit vectors after (2.4.2-144)-(2.4.2-146). With the DOS after (2.4.2-156) the divergence (2.4.2-154) becomes

$$\begin{aligned}
\text{div}_{\vec{\bar{k}}^\nu}\left(\vec{F}\right) &= \frac{1}{\text{dos}^\nu(\vec{r}, \epsilon^\nu)}\left[\frac{\partial}{\partial \epsilon^\nu}\left(\frac{\text{dos}^\nu(\vec{r}, \epsilon^\nu)}{h_{\epsilon^\nu}}F_{\epsilon^\nu}\right) + \frac{\partial}{\partial \mu}\left(\frac{\text{dos}^\nu(\vec{r}, \epsilon^\nu)}{h_\mu}F_\mu\right)\right. \\
&\left. + \frac{\partial}{\partial \varphi}\left(\frac{\text{dos}^\nu(\vec{r}, \epsilon^\nu)}{h_\varphi}F_\varphi\right)\right],
\end{aligned} \tag{2.4.2-157}$$

which will be used further on.

2.4.3 Transformed BTE

Based on the coordinate transformation given in section 2.4.2 and its sub-sections, the BTE after (2.4.0-111) is transformed into the new coordinate system. For convenience, each term of the BTE is examined separately and the arguments of the involved quantities are omitted (except for the transient term).

Transient term

Since the coordinate transformation described in section 2.4.2 does not affect the time dependence of the BTE, the transient term is directly transformed by

$$\frac{\partial f^\nu(\vec{r}, \vec{\bar{k}}^\nu, t)}{\partial t} \rightarrow \frac{\partial f^\nu(\vec{r}_\text{T}, \epsilon^\nu, \mu, \varphi, t)}{\partial t}. \tag{2.4.3-158}$$

Diffusion term

The transformed diffusion term reads

$$\text{div}_{\vec{r}}\left(T^{\text{HV},\nu}\vec{v}_{\text{g}}^{\nu}f^{\nu}\right) \rightarrow \frac{1}{\hbar}\text{div}_{\vec{r}_{\text{T}}}\left(\frac{f^{\nu}}{\frac{\partial|\vec{k}^{\nu}|}{\partial\epsilon^{\nu}}}\begin{pmatrix}T^{\text{HV},\nu}_{1,1}\mu\\T^{\text{HV},\nu}_{2,2}\sqrt{1-\mu^2}\cos(\varphi)\\T^{\text{HV},\nu}_{3,3}\sqrt{1-\mu^2}\sin(\varphi)\end{pmatrix}\right)$$

$$+\frac{1}{\hbar}\frac{\partial}{\partial\epsilon^{\nu}}\left(\frac{f^{\nu}}{\frac{\partial|\vec{k}^{\nu}|}{\partial\epsilon^{\nu}}}\right)\left[\begin{pmatrix}T^{\text{HV},\nu}_{1,1}\mu\\T^{\text{HV},\nu}_{2,2}\sqrt{1-\mu^2}\cos(\varphi)\\T^{\text{HV},\nu}_{3,3}\sqrt{1-\mu^2}\sin(\varphi)\end{pmatrix}\cdot\begin{pmatrix}\frac{\partial\epsilon^{\nu}}{\partial x}\\\frac{\partial\epsilon^{\nu}}{\partial y}\\\frac{\partial\epsilon^{\nu}}{\partial z}\end{pmatrix}\right]$$

$$+\frac{1}{\hbar}\frac{1}{\frac{\partial|\vec{k}^{\nu}|}{\partial\epsilon^{\nu}}}\left[\begin{pmatrix}T^{\text{HV},\nu}_{1,1}(\frac{\partial f^{\nu}}{\partial\mu}\mu+f^{\nu})\\T^{\text{HV},\nu}_{2,2}\sqrt{1-\mu^2}\cos(\varphi)(\frac{\partial f^{\nu}}{\partial\mu}-f^{\nu}\frac{\mu}{1-\mu^2})\\T^{\text{HV},\nu}_{3,3}\sqrt{1-\mu^2}\sin(\varphi)(\frac{\partial f^{\nu}}{\partial\mu}-f^{\nu}\frac{\mu}{1-\mu^2})\end{pmatrix}\cdot\begin{pmatrix}\frac{\partial\mu}{\partial x}\\\frac{\partial\mu}{\partial y}\\\frac{\partial\mu}{\partial z}\end{pmatrix}\right]$$

$$+\frac{1}{\hbar}\frac{1}{\frac{\partial|\vec{k}^{\nu}|}{\partial\epsilon^{\nu}}}\left[\begin{pmatrix}T^{\text{HV},\nu}_{1,1}\mu\frac{\partial f^{\nu}}{\partial\varphi}\\T^{\text{HV},\nu}_{2,2}\sqrt{1-\mu^2}(\frac{\partial f^{\nu}}{\partial\varphi}\cos(\varphi)-f^{\nu}\sin(\varphi))\\T^{\text{HV},\nu}_{3,3}\sqrt{1-\mu^2}(\frac{\partial f^{\nu}}{\partial\varphi}\sin(\varphi)+f^{\nu}\cos(\varphi))\end{pmatrix}\cdot\begin{pmatrix}\frac{\partial\varphi}{\partial x}\\\frac{\partial\varphi}{\partial y}\\\frac{\partial\varphi}{\partial z}\end{pmatrix}\right]. \tag{2.4.3-159}$$

In the transformed diffusion term (2.4.3-159), the transformed real space divergence (2.4.2-125), the group-velocity after (2.4.2-139) and the Herring-Vogt transformation have already been included. Note, the last three terms on the right-hand side of (2.4.3-159) are only nonzero, if a position dependent dispersion relation is assumed. For the sake of clarity, the analytic expressions of the derivatives $\frac{\partial\epsilon^{\nu}}{\partial\{x,y,z\}}$, $\frac{\partial\mu}{\partial\{x,y,z\}}$ and $\frac{\partial\varphi}{\partial\{x,y,z\}}$ have not been inserted and can be found in section 2.4.2.2, equation (2.4.2-140), (2.4.2-142) and (2.4.2-143), respectively.

Drift term

For the drift term, the transformed k-space divergence (2.4.2-157) is employed. The Herring-Vogt transformation matrix (2.2.0-46) is applied to the driving force $E^{\vec{\nu}}_{\text{eff.}}$, and the resulting vector field is projected onto the $(\epsilon^\nu, \mu, \varphi)$ frame for consistency with (2.4.2-157). The transformed drift term of the BTE (2.4.0-111) becomes

$$
-\frac{q}{\hbar}\mathrm{div}_{\vec{k}^\nu}\left(T^{\mathrm{HV},\nu}E^{\vec{\nu}}_{\text{eff.}}f^\nu\right) \to -\frac{q}{\hbar\,\mathrm{dos}^\nu}\Bigg\{
$$

$$
\frac{\partial}{\partial\epsilon^\nu}\left(\frac{\mathrm{dos}^\nu f^\nu}{\frac{\partial|\vec{k}^\nu|}{\partial\epsilon^\nu}}\begin{pmatrix}\mu\\\sqrt{1-\mu^2}\cos(\varphi)\\\sqrt{1-\mu^2}\sin(\varphi)\end{pmatrix}\cdot\begin{pmatrix}T^{\mathrm{HV},\nu}_{1,1}E^{\nu,x}_{\text{eff.}}\\T^{\mathrm{HV},\nu}_{2,2}E^{\nu,y}_{\text{eff.}}\\T^{\mathrm{HV},\nu}_{3,3}E^{\nu,z}_{\text{eff.}}\end{pmatrix}\right)
$$

$$
+\frac{\partial}{\partial\mu}\left(\frac{\mathrm{dos}^\nu f^\nu\sqrt{1-\mu^2}}{|\vec{k}^\nu|}\begin{pmatrix}\sqrt{1-\mu^2}\\-\mu\cos(\varphi)\\-\mu\sin(\varphi)\end{pmatrix}\cdot\begin{pmatrix}T^{\mathrm{HV},\nu}_{1,1}E^{\nu,x}_{\text{eff.}}\\T^{\mathrm{HV},\nu}_{2,2}E^{\nu,y}_{\text{eff.}}\\T^{\mathrm{HV},\nu}_{3,3}E^{\nu,z}_{\text{eff.}}\end{pmatrix}\right)
$$

$$
+\frac{\partial}{\partial\varphi}\left(\frac{\mathrm{dos}^\nu f^\nu}{|\vec{k}^\nu|\sqrt{1-\mu^2}}\begin{pmatrix}0\\-\sin(\varphi)\\\cos(\varphi)\end{pmatrix}\cdot\begin{pmatrix}T^{\mathrm{HV},\nu}_{1,1}E^{\nu,x}_{\text{eff.}}\\T^{\mathrm{HV},\nu}_{2,2}E^{\nu,y}_{\text{eff.}}\\T^{\mathrm{HV},\nu}_{3,3}E^{\nu,z}_{\text{eff.}}\end{pmatrix}\right)\Bigg\} \quad.(2.4.3\text{-}160)
$$

Note, the analytic expressions of h_{ϵ^ν}, h_μ and h_φ (cf. (2.4.2-147)-(2.4.2-149)), involved in the transformed divergence (2.4.2-157), have already been inserted in (2.4.3-160).

Collision term

In the new coordinate system, the infinitesimal volume $d\vec{k}^{\nu'}$ is given by the equations (2.4.2-155)-(2.4.2-156) in section 2.4.2.3. Hence, the collision term (2.4.1-115) transforms to:

$$
C\left(f^\nu, f^{\nu'}\right) = \sum_X \frac{\Omega}{8\pi^3}\int_{\mathrm{BZ}}\Bigg(S^X(\{\acute{\epsilon}^{\nu'},\mu',\varphi'\},\{\epsilon^\nu,\mu,\varphi\})f^{\nu'}\left(1-f^\nu\right)
$$

$$
-S^X(\{\epsilon^\nu,\mu,\varphi\},\{\acute{\epsilon}^{\nu'},\mu',\varphi'\})f^\nu\left(1-f^{\nu'}\right)\Bigg)\mathrm{dos}^{\nu'}d\acute{\epsilon}^{\nu'}d\mu'd\varphi' \quad. (2.4.3\text{-}161)
$$

2.4.4 Conclusions and remarks

In section 2.4.2, a rotated spherical coordinate system is employed for expressing the k-vectors $\vec{k}^\nu$ (cf. equation (2.4.2-122)). The advantage of the rotation can directly be deduced from the transformed drift term (2.4.3-160). For the simulation of a 1D carrier transport in a 3D material in x-direction (in real space), homogeneity in the remaining real space directions (y- and z-direction) is assumed. Nevertheless, a 3D k-space must be considered, when the transport direction is confined only within the semiconducting crystal. The dimension of the crystal and thus of its reciprocal space is unaffected. Thus, a direct discretization of the k-space leads in total to a 4D simulation domain. However, by using the transformation described in section 2.4.2 and its subsections, $(\epsilon^\nu, \mu, \varphi)$-space is used instead of $\vec{k}^\nu$-space. Therefore, in the case of a 1D carrier transport in x-direction the φ-dimension becomes redundant. First, due to the assumed homogeneity for a pure 1D transport, the driving force $\vec{E}^\nu_{\text{eff.}}$ consists only of a nonzero x-component:

$$
\vec{E}^\nu_{\text{eff.}} = \begin{pmatrix} E^{\nu,x}_{\text{eff.}} \\ 0 \\ 0 \end{pmatrix}. \tag{2.4.4-162}
$$

Inserting (2.4.4-162) into the transformed drift term (2.4.3-160) results in

$$
-\frac{q}{\hbar}\operatorname{div}_{\vec{k}^\nu}\left(T^{\text{HV},\nu}\vec{E}^\nu_{\text{eff.}}f^\nu\right) \xrightarrow{\text{1D, }x} -\frac{q}{\hbar\,\text{dos}^\nu}\left\{\frac{\partial}{\partial\epsilon^\nu}\left(\frac{\text{dos}^\nu f^\nu}{\frac{\partial|\vec{k}^\nu|}{\partial\epsilon^\nu}}\mu T^{\text{HV},\nu}_{1,1}E^{\nu,x}_{\text{eff.}}\right) \right.
$$
$$
\left. +\frac{\partial}{\partial\mu}\left(\frac{\text{dos}^\nu f^\nu(1-\mu^2)}{|\vec{k}^\nu|}T^{\text{HV},\nu}_{1,1}E^{\nu,x}_{\text{eff.}}\right)\right\}. \tag{2.4.4-163}
$$

Analogously, the diffusion term (2.4.3-159) of the BTE consists only of derivatives w.r.t. the x-direction:

$$\text{div}_{\vec{r}}\left(T^{\text{HV},\nu}\vec{v}_g^\nu f^\nu\right) \overset{\text{1D, x}}{\longrightarrow} \frac{1}{\hbar}\frac{\partial}{\partial x_{\text{T}}}\left(\frac{f^\nu}{\frac{\partial |\vec{k}^\nu|}{\partial \epsilon^\nu}}T_{1,1}^{\text{HV},\nu}\mu\right) + \frac{1}{\hbar}T_{1,1}^{\text{HV},\nu}\Bigg\{$$

$$\left[\mu\frac{\partial \epsilon^\nu}{\partial x}\right]\frac{\partial}{\partial \epsilon^\nu}\left(\frac{f^\nu}{\frac{\partial |\vec{k}^\nu|}{\partial \epsilon^\nu}}\right) + \frac{1}{\frac{\partial |\vec{k}^\nu|}{\partial \epsilon^\nu}}\left[\left(\frac{\partial f^\nu}{\partial \mu}\mu + f^\nu\right)\frac{\partial \mu}{\partial x}\right] + \frac{1}{\frac{\partial |\vec{k}^\nu|}{\partial \epsilon^\nu}}\left[\mu\overbrace{\frac{\partial f^\nu}{\partial \varphi}}^{=0}\frac{\partial \varphi}{\partial x}\right]\Bigg\}$$

$$= \frac{1}{\hbar}\frac{\partial}{\partial x_{\text{T}}}\left(\frac{f^\nu}{\frac{\partial |\vec{k}^\nu|}{\partial \epsilon^\nu}}T_{1,1}^{\text{HV},\nu}\mu\right) + \frac{1}{\hbar}T_{1,1}^{\text{HV},\nu}\Bigg\{\left[\mu\frac{\partial \epsilon^\nu}{\partial x}\right]\frac{\partial}{\partial \epsilon^\nu}\left(\frac{f^\nu}{\frac{\partial |\vec{k}^\nu|}{\partial \epsilon^\nu}}\right)$$

$$+ \frac{1}{\frac{\partial |\vec{k}^\nu|}{\partial \epsilon^\nu}}\left[\left(\frac{\partial f^\nu}{\partial \mu}\mu + f^\nu\right)\frac{\partial \mu}{\partial x}\right]\Bigg\}\,. \tag{2.4.4-164}$$

In the case of the transformed drift-term (2.4.4-163) the φ-dependence vanishes due to the transformation of the driving force (including Herring-Vogt) into the $(\epsilon^\nu, \mu, \varphi)$-frame (cf. equation (2.4.2-150) and (2.4.3-160))

$$\begin{pmatrix} T_{1,1}^{\text{HV},\nu}E_{\text{eff.}}^{\nu,x} \\ 0 \\ 0 \end{pmatrix} = \mu T_{1,1}^{\text{HV},\nu}E_{\text{eff.}}^{\nu,x}\,\vec{e}_{\epsilon^\nu} + \sqrt{1-\mu^2}T_{1,1}^{\text{HV},\nu}E_{\text{eff.}}^{\nu,x}\,\vec{e}_\mu + 0\,\vec{e}_\varphi\,. \tag{2.4.4-165}$$

Thus, the φ-dependence vanishes due to the equations involved. However, the φ-dependence of the transformed diffusion term (2.4.4-164) vanishes due to $\frac{\partial f^\nu}{\partial \varphi} = 0$, which originates from physical considerations. For a driving force after (2.4.4-162) with a negative x-component, the distribution function f^ν is pushed in positive $\tilde{k}_x^\nu$ direction. Since $E_{\text{eff.}}^{\vec{\nu}}$ consists only of a nonzero x-component, the distribution of f^ν among any semi-circle with constant $|\vec{k}^\nu|$ (or constant kinetic energy ϵ^ν) is independent of the actual φ. Thus, the omission of the φ-dependence is justified by the rotational symmetry of f^ν. Figure 2.21 illustrates schematically the situation. It should be noted that the term h_x (see equation (2.4.2-141) for $d = x$ in section 2.4.2.2) is function of φ and thus $E_{\text{eff.}}^{\nu,x}$ (2.4.0-108) (via $\frac{\partial \epsilon^\nu}{\partial x}$ (2.4.2-140)) and $\frac{\partial \mu}{\partial x}$ (2.4.2-142). However, for the 1D carrier transport in x-direction in combination with the used discretization scheme (see section 3), the term h_x transforms effectively to a φ-independent formulation $\overline{h_x}$ (see equation (B.3.0-28) in appendix B.3) and therefore $E_{\text{eff.}}^{\nu,x}$ and $\frac{\partial \mu}{\partial x}$ become for the 1D carrier transport in x-direction φ-independent, too. Table 2.4 summarizes the total number of dimensions

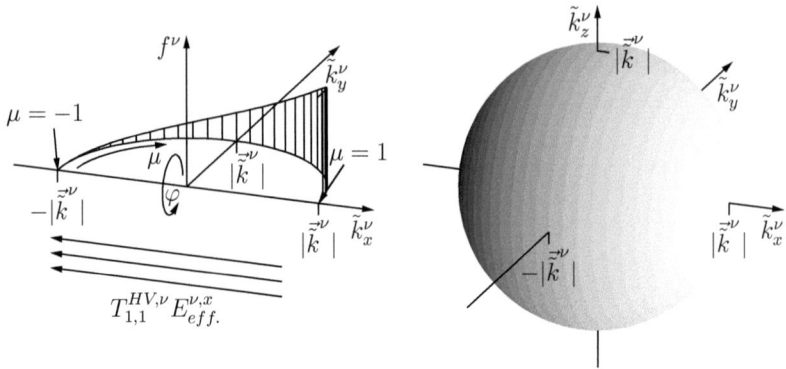

Figure 2.21: Schematic illustration of the rotational symmetry of f^ν for a 1D (real space) simulation (negative $E_{\text{eff.}}^{\nu,x}$). On the left-hand side, f^ν is shown on the semi-circle for a constant value of $|\vec{k}^\nu|$ (ϵ^ν). The course of the μ-variable used for the coordinate transformation is additionally sketched. On the right-hand side, the sphere with identical $|\vec{k}^\nu|$ is illustrated. The shading of the sphere corresponds to f^ν on the right-hand side, where small/large values of f^ν are indicated by a dark/bright shading, respectively.

needed for solving the BTE as function of the considered real space dimensions. The number of dimensions are given for both the Cartesian and the transformed $(\epsilon^\nu, \mu, \varphi)$ coordinate system of the reciprocal space. Thus, the employed coordinate transformation is very efficient when only a 1D transport of carriers is considered.

Transport regime (real space)	1D	2D	3D
Cartesian system $(\tilde{k}_x^\nu, \tilde{k}_y^\nu, \tilde{k}_z^\nu)$	4 (+1)	5 (+1)	6 (+1)
Tranformed system $(\epsilon^\nu, \mu, \varphi)$	3 (+1)	5 (+1)	6 (+1)

Table 2.4: Dimensional size of the considered simulation domain for the BTE. Note, the (+1) reflects the time domain, where only a few time slices need to be saved for approximating the transient term.

CHAPTER 3

Numerical treatment of the Boltzmann transport equation

In this chapter, the numerical treatment of the Boltzmann transport equation is described. Here, the main focus is put on the 1D transport of carriers, since the extension for 2D/3D transport can be deduced in a straight-forward manner from the 1D case. In addition, the computational burden (simulation time and memory requirements) is the smallest one for the 1D case due to the rotational symmetry of the carrier distribution function accomplished by the coordinate transformation (see section 2.4.2 and 2.4.4). Regarding the physical significance, 1D simulations are still of importance. For most semiconductor devices the main component of the carrier flux is aligned to one spatial direction. Thus, the main transport direction of carriers can be investigated by 1D BTE simulations, whereas simpler simulation methods, e.g. drift-diffusion (DD) or hydrodynamic (HD) can be used to account for 2D/3D effects. Having performed the coordinate transformation, the 1D BTE reads

(cf. section 2.4.3)

$$
\frac{\partial f^\nu}{\partial t} + \frac{1}{\hbar}\frac{\partial}{\partial x_\mathrm{T}}\left(\frac{f^\nu}{\frac{\partial |\vec{k}^\nu|}{\partial \epsilon^\nu}}T_{1,1}^{\mathrm{HV},\nu}\mu\right) + \frac{1}{\hbar}\left[T_{1,1}^{\mathrm{HV},\nu}\mu\frac{\partial \epsilon^\nu}{\partial x}\right]\frac{\partial}{\partial \epsilon^\nu}\left(\frac{f^\nu}{\frac{\partial |\vec{k}^\nu|}{\partial \epsilon^\nu}}\right)
$$

$$
+\frac{T_{1,1}^{\mathrm{HV},\nu}}{\hbar\frac{\partial |\vec{k}^\nu|}{\partial \epsilon^\nu}}\left(\frac{\partial f^\nu}{\partial \mu}\mu + f^\nu\right)\frac{\partial \mu}{\partial x} - \frac{q}{\hbar\,\mathrm{dos}^\nu}\left\{\frac{\partial}{\partial \epsilon^\nu}\left(\frac{\mathrm{dos}^\nu f^\nu}{\frac{\partial |\vec{k}^\nu|}{\partial \epsilon^\nu}}\mu T_{1,1}^{\mathrm{HV},\nu}E_{\mathrm{eff.}}^{\nu,x}\right)\right.
$$

$$
\left.+\frac{\partial}{\partial \mu}\left(\frac{\mathrm{dos}^\nu f^\nu(1-\mu^2)}{|\vec{k}^\nu|}T_{1,1}^{\mathrm{HV},\nu}E_{\mathrm{eff.}}^{\nu,x}\right)\right\} = C\ , \tag{3.0.0-1}
$$

in which for convenience the dependencies of the quantities involved have been omitted (e.g. $f^\nu(x,\epsilon^\nu,\mu,\varphi)$, $T_{1,1}^{\mathrm{HV},\nu}(x)$). For the further treatment of (3.0.0-1), a new solution variable is introduced [19, 20, 48]:

$$
F^\nu(x_\mathrm{T},\epsilon^\nu,\mu,\varphi) = f^\nu(x_\mathrm{T},\epsilon^\nu,\mu,\varphi)\,\mathrm{dos}^\nu(x_\mathrm{T},\epsilon^\nu)\ . \tag{3.0.0-2}
$$

With F^ν after (3.0.0-2) numerical issues due to the range of values of f^ν are damped and the integration over the reciprocal space is prepared. As depicted in section 2.4.2.3, the DOS is a remnant of the coordinate transformation in the reciprocal space and transforms the infinitesimal volume $d\vec{k}^\nu$ in the new frame (see (2.4.2-155)). Thus, when a **finite volume** or **box integration method** is employed for the discretization of (3.0.0-1), the BTE is integrated over a discretized control volume in the reciprocal space. Thus, the BTE (3.0.0-1) is multiplied first by the DOS in order to integrate over the transformed reciprocal space and reads with (3.0.0-2)

$$
\frac{\partial F^\nu}{\partial t} + \frac{\mathrm{dos}^\nu}{\hbar}\frac{\partial}{\partial x_\mathrm{T}}\left(\frac{f^\nu}{\frac{\partial |\vec{k}^\nu|}{\partial \epsilon^\nu}}T_{1,1}^{\mathrm{HV},\nu}\mu\right) + \frac{\mathrm{dos}^\nu}{\hbar}\left[T_{1,1}^{\mathrm{HV},\nu}\mu\frac{\partial \epsilon^\nu}{\partial x}\right]\frac{\partial}{\partial \epsilon^\nu}\left(\frac{F^\nu}{\mathrm{dos}^\nu\frac{\partial |\vec{k}^\nu|}{\partial \epsilon^\nu}}\right)
$$

$$
+\frac{T_{1,1}^{\mathrm{HV},\nu}}{\hbar\frac{\partial |\vec{k}^\nu|}{\partial \epsilon^\nu}}\left(\frac{\partial F^\nu}{\partial \mu}\mu + F^\nu\right)\frac{\partial \mu}{\partial x} - \frac{q}{\hbar}\left\{\frac{\partial}{\partial \epsilon^\nu}\left(\frac{F^\nu}{\frac{\partial |\vec{k}^\nu|}{\partial \epsilon^\nu}}\mu T_{1,1}^{\mathrm{HV},\nu}E_{\mathrm{eff.}}^{\nu,x}\right)\right.
$$

$$
\left.+\frac{\partial}{\partial \mu}\left(\frac{F^\nu(1-\mu^2)}{|\vec{k}^\nu|}T_{1,1}^{\mathrm{HV},\nu}E_{\mathrm{eff.}}^{\nu,x}\right)\right\} = \mathrm{dos}^\nu\,C\ . \tag{3.0.0-3}
$$

Rewriting the derivative $\frac{\partial}{\partial x_T}$ associated with the transformed diffusion term (see appendix B.2, equation (B.2.0-15)) leads to

$$
\frac{\partial F^\nu}{\partial t} + \frac{1}{\hbar}\frac{\partial}{\partial x_T}\left(\frac{F^\nu}{\frac{\partial |\vec{k}^\nu|}{\partial \epsilon^\nu}}T_{1,1}^{\mathrm{HV},\nu}\mu\right) - \frac{1}{\hbar}\frac{F^\nu}{\frac{\partial |\vec{k}^\nu|}{\partial \epsilon^\nu}}T_{1,1}^{\mathrm{HV},\nu}\mu\frac{1}{\mathrm{dos}^\nu}\frac{\partial \mathrm{dos}^\nu}{\partial x_T}
$$

$$
+ \frac{\mathrm{dos}^\nu}{\hbar}\left[T_{1,1}^{\mathrm{HV},\nu}\mu\frac{\partial \epsilon^\nu}{\partial x}\right]\frac{\partial}{\partial \epsilon^\nu}\left(\frac{F^\nu}{\mathrm{dos}^\nu\frac{\partial |\vec{k}^\nu|}{\partial \epsilon^\nu}}\right) + \frac{T_{1,1}^{\mathrm{HV},\nu}}{\hbar\frac{\partial |\vec{k}^\nu|}{\partial \epsilon^\nu}}\left(\frac{\partial F^\nu}{\partial \mu}\mu + F^\nu\right)\frac{\partial \mu}{\partial x}
$$

$$
- \frac{q}{\hbar}\left\{\frac{\partial}{\partial \epsilon^\nu}\left(\frac{F^\nu}{\frac{\partial |\vec{k}^\nu|}{\partial \epsilon^\nu}}\mu T_{1,1}^{\mathrm{HV},\nu}E_{\mathrm{eff.}}^{\nu,x}\right) + \frac{\partial}{\partial \mu}\left(\frac{F^\nu(1-\mu^2)}{|\vec{k}^\nu|}T_{1,1}^{\mathrm{HV},\nu}E_{\mathrm{eff.}}^{\nu,x}\right)\right\} = \mathrm{dos}^\nu\, C\ .
$$

$$(3.0.0\text{-}4)$$

For further treatment of the BTE, the dispersion relation is needed. Here, the non-parabolic dispersion relation (2.2.0-43) is assumed, where a possible anisotropy is already accounted for in (3.0.0-1) via the Herring-Vogt transformation (2.2.0-46). Although the scheme described so far is capable of more complex descriptions of the dispersion relation $\epsilon^\nu(|\vec{k}^\nu|)$, the non-parabolic one (2.2.0-43) is used further on for comparison purposes with other simulation techniques and published results. With (2.2.0-43) the missing quantities in (3.0.0-4) read

$$
|\vec{k}^\nu| = \frac{\sqrt{2m_0 m_\nu^*}}{\hbar}\sqrt{\epsilon^\nu\left(1 + \alpha_\nu \epsilon^\nu\right)}\ , \tag{3.0.0-5}
$$

$$
\frac{\partial |\vec{k}^\nu|}{\partial \epsilon^\nu} = \frac{\sqrt{2m_0 m_\nu^*}}{\hbar}\frac{1}{2}\frac{1 + 2\alpha_\nu \epsilon^\nu}{\sqrt{\epsilon^\nu\left(1 + \alpha_\nu \epsilon^\nu\right)}}\ , \tag{3.0.0-6}
$$

$$
\frac{\partial |\vec{k}^\nu|}{\partial x} = \frac{\sqrt{2m_0 m_\nu^*}}{\hbar}\sqrt{\epsilon^\nu\left(1 + \alpha_\nu \epsilon^\nu\right)}\frac{1}{2}\left(\frac{1}{m_\nu^*}\frac{\partial m_\nu^*}{\partial x} + \frac{(\epsilon^\nu)^2}{\epsilon^\nu\left(1 + \alpha_\nu \epsilon^\nu\right)}\frac{\partial \alpha_\nu}{\partial x}\right)\ , \tag{3.0.0-7}
$$

$$
\mathrm{dos}^\nu = \frac{1}{2}\left(\frac{\sqrt{2m_0 m_\nu^*}}{\hbar}\right)^3\sqrt{\epsilon^\nu\left(1 + \alpha_\nu \epsilon^\nu\right)}\left(1 + 2\alpha_\nu \epsilon^\nu\right)\ . \tag{3.0.0-8}
$$

3.1 Normalization of the BTE

The numerical treatment demands a dimensionless equation, which is obtained by dividing the BTE by its normalization factor. With a suitable choice of normalization factors the BTE can additionally be scaled in such a way that numerical issues are prevented (numerical under- or overflow). Thus, it is useful to construct the normalization factors in such a way that they either consist of physical constants or physical quantities involved in the BTE. For potentials and voltages it is a natural choice to use the thermal voltage as normalization factor:

$$V_T = \frac{k_b T_L}{q} = V_N \ , \tag{3.1.0-9}$$

where T_L is the lattice temperature and V_N labels the normalization factor for potentials or voltages. Based on (3.1.0-9), the normalization factor for energies can directly be derived by a multiplication with elementary charge q:

$$E_N = qV_N = k_b T_L \ . \tag{3.1.0-10}$$

With (3.1.0-10) and the $|\vec{k}^\nu|$ after (3.0.0-5) the normalization factor for k-vectors reads

$$k_N = \frac{\sqrt{2m_0 k_b T_L}}{\hbar} \ . \tag{3.1.0-11}$$

With k_N, the carrier and doping densities are normalized by

$$D_N = k_N^3 = \left(\frac{\sqrt{2m_0 E_N}}{\hbar} \right)^3 = \left(\frac{\sqrt{2m_0 k_b T_L}}{\hbar} \right)^3 \ . \tag{3.1.0-12}$$

Based on the obtained normalization factor D_N, the DOS after (3.0.0-8) turns into its normalized form to

$$
\begin{aligned}
\mathrm{dos}^\nu \ &= \ \frac{1}{2} \left(\frac{\sqrt{2m_0 m_\nu^*}}{\hbar} \right)^3 \sqrt{\epsilon^\nu \left(1 + \alpha_\nu \epsilon^\nu\right)} \left(1 + 2\alpha_\nu \epsilon^\nu\right) \ , \\
&= \ \frac{D_N}{E_N} \frac{\left(\sqrt{m_\nu^*}\right)^3}{2} s^\nu(w^\nu) \ ,
\end{aligned}
\tag{3.1.0-13}
$$

with the energy dependent function $s^\nu(w^\nu)$ associated to the DOS

$$s^\nu(w^\nu) = \sqrt{w^\nu(1 + a_\nu w^\nu)}\,(1 + 2a_\nu w^\nu) \qquad (3.1.0\text{-}14)$$

where $w^\nu = \frac{\epsilon^\nu}{E_N}$ and the normalized non-parabolicity factor

$$a_\nu = \alpha_\nu E_N \ . \qquad (3.1.0\text{-}15)$$

With (3.1.0-13) and (3.1.0-14) the new solution variable after (3.0.0-2) reads

$$
\begin{aligned}
F^\nu &= \frac{1}{2}\frac{D_N}{E_N}(\sqrt{m_\nu^*})^3 f^\nu s^\nu(w^\nu) \\
&= \frac{1}{2}\frac{D_N}{E_N}\Phi^\nu \ , \\
\Phi^\nu &= (\sqrt{m_\nu^*})^3 f^\nu s^\nu(w^\nu) \ , \qquad (3.1.0\text{-}16)
\end{aligned}
$$

where Φ^ν is the normalized version of F^ν after (3.0.0-2) including the term $(\sqrt{m_\nu^*})^3$. The inclusion of the effective mass m_ν^* in (3.1.0-16) facilitates the discretization of the BTE, especially for a position dependent m_ν^*. It will be shown later on that the inclusion of m_ν^* in the new solution variable (3.1.0-16) prevents corrective terms associated with a position dependence of m_ν^*. For the BTE (3.0.0-4), normalization factors for the energy and the k-vectors are given by (3.1.0-10) and (3.1.0-11), respectively. The variables μ and φ are given in radians and hence are dimensionless. However, normalization factors for the spatial dimensions x, x_T and the time t are missing. For the normalization factor of the time t it is focused on the collision term of the BTE, since its unit is $1/s$. The unit $1/s$ originates from transition rates in conjunction with integration over the volume of the reciprocal space (the distribution functions f^ν, $f^{\nu'}$ are dimensionless). The following normalization factor is found for the time t:

$$t_N = \frac{E_N}{D_N K_N} \ . \qquad (3.1.0\text{-}17)$$

K_N is the normalization factor for all transition rates:

$$K_N = \max_X \left(K_N^X \right) = \max_X \left(\frac{\Omega}{8\pi^3} A^X \right) \, , \tag{3.1.0-18}$$

where X labels the considered scattering mechanisms with transition rates of the form

$$S^X \quad = \quad A^X \delta \left(\epsilon^{\nu'} - \epsilon^\nu \pm \Delta\epsilon^X \right) \, . \tag{3.1.0-19}$$

$\Delta\epsilon^X$ accounts for an inelastic scattering process and equals zero for an elastic one. The last remaining normalization factor is the one for lengths labeled by L_N. For L_N the maximum length of the real space simulation domain is usually employed. However, for BULK simulations – a single point within an infinite homogeneous simulation domain (semiconducting crystal) – L_N is not defined. Nevertheless, L_N is needed in order to normalize the electric fields (by V_N/L_N). An alternative definition of L_N valid for both BULK and device simulations reads

$$L_N = \frac{E_N^2}{\hbar D_N k_N K_N} \, . \tag{3.1.0-20}$$

For more information about the normalization factors and their derivations the reader is referred to appendix B.1. Table 3.1 summarizes the independent and table 3.2 the dependent normalization factors, respectively.

Quantity	Symbol	Normal. factor	Unit
Energy	ϵ	$E_N = k_b T_L$	eV
Recip. vectors	$\vec{k}, \vec{q}_d, ...$	$k_N = \frac{\sqrt{2m_0 k_b T_L}}{\hbar}$	m^{-1}
Transition rates	$\frac{\Omega}{8\pi^3} S^X(\vec{k}, \vec{k}')$	$K_N = \max_X \left(\frac{\Omega}{8\pi^3} A^X \right)$	eVm^3/s

Table 3.1: Independent normalization factors.

Quantity	Symbol	Normal. factor	Unit
Potentials	V, ψ	$V_N = \frac{k_b T_L}{q} = \frac{E_N}{q}$	V
Densities	$n, N_D,...$	$D_N = k_N^3$	m^{-3}
Lengths	$x, \Delta x, y,...$	$L_N = \frac{E_N^2}{\hbar D_N k_N K_N}$	m
Electric fields	$\vec{E}$	$F_N = \frac{V_N}{L_N}$	V/m
Time	t	$t_N = \frac{E_N}{D_N K_N}$	s
Velocities	$\vec{v_g}$	$v_N = \frac{L_N}{t_N}$	m/s

Table 3.2: Dependent normalization factors.

Transient term

First, the transient term in equation (3.0.0-4) is considered:

$$\frac{\partial F^\nu}{\partial t} = \frac{1}{t_N}\frac{\partial F^\nu}{\partial \overline{t}} = \frac{D_N}{E_N t_N}\frac{1}{2}\frac{\partial \Phi^\nu}{\partial \overline{t}}$$

$$= \frac{1}{2}\frac{D_N^2 K_N}{E_N^2}\frac{\partial \Phi^\nu}{\partial \overline{t}} \, , \tag{3.1.0-21}$$

with Φ^ν after (3.1.0-16). At this point, the normalization factor for the BTE (3.0.0-4) can be derived from (3.1.0-21):

$$BTE_N = \frac{1}{2}\frac{D_N^2 K_N}{E_N^2} \, . \tag{3.1.0-22}$$

The unit of the normalization factor (3.1.0-22) equals $\frac{m^{-3}}{eV\,s}$ instead of s^{-1}, since both the BTE (3.0.0-4) and the the new solution variable (3.0.0-2) have been multiplied by the DOS. Note, the effective mass m_ν^* of the currently considered valley ν is not included in the normalization factor, but is considered in the normalized solution variable (see equation (3.1.0-16)). Thus, the normalization factor of the BTE after (3.1.0-22) is position independent.

Diffusion term

The normalization of the diffusion term in conjunction with the factor (3.1.0-22) is more elaborate due to the derivatives w.r.t. the real space dimension x_T. The first two terms of the transformed diffusion term in (3.0.0-4) become with (3.0.0-6) and (3.0.0-8)

$$
\frac{1}{\hbar} \frac{\partial}{\partial x_T} \left(\frac{F^\nu}{\frac{\partial |\vec{k}^\nu|}{\partial \epsilon^\nu}} T_{1,1}^{\mathrm{HV},\nu} \mu \right) - \frac{1}{\hbar} \frac{F^\nu}{\frac{\partial |\vec{k}^\nu|}{\partial \epsilon^\nu}} T_{1,1}^{\mathrm{HV},\nu} \mu \frac{1}{\mathrm{dos}^\nu} \frac{\partial \mathrm{dos}^\nu}{\partial x_T}
$$

$$
= \frac{1}{\hbar} \frac{D_\mathrm{N}}{k_\mathrm{N} L_\mathrm{N}} \frac{1}{2} \left[\frac{\partial}{\partial \overline{x_T}} \left(\frac{\Phi^\nu}{\sqrt{m_\nu^*}} \frac{2\sqrt{w^\nu (1 + a_\nu w^\nu)}}{1 + 2a_\nu w^\nu} T_{1,1}^{\mathrm{HV},\nu} \mu \right) \right.
$$

$$
\left. - \frac{\Phi^\nu T_{1,1}^{\mathrm{HV},\nu} \mu}{\sqrt{m_\nu^*}(1 + 2a_\nu w^\nu)^2} \left(\frac{3}{m_\nu^*} \frac{\partial m_\nu^*}{\partial x_T} s^\nu(w^\nu) + \frac{w^{\nu 2}(6a_\nu w^\nu + 5)}{\sqrt{w^\nu (1 + a_\nu w^\nu)}} \frac{\partial a_\nu}{\partial \overline{x_T}} \right) \right],
$$

$$\tag{3.1.0-23}$$

with $\overline{x_T} = x_T / L_\mathrm{N}$. Note, due to the definition of L_N after (3.1.0-20), the prefactor of (3.1.0-23) is identical to the normalization factor (3.1.0-22) for the whole BTE:

$$
\frac{1}{\hbar} \frac{D_\mathrm{N}}{k_\mathrm{N} L_\mathrm{N}} \frac{1}{2} = \frac{1}{2} \frac{D_\mathrm{N}^2 K_\mathrm{N}}{E_\mathrm{N}^2} .
$$

$$\tag{3.1.0-24}$$

Inserting (3.0.0-6)-(3.0.0-8) into the remaining two parts of the transformed diffusion term leads after normalizing to

$$
\frac{\mathrm{dos}^\nu}{\hbar} \left[T_{1,1}^{\mathrm{HV},\nu} \mu \frac{\partial \epsilon^\nu}{\partial x} \right] \frac{\partial}{\partial \epsilon^\nu} \left(\frac{F^\nu}{\mathrm{dos}^\nu \frac{\partial |\vec{k}^\nu|}{\partial \epsilon^\nu}} \right)
$$

$$
= BTE_\mathrm{N} \left[\frac{\partial}{\partial w^\nu} \left(\frac{2\sqrt{w^\nu(1 + a_\nu w^\nu)}}{\sqrt{m_\nu^*}(1 + 2a_\nu w^\nu)} \Phi^\nu T_{1,1}^{\mathrm{HV},\nu} \mu \frac{\partial w^\nu}{\partial \overline{x}} \right) \right.
$$

$$
\left. - \frac{2\Phi^\nu T_{1,1}^{\mathrm{HV},\nu} \mu}{\sqrt{m_\nu^*}(1 + 2a_\nu w^\nu)^2} \frac{\partial}{\partial w^\nu} \left(s^\nu(w^\nu) \frac{\partial w^\nu}{\partial \overline{x}} \right) \right]
$$

$$\tag{3.1.0-25}$$

and

$$\frac{T_{1,1}^{\mathrm{HV},\nu}}{\hbar\frac{\partial|\vec{k}^\nu|}{\partial\epsilon^\nu}}\left(\frac{\partial F^\nu}{\partial\mu}\mu + F^\nu\right)\frac{\partial\mu}{\partial x} = BTE_{\mathrm{N}}\left[\frac{\partial}{\partial\mu}\left(\frac{2\sqrt{w^\nu(1+a_\nu w^\nu)}}{\sqrt{m_\nu^*}(1+2a_\nu w^\nu)}\Phi^\nu T_{1,1}^{\mathrm{HV},\nu}\mu\frac{\partial\mu}{\partial\overline{x}}\right)\right.$$

$$\left.-\frac{2\sqrt{w^\nu(1+a_\nu w^\nu)}}{\sqrt{m_\nu^*}(1+2a_\nu w^\nu)}\Phi^\nu\mu T_{1,1}^{\mathrm{HV},\nu}\frac{\partial}{\partial\mu}\left(\frac{\partial\mu}{\partial\overline{x}}\right)\right]$$

$$(3.1.0\text{-}26)$$

with $\overline{x} = x/L_{\mathrm{N}}$. The matrix elements $T_{1,1}^{\mathrm{HV},\nu} - T_{3,3}^{\mathrm{HV},\nu}$ of the Herring-Vogt transformation matrix (2.2.0-46), where $T_{2,2}^{\mathrm{HV},\nu}$ and $T_{3,3}^{\mathrm{HV},\nu}$ are involved in $\frac{\partial w^\nu}{\partial\overline{x}}$ (3.1.0-25) and $\frac{\partial\mu}{\partial\overline{x}}$ (3.1.0-26) (see section 2.4.2.2), are intentionally not replaced by their analytic expressions. The reason is to emphasize the contribution of an anisotropic dispersion relation. Thus, for an isotropic dispersion relation equations (3.1.0-23)-(3.1.0-26) are also valid, but the matrix elements are all set to one. In this case however, the effective mass m_ν^* (and the normalized non-parabolicity factor a_ν) might still be a function of the real space position (position dependent isotropic dispersion relation), which is also captured by the equations above. For a more detailed derivation of (3.1.0-23)-(3.1.0-26), the reader is referred to appendix B.2.

Drift term

The drift term of the BTE (3.0.0-4) becomes

$$-\frac{q}{\hbar}\left\{\frac{\partial}{\partial\epsilon^\nu}\left(\frac{F^\nu}{\partial|\vec{k}^\nu|}\mu T_{1,1}^{\mathrm{HV},\nu}E_{\mathrm{eff.}}^{\nu,x}\right)+\frac{\partial}{\partial\mu}\left(\frac{F^\nu(1-\mu^2)}{|\vec{k}^\nu|}T_{1,1}^{\mathrm{HV},\nu}E_{\mathrm{eff.}}^{\nu,x}\right)\right\}$$

$$=-\frac{q}{\hbar}\frac{D_{\mathrm{N}}}{E_{\mathrm{N}}}\frac{1}{2k_{\mathrm{N}}}F_{\mathrm{N}}\left\{\frac{\partial}{\partial w^\nu}\left(\frac{2\Phi^\nu\sqrt{w^\nu(1+a_\nu w^\nu)}}{\sqrt{m_\nu^*}(1+2a_\nu w^\nu)}\mu T_{1,1}^{\mathrm{HV},\nu}\overline{E_{\mathrm{eff.}}^{\nu,x}}\right)\right.$$

$$\left.+\frac{\partial}{\partial\mu}\left(\frac{\Phi^\nu(1-\mu^2)}{\sqrt{m_\nu^*}\sqrt{w^\nu(1+a_\nu w^\nu)}}T_{1,1}^{\mathrm{HV},\nu}\overline{E_{\mathrm{eff.}}^{\nu,x}}\right)\right\}, \quad (3.1.0\text{-}27)$$

with F_N from table 3.2 and the normalized field $\overline{E_{\text{eff.}}^{\nu,x}} = E_{\text{eff.}}^{\nu,x}/F_N$. Again, the prefactor in (3.1.0-27) is equal to the normalization factor BTE_N (3.1.0-22):

$$\frac{q}{\hbar}\frac{D_N}{E_N}\frac{1}{2k_N}F_N = \frac{1}{2}\frac{D_N^2 K_N}{E_N^2} \; . \tag{3.1.0-28}$$

Collision term

Based on the transformed collision term (see section 2.4.3, equation (2.4.3-161)) the collision term in (3.0.0-4) becomes

$$\text{dos}^\nu C = \text{dos}^\nu \sum_X \frac{\Omega}{8\pi^3} \int_{\text{BZ}} \left(S^X(\{\acute{\epsilon}^{\nu'},\mu',\varphi'\},\{\epsilon^\nu,\mu,\varphi\}) f^{\nu'} (1 - f^\nu) \right.$$

$$\left. - S^X(\{\epsilon^\nu,\mu,\varphi\},\{\acute{\epsilon}^{\nu'},\mu',\varphi'\}) f^\nu \left(1 - f^{\nu'}\right) \right) \text{dos}^{\nu'} d\acute{\epsilon}^{\nu'} d\mu' d\varphi'$$

$$= \frac{(\sqrt{m_\nu^*})^3}{2}\frac{D_N^2 K_N}{E_N^2} s^\nu(w^\nu) \left[\sum_X \int_{\text{BZ}} \left(\overline{S^X}(\{\acute{w}^{\nu'},\mu',\varphi'\},\{w^\nu,\mu,\varphi\}) f^{\nu'} (1 - f^\nu) \right. \right.$$

$$\left. \left. - \overline{S^X}(\{w^\nu,\mu,\varphi\},\{\acute{w}^{\nu'},\mu',\varphi'\}) f^\nu \left(1 - f^{\nu'}\right) \right) \frac{(\sqrt{m_{\nu'}^*})^3 s^{\nu'}(\acute{w}^{\nu'})}{2} d\acute{w}^{\nu'} d\mu' d\varphi' \right]$$

$$\tag{3.1.0-29}$$

$$= BTE_N (\sqrt{m_\nu^*})^3 s^\nu(w^\nu)\overline{C} \; , \tag{3.1.0-30}$$

with $s^\nu(w^\nu)$ and $s^{\nu'}(\acute{w}^{\nu'})$ after (3.1.0-14) and the normalized transition rate

$$\overline{S^X}(\{w^\nu,\mu,\varphi\},\{\acute{w}^{\nu'},\mu',\varphi'\}) = \frac{\frac{\Omega}{8\pi^3} S^X(\{\epsilon^\nu,\mu,\varphi\},\{\acute{\epsilon}^{\nu'},\mu',\varphi'\})}{K_N} \; , \tag{3.1.0-31}$$

where superscript X labels the considered scattering mechanisms. Note, in the course of the normalization the factor E_N originating from the normalization of $d\acute{\epsilon}^{\nu'}$ cancels out with $1/E_N$ originating from ϵ^ν, $\acute{\epsilon}^{\nu'}$ in the argument of the transition rates $S^X(\dots)$. The cancellation of E_N stems from the normalization of the δ-functions involved in the transition rates, describing the energy conservation (see section 2.3). An example for the cancellation of E_N is also presented in appendix B.1 by equation (B.1.0-4) and (B.1.0-5).

3.2 Discretization of the 1D BULK BTE

Based on normalized terms of the BTE obtained in the previous section, their discretization is described in this section. Here, the BULK case is considered first, which represents the simulation of one point within an infinitely large and homogeneous semiconductor. Thus, in the BULK case the diffusion term of the BTE is neglected due to the assumed homogeneity in real space. With the normalized transient (3.1.0-21), drift (3.1.0-27) and collision term (3.1.0-30), the BULK BTE reads

$$
\frac{\partial \Phi^\nu}{\partial \overline{t}} - \left\{ \frac{\partial}{\partial w^\nu} \left(\frac{2\Phi^\nu \sqrt{w^\nu (1 + a_\nu w^\nu)}}{\sqrt{m_\nu^*} (1 + 2a_\nu w^\nu)} \mu T_{1,1}^{\mathrm{HV},\nu} \overline{E_{\mathrm{eff.}}^{\nu,x}} \right) \right.
$$
$$
\left. + \frac{\partial}{\partial \mu} \left(\frac{\Phi^\nu (1 - \mu^2)}{\sqrt{m_\nu^*} \sqrt{w^\nu (1 + a_\nu w^\nu)}} T_{1,1}^{\mathrm{HV},\nu} \overline{E_{\mathrm{eff.}}^{\nu,x}} \right) \right\} = (\sqrt{m_\nu^*})^3 s^\nu (w^\nu) \overline{C} \ , \quad (3.2.0\text{-}32)
$$

where the BULK BTE (3.2.0-32) is divided by the normalization factor BTE_N (3.1.0-22). Note, since (3.2.0-32) is only capable of considering the effective field $\overline{E_{\mathrm{eff.}}^{\nu,x}}$ in x-direction, (3.2.0-32) is referred to as 1D BULK BTE. In addition, for the 1D BULK BTE the effective field $\overline{E_{\mathrm{eff.}}^{\nu,x}}$ equals the electric field – due to the assumed homogeneity in real space – and is specified by the user.

The BTE (3.2.0-32) depends primarily on the variables w^ν and μ. Thus, a grid needs to be defined for the w^ν-μ simulation domain. The w^ν-μ domain ($w^\nu \in [0, w_{\max}]$, $\mu \in [-1, 1]$) is portioned into equidistant sections Δw and $\Delta \mu$, respectively. The discretized energy and angle is given by

$$
w_i^\nu = \frac{\Delta w}{2} + (i-1)\Delta w \ , \ i = 1, ..., N_w = \frac{w_{\max}}{\Delta w} \ , \quad (3.2.0\text{-}33)
$$
$$
\mu_j = -1 + \frac{\Delta \mu}{2} + (j-1)\Delta \mu \ , \ j = 1, ..., N_\mu = \frac{2}{\Delta \mu} \ , \quad (3.2.0\text{-}34)
$$

where N_w and N_μ are the number of grid points in w^ν and μ direction, respectively. Note, in contrast to common grids, a so called staggered grid is employed here. Therefore, the boundaries of the simulation domain are not explicitly discretized by grid points. For each point denoted by i, j, equation (3.2.0-32) is integrated over the control volume $[w_{i-\frac{1}{2}}, w_{i+\frac{1}{2}}] \times [\mu_{j-\frac{1}{2}}, \mu_{j+\frac{1}{2}}]$.

The integration can directly be applied to (3.2.0-32), since the BTE is already multiplied by DOS (cf. conversion of (3.0.0-1) to (3.0.0-3), (3.0.0-4)). Within the framework of the BIM the following assumptions/approximations are made:

i.) scalar quantities associated with a discretized grid point are assumed to be constant within the control volume spanned around the grid point and

ii.) vector fields/fluxes between two adjacent grid points are assumed to be constant between those two points and are assigned to the bounding surface of the control volume(s).

For the first assumption, a change of the variable Φ^ν is performed [19, 20, 48]. Instead of solving (3.2.0-32) directly for Φ^ν, the average value of Φ^ν within the control volume is used:

$$\langle \Phi^\nu \left(w^\nu, \mu, \bar{t} \right) \rangle = \frac{1}{2\pi \Delta w \Delta \mu} \int_0^{2\pi} \int_{\mu_{j-\frac{1}{2}}}^{\mu_{j+\frac{1}{2}}} \int_{w_{i-\frac{1}{2}}^\nu}^{w_{i+\frac{1}{2}}^\nu} \Phi^\nu \left(w^\nu, \mu, \bar{t} \right) dw^\nu \, d\mu \, d\varphi$$

$$= \frac{(\sqrt{m_\nu^*})^3}{\Delta w \Delta \mu} \int_{\mu_{j-\frac{1}{2}}}^{\mu_{j+\frac{1}{2}}} \int_{w_{i-\frac{1}{2}}^\nu}^{w_{i+\frac{1}{2}}^\nu} f^\nu \left(w^\nu, \mu, \bar{t} \right) s^\nu (w^\nu) dw^\nu d\mu = \frac{n_{i,j}^\nu(\bar{t})}{\Delta w \Delta \mu} \quad . \quad (3.2.0\text{-}35)$$

From the physical point of view, $n_{i,j}^\nu(\bar{t})$ equals the normalized electron density within the control volume of the valley ν in the reciprocal space. By using $n_{i,j}^\nu(\bar{t})$ instead of $f^\nu s^\nu$ or Φ^ν no additional errors/assumptions have to be made for the integration over the control volume. Note, the integration over φ in (3.2.0-35) results in a multiplication by 1 due to the coordinate transformation and rotational symmetry of f^ν for a 1D (real space) simulation (cf. section 2.4.3 and figure 2.21). Nevertheless, for a 2D/3D real space domain, the control volume also spans in φ-direction by a $\Delta\varphi$. In this case, the φ-dependence needs to be considered analogously to w^ν-, μ-dependencies in (3.2.0-35).

The above listed assumption ii.) is advantageous, if a volume integral over the divergence of a vector field needs to be evaluated. By using *Gauss's* theorem such an integral can be converted into a surface integral [8]:

$$\int \text{div}\,(\vec{v})\ dV = \oint \vec{v}\cdot\vec{n}\ dS\ , \qquad (3.2.0\text{-}36)$$

with a vector field $\vec{v}$ and the outward pointing unit normal vector $\vec{n}$ of the bounding surface. Since a vector field is assigned to the bounding surface of the control volume, the right-hand side of (3.2.0-36) can directly be applied:

$$\oint_{d\Omega_{i,j}} \vec{v}\cdot\vec{n}\ dS = \int_{w^{\nu}_{i-\frac{1}{2}}}^{w^{\nu}_{i+\frac{1}{2}}} v^{\mu}_{i,j+\frac{1}{2}}\,dw^{\nu} + \int_{\mu_{j-\frac{1}{2}}}^{\mu_{j+\frac{1}{2}}} v^{w^{\nu}}_{i+\frac{1}{2},j}\,d\mu$$

$$- \left(\int_{w^{\nu}_{i-\frac{1}{2}}}^{w^{\nu}_{i+\frac{1}{2}}} v^{\mu}_{i,j-\frac{1}{2}}\,dw^{\nu} + \int_{\mu_{j-\frac{1}{2}}}^{\mu_{j+\frac{1}{2}}} v^{w^{\nu}}_{i-\frac{1}{2},j}\,d\mu \right)\ , (3.2.0\text{-}37)$$

where $d\Omega_{i,j}$ is the bounding surface of the control volume $\Omega_{i,j}$. Therefore, it is advantageous to use the conservative form of the BTE (cf. section 2.4, (2.4.0-107) to (2.4.0-110)). Figure 3.1 illustrates the applied discretization and the components of a vector field $\vec{v}$ assigned to the bounding surface of a control volume. Note, the minus in front of the second part of (3.2.0-37) stems from the assumed direction of $\vec{v}$ as also shown in figure 3.1. Since $v^{\mu}_{i,j-\frac{1}{2}}$ and $v^{w^{\nu}}_{i-\frac{1}{2},j}$ are inward pointing, the direction is corrected by the minus.

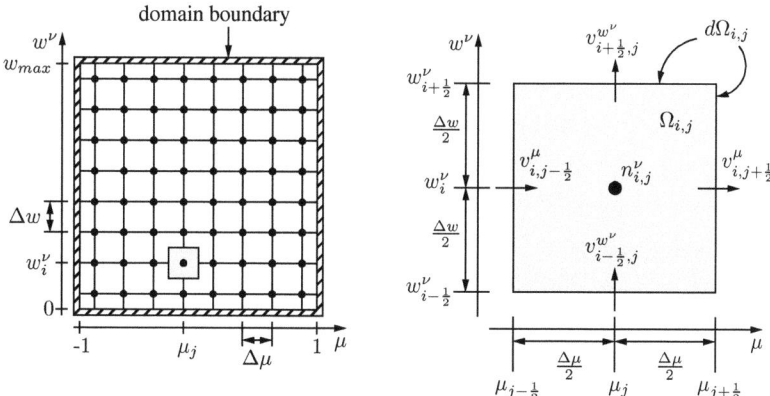

Figure 3.1: Discretization of the transformed reciprocal space (left) and a control volume together with the assumed directions of a flux $\vec{v}$ (right).

3.2.1 Transient term

The transient term of (3.2.0-32) can directly be integrated over the control volume $\Omega_{i,j}$ and becomes

$$\int_{w_{i-\frac{1}{2}}^{\nu}}^{w_{i+\frac{1}{2}}^{\nu}} \int_{\mu_{j-\frac{1}{2}}}^{\mu_{j+\frac{1}{2}}} \frac{\partial \Phi^\nu}{\partial \bar{t}} d\mu \, dw^\nu \approx \int_{w_{i-\frac{1}{2}}^{\nu}}^{w_{i+\frac{1}{2}}^{\nu}} \int_{\mu_{j-\frac{1}{2}}}^{\mu_{j+\frac{1}{2}}} \frac{\partial \langle \Phi^\nu \rangle}{\partial \bar{t}} d\mu \, dw^\nu$$

$$= \frac{\partial n_{i,j}^\nu}{\partial \bar{t}} \,, \qquad (3.2.1\text{-}38)$$

where (3.2.0-35) has been employed. For the numerical treatment of the time derivative, one can for example use the *implicit Euler* method:

$$\frac{\partial n_{i,j}^\nu}{\partial \bar{t}} \approx \frac{n_{i,j}^{\nu,\tau} - n_{i,j}^{\nu,\tau-1}}{\Delta \bar{t}} \,, \qquad (3.2.1\text{-}39)$$

with the currently observed time slice τ and the previous time slice $\tau - 1$ being separated by the normalized time step $\Delta \bar{t}$. According to (3.2.1-39) only one additional time slice needs to be known (here $\tau - 1$). Also higher order schemes applied to the time derivative do not need the results of all considered time slices. Thus, the discretization of the total considered time range for the simulation can be omitted, which is reflected by (+1) in table 2.4, section 2.4.3.

3.2.2 Drift term

Focusing on the drift part of (3.2.0-32), derivatives $\frac{\partial}{\partial w^\nu}$ and $\frac{\partial}{\partial \mu}$ are present. As shown in section 2.4.2.3, equation (2.4.2-157), these derivatives represent the transformed k-space divergence in the new coordinate system $(\epsilon^\nu, \mu, \varphi)$, or in normalized form (w^ν, μ, φ). Thus, the arguments of the derivatives of $\frac{\partial}{\partial w^\nu}$ and $\frac{\partial}{\partial \mu}$ are the fluxes v^{w^ν} and v^μ, respectively (cf. figure 3.1):

$$v^{w^\nu} = -\frac{2\Phi^\nu \sqrt{w^\nu(1 + a_\nu w^\nu)}}{\sqrt{m_\nu^*}(1 + 2a_\nu w^\nu)} \mu T_{1,1}^{\text{HV},\nu} \overline{E_{\text{eff.}}^{\nu,x}} \,, \qquad (3.2.2\text{-}40)$$

$$v^\mu = -\frac{\Phi^\nu(1 - \mu^2)}{\sqrt{m_\nu^*}\sqrt{w^\nu(1 + a_\nu w^\nu)}} T_{1,1}^{\text{HV},\nu} \overline{E_{\text{eff.}}^{\nu,x}} \,, \qquad (3.2.2\text{-}41)$$

where the negative signs stem from the minus sign in front of the braces enclosing the drift term in (3.2.0-32). For Gauss's theorem, the fluxes have to be evaluated first at the bounding surfaces of the control volume (see figure 3.1). Proceeding for Φ^ν analogously to (3.2.0-35) one obtains

$$v^{w^\nu}_{i\pm\frac{1}{2},j} = -\frac{2\sqrt{w^\nu_{i\pm\frac{1}{2}}(1+a_\nu w^\nu_{i\pm\frac{1}{2}})}}{\sqrt{m^*_\nu}\left(1+2a_\nu w^\nu_{i\pm\frac{1}{2}}\right)}\mu T^{\mathrm{HV},\nu}_{1,1}\overline{E^{\nu,x}_{\mathrm{eff.}}}\frac{n^{\nu,\tau}_{i\pm\frac{1}{2},j}}{\Delta w\Delta\mu}\ , \quad (3.2.2\text{-}42)$$

$$v^\mu_{i,j\pm\frac{1}{2}} = -\frac{(1-\mu^2_{j\pm\frac{1}{2}})}{\sqrt{m^*_\nu}\sqrt{w^\nu(1+a_\nu w^\nu)}}T^{\mathrm{HV},\nu}_{1,1}\overline{E^{\nu,x}_{\mathrm{eff.}}}\frac{n^{\nu,\tau}_{i,j\pm\frac{1}{2}}}{\Delta w\Delta\mu}\ . \quad (3.2.2\text{-}43)$$

The integrals on the right-hand side of (3.2.0-37) evaluate with (3.2.2-42) and (3.2.2-43) to

$$\int_{\mu_{j-\frac{1}{2}}}^{\mu_{j+\frac{1}{2}}} v^{w^\nu}_{i\pm\frac{1}{2},j}\,d\mu = -\frac{1}{\sqrt{m^*_\nu}}T^{\mathrm{HV},\nu}_{1,1}\frac{\overline{E^{\nu,x}_{\mathrm{eff.}}}}{\Delta w\Delta\mu}\Bigg\{$$

$$\frac{\sqrt{w^\nu_{i\pm\frac{1}{2}}(1+a_\nu w^\nu_{i\pm\frac{1}{2}})}}{\left(1+2a_\nu w^\nu_{i\pm\frac{1}{2}}\right)}\left(\mu^2_{j+\frac{1}{2}}-\mu^2_{j-\frac{1}{2}}\right)\Bigg\}n^{\nu,\tau}_{i\pm\frac{1}{2},j}$$

$$= \langle a^{\nu,\tau}_{i\pm\frac{1}{2},j}\rangle n^{\nu,\tau}_{i\pm\frac{1}{2},j}\ , \quad (3.2.2\text{-}44)$$

$$\int_{w^\nu_{i-\frac{1}{2}}}^{w^\nu_{i+\frac{1}{2}}} v^\mu_{i,j\pm\frac{1}{2}}\,dw^\nu = -\frac{1}{\sqrt{m^*_\nu}}T^{\mathrm{HV},\nu}_{1,1}\frac{\overline{E^{\nu,x}_{\mathrm{eff.}}}}{\Delta w\Delta\mu}\Bigg\{$$

$$\frac{(1-\mu^2_{j\pm\frac{1}{2}})}{\sqrt{a_\nu}}\log\left(\frac{2\sqrt{a_\nu}\sqrt{w^\nu_{i+\frac{1}{2}}(1+a_\nu w^\nu_{i+\frac{1}{2}})}+(1+2a_\nu w^\nu_{i+\frac{1}{2}})}{2\sqrt{a_\nu}\sqrt{w^\nu_{i-\frac{1}{2}}(1+a_\nu w^\nu_{i-\frac{1}{2}})}+(1+2a_\nu w^\nu_{i-\frac{1}{2}})}\right)\Bigg\}n^{\nu,\tau}_{i,j\pm\frac{1}{2}}$$

$$= \langle a^{\nu,\tau}_{i,j\pm\frac{1}{2}}\rangle n^{\nu,\tau}_{i,j\pm\frac{1}{2}}\ . \quad (3.2.2\text{-}45)$$

Note, the densities $n^{\nu,\tau}_{i\pm\frac{1}{2},j}$, $n^{\nu,\tau}_{i,j\pm\frac{1}{2}}$ at the bounding surfaces of the control volume used in (3.2.2-42)-(3.2.2-45) are unspecified and need to be approximated. In [31], a *Scharfetter-Gummel* discretization scheme in conjunction with the *maximum entropy dissipation scheme* is used. For this scheme, approximations at bounding surfaces are not needed, but the values of $n^{\nu,\tau}$ at the discretized points must be provided. However, in [19, 20] a less elaborate method based on an upwind scheme in combination with a slope limiter is successfully applied. This approach leads to simple and therefore attractive approximations of $n_{i\pm\frac{1}{2},j}^{\nu,\tau}$ and $n_{i,j\pm\frac{1}{2}}^{\nu,\tau}$. The main idea behind the upwind scheme is to evaluate the flux-direction (wind-direction) and to incorporate it into the interpolation of $n^{\nu,\tau}$ at the bounding surfaces. The wind-direction is evaluated by the sign of $\langle a_{i\pm\frac{1}{2},j}^{\nu,\tau}\rangle$ in (3.2.2-44) for the w^ν and the sign of $\langle a_{i,j\pm\frac{1}{2}}^{\nu,\tau}\rangle$ in (3.2.2-45) for the μ-direction, respectively. The sign of $\langle a_{i\pm\frac{1}{2},j\pm\frac{1}{2}}^{\nu,\tau}\rangle$ is inverted w.r.t. the effective field $\overline{E_{\text{eff}}^{\nu,x}}$ and thus the carriers move in the opposite direction of $\overline{E_{\text{eff}}^{\nu,x}}$. As slope limiter, a *minmod* limiter is used in [19,20], which is defined as [42]

$$
\text{MM}(a,b) = \begin{cases} \text{sign}(a)\min(|a|,|b|), & \text{if } ab > 0 \text{ ,} \\ 0, & \text{else,} \end{cases} \tag{3.2.2-46}
$$

where a and b are two slopes to be selected. Based on the wind direction and the *minmod* limiter, $n_{\iota+\frac{1}{2}}^{\nu,\tau}$ at the bounding surface is linearly interpolated by

- $\langle a_{\iota+\frac{1}{2}}^{\nu,\tau}\rangle > 0$:

$$
\begin{aligned}
n_{\iota+\frac{1}{2}}^{\nu,\tau} &= n_\iota^{\nu,\tau} + \frac{\text{MM}(\overbrace{n_{\iota+1}^{\nu,\tau}-n_\iota^{\nu,\tau}}^{m_2},\overbrace{n_\iota^{\nu,\tau}-n_{\iota-1}^{\nu,\tau}}^{m_1})}{2} \\
&= \begin{cases} \frac{1}{2}(n_{\iota+1}^{\nu,\tau}+n_\iota^{\nu,\tau}) \text{ ,} & |m_1| > |m_2| \text{ , } m_1 m_2 > 0 \text{ ,} \\ \frac{1}{2}(3n_\iota^{\nu,\tau}-n_{\iota-1}^{\nu,\tau}) \text{ ,} & |m_1| < |m_2| \text{ , } m_1 m_2 > 0 \text{ ,} \\ n_\iota^{\nu,\tau} \text{ , else (local minimum/maximum),} \end{cases} \tag{3.2.2-47}
\end{aligned}
$$

- $\langle a_{\iota+\frac{1}{2}}^{\nu,\tau}\rangle < 0$:

$$
\begin{aligned}
n_{\iota+\frac{1}{2}}^{\nu,\tau} &= n_{\iota+1}^{\nu,\tau} - \frac{\text{MM}(\overbrace{n_{\iota+2}^{\nu,\tau}-n_{\iota+1}^{\nu,\tau}}^{m_2},\overbrace{n_{\iota+1}^{\nu,\tau}-n_\iota^{\nu,\tau}}^{m_1})}{2} \\
&= \begin{cases} \frac{1}{2}(3n_{\iota+1}^{\nu,\tau}-n_{\iota+2}^{\nu,\tau}) \text{ ,} & |m_1| > |m_2| \text{ , } m_1 m_2 > 0 \text{ ,} \\ \frac{1}{2}(n_{\iota+1}^{\nu,\tau}+n_\iota^{\nu,\tau}) \text{ ,} & |m_1| < |m_2| \text{ , } m_1 m_2 > 0 \text{ ,} \\ n_{\iota+1}^{\nu,\tau} \text{ , else (local minimum/maximum),} \end{cases} \tag{3.2.2-48}
\end{aligned}
$$

where $\langle a_{\iota+\frac{1}{2}}^{\nu,\tau} \rangle$ is a placeholder for $\langle a_{i+\frac{1}{2},j}^{\nu,\tau} \rangle$, $\langle a_{i,j+\frac{1}{2}}^{\nu,\tau} \rangle$ and $n_{\iota+\frac{1}{2}}^{\nu,\tau}$ for $n_{i+\frac{1}{2},j}^{\nu,\tau}$, $n_{i,j+\frac{1}{2}}^{\nu,\tau}$, respectively. Figure 3.2 illustrates the Upwind-MinMod interpolation scheme schematically. The interpolation formula for $n_{i-\frac{1}{2},j}^{\nu,\tau}$ or $n_{i,j-\frac{1}{2}}^{\nu,\tau}$ can be obtained

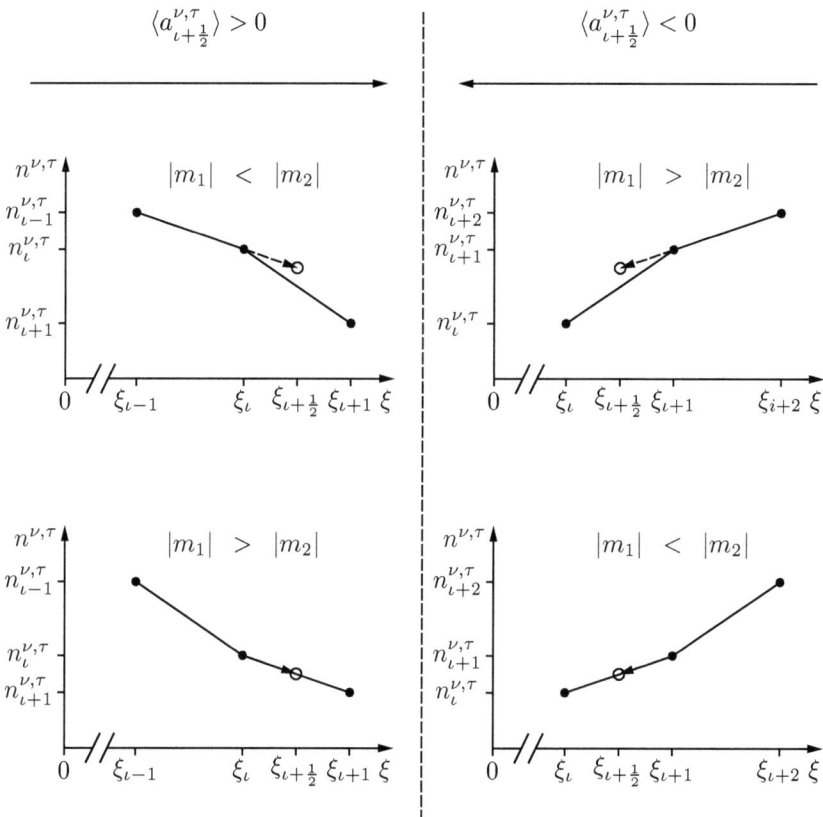

Figure 3.2: Schematic illustration of the Upwind-MinMod scheme for interpolating the values of $n_{\iota+\frac{1}{2}}^{\nu,\tau}$ at the bounding surfaces (here labeled by $\xi_{\iota+\frac{1}{2}}$) of a control volume.

by replacing ι with $\iota - 1$ in (3.2.2-47) and (3.2.2-48). The major advantage of the Upwind-MinMod scheme, aside from its simplicity, is that nonphysical negative values of $n^{\nu,\tau}$ at the bounding surfaces and thus oscillations during the subsequent solution process are suppressed by the *minmod* limiter [42].

3.2.3 Collision term

Starting point for the discretized collision term is its normalized version (3.1.0-29). For consistency with (3.2.0-32), the collision term (3.1.0-29) is divided by BTE_N (3.1.0-22) and reads

$$
(\sqrt{m_\nu^*})^3 s^\nu(w^\nu)\overline{C} = \left[\sum_X \int_{\mathrm{BZ}} \frac{1}{2} \left(\underbrace{\overline{S^X}(\{\acute{w}^{\nu'},\mu',\varphi'\},\{w^\nu,\mu,\varphi\})}_{\overline{S^{X,\mathrm{in}}}} f^{\nu'}(1-f^\nu) \right.\right.
$$

$$
\left.\left. - \underbrace{\overline{S^X}(\{w^\nu,\mu,\varphi\},\{\acute{w}^{\nu'},\mu',\varphi'\})}_{\overline{S^{X,\mathrm{out}}}} f^\nu \left(1-f^{\nu'}\right) \right) (\sqrt{m_{\nu'}^*})^3 s^{\nu'}(\acute{w}^{\nu'}) d\acute{w}^{\nu'} d\mu' d\varphi' \right]
$$

$$
\times (\sqrt{m_\nu^*})^3 s^\nu(w^\nu) \, . \tag{3.2.3-49}
$$

Using the normalized solution variable (3.1.0-16) for the valley ν' ($\Phi^{\nu'}$) and splitting the integration over the first *Brillouin-zone* into an integration over $(\acute{w}^{\nu'},\mu',\varphi')$, summed over all considered valleys, leads to

$$
(\sqrt{m_\nu^*})^3 s^\nu(w^\nu)\overline{C} = (\sqrt{m_\nu^*})^3 s^\nu(w^\nu) \left[\sum_X \sum_{\nu'=1}^{N_{\mathrm{val.}}} \frac{1}{2} \int_{\varphi'=0}^{2\pi} \int_{\mu'=-1}^{1} \int_{\acute{w}^{\nu'}=0}^{w_{\max}} \right.
$$

$$
\left. \left(\overline{S^{X,\mathrm{in}}}\, \Phi^{\nu'}(1-f^\nu) - \overline{S^{X,\mathrm{out}}} f^\nu \left((\sqrt{m_{\nu'}^*})^3 s^{\nu'}(\acute{w}^{\nu'}) - \Phi^{\nu'}\right) \right) d\acute{w}^{\nu'} d\mu' d\varphi' \right] \, , \tag{3.2.3-50}
$$

with the total number of considered valleys $N_{\mathrm{val.}}$ and X representing all scattering mechanisms to be accounted for. Next, the collision term (3.2.3-50) is integrated over the currently considered control volume $\Omega_{i,j}$ (BIM):

$$
\int_{w_{i-\frac{1}{2}}^\nu}^{w_{i+\frac{1}{2}}^\nu} \int_{\mu_{j-\frac{1}{2}}}^{\mu_{j+\frac{1}{2}}} (\sqrt{m_\nu^*})^3 s^\nu(w^\nu)\overline{C} d\mu\, dw^\nu = \int_{w_{i-\frac{1}{2}}^\nu}^{w_{i+\frac{1}{2}}^\nu} \int_{\mu_{j-\frac{1}{2}}}^{\mu_{j+\frac{1}{2}}} \left[\sum_X \sum_{\nu'=1}^{N_{\mathrm{val.}}} \frac{1}{2} \right.
$$

$$
\int_0^{2\pi} \int_{-1}^{1} \int_0^{w_{\max}} \left(\overline{S^{X,\mathrm{in}}}\, \Phi^{\nu'}(1-f^\nu) - \overline{S^{X,\mathrm{out}}} f^\nu \left((\sqrt{m_{\nu'}^*})^3 s^{\nu'}(\acute{w}^{\nu'}) - \Phi^{\nu'}\right) \right)
$$

$$
\left. d\acute{w}^{\nu'} d\mu' d\varphi' \right] (\sqrt{m_\nu^*})^3 s^\nu(w^\nu) d\mu\, dw^\nu \, . \tag{3.2.3-51}
$$

The physical meaning of (3.2.3-51) becomes more clear, if (3.2.3-50) is considered first. For each scattering mechanism, all contributing states are specified for the evaluation of the collision term by the integration over the *Brillouin-zone*. The states involved in the collision term are given for each scattering mechanism by its transition rate for in- and out-scattering. Since for BIM a integration over the control volume is performed, equation (3.2.3-50) turns into (3.2.3-51). Thus, for all states within the control volume the scattering interaction is considered instead of for the single state at the grid point of the control volume.

3.2.3.1 Acoustic deformation potential scattering

The main steps for calculating the discretized collision term (3.2.3-51) – integrated over the control volume $\Omega_{i,j}$ – are exemplary shown for acoustic deformation potential scattering. The transition rate for acoustic deformation potential scattering (2.3.1-62) (section 2.3.1.1) is normalized by (3.1.0-31) and reads

$$\overline{S^{\mathrm{ADP,out}}} = K^{\mathrm{ADP}} \delta \left(W_0^{\nu} + w^{\nu} - W_0^{\nu'} - \acute{w}^{\nu'} \right) \delta_{\nu,\nu'} \ , \quad (3.2.3\text{-}52)$$

$$K^{\mathrm{ADP}} = \frac{1}{K_{\mathrm{N}}} \frac{D_{\mathrm{A}}^2 k_{\mathrm{b}} T_{\mathrm{L}}}{4\pi^2 \hbar \rho v_{\mathrm{s}}^2} \ , \quad (3.2.3\text{-}53)$$

with the normalized band edge energies W_0^{ν} and $W_0^{\nu'}$ of the valleys ν and ν', respectively. Inserting (3.2.3-52) in the collision term (3.2.3-50) gives

$$(\sqrt{m_{\nu}^*})^3 s^{\nu}(w^{\nu}) \overline{C^{\mathrm{ADP}}} = (\sqrt{m_{\nu}^*})^3 s^{\nu}(w^{\nu}) K^{\mathrm{ADP}} \pi \Bigg[$$

$$\int_{\mu'=-1}^{1} \int_{\acute{w}^{\nu}=0}^{w_{\max}} \Bigg(\delta \left(\acute{w}^{\nu} - w^{\nu} \right) \Phi^{\nu}(\acute{w}^{\nu}, \mu') \left(1 - f^{\nu} \right)$$

$$-\delta \left(w^{\nu} - \acute{w}^{\nu} \right) f^{\nu} \left((\sqrt{m_{\nu}^*})^3 s^{\nu}(\acute{w}^{\nu}) - \Phi^{\nu}(\acute{w}^{\nu}, \mu') \right) \Bigg) d\acute{w}^{\nu} \, d\mu' \Bigg] \ . \quad (3.2.3\text{-}54)$$

The Kronecker delta $\delta_{\nu,\nu'}$ in (3.2.3-52) permits only a contribution for $\nu' = \nu$ (intravalley scattering). Thus, the summation over ν' in (3.2.3-50) can be omitted. The integration over φ' results in a multiplication of 2π. Furthermore, the integration over $\acute{w}^\nu$ gives $\acute{w}^\nu = w^\nu$ due to the Dirac function

$$\int_{x_1}^{x_u} \delta(x - x_0) f(x) dx = \begin{cases} f(x_0) \text{ , for } x_1 < x_0 < x_u \text{ ,} \\ 0, \text{ else.} \end{cases} \quad (3.2.3\text{-}55)$$

Thus, one obtains

$$(\sqrt{m_\nu^*})^3 s^\nu(w^\nu) \overline{C^{\mathrm{ADP}}} = s^\nu(w^\nu) K^{\mathrm{ADP}} \pi (\sqrt{m_\nu^*})^3 \left[\int_{-1}^{1} d\mu' \left(\Phi^\nu(w^\nu, \mu') \right. \right.$$

$$\left. \left. - \underbrace{f^\nu (\sqrt{m_\nu^*})^3 s^\nu(w^\nu)}_{\Phi^\nu(w^\nu, \mu)} \right) \right] . \quad (3.2.3\text{-}56)$$

At this point, the solution variables Φ^ν are replaced by their average values across the considered control volumes after (3.2.0-35). This corresponds to the transition from the analytic description to a discretized version. Evaluating the integral over μ' in (3.2.3-56) results in

$$(\sqrt{m_\nu^*})^3 s^\nu(w^\nu) \overline{C^{\mathrm{ADP}}} =$$

$$s^\nu(w^\nu) K^{\mathrm{ADP}} \pi (\sqrt{m_\nu^*})^3 \left[\sum_{j'=1}^{N_\mu} \frac{n_{i,j'}^{\nu,\tau}}{\Delta w} - 2 \frac{n_{i,j}^{\nu,\tau}}{\Delta w \Delta \mu} \right] . \quad (3.2.3\text{-}57)$$

According to (3.2.3-51), equation (3.2.3-57) needs to be integrated over the control volume $\Omega_{i,j}$ and leads to

$$\int_{w_{i-\frac{1}{2}}^\nu}^{w_{i+\frac{1}{2}}^\nu} \int_{\mu_{j-\frac{1}{2}}}^{\mu_{j+\frac{1}{2}}} (\sqrt{m_\nu^*})^3 s^\nu(w^\nu) \overline{C^{\mathrm{ADP}}} d\mu \, dw^\nu =$$

$$S^\nu(w^\nu)\big|_{w_{i-\frac{1}{2}}^\nu}^{w_{i+\frac{1}{2}}^\nu} \Delta\mu K^{\mathrm{ADP}} \pi (\sqrt{m_\nu^*})^3 \left[\sum_{j'=1}^{N_\mu} \frac{n_{i,j'}^{\nu,\tau}}{\Delta w} - 2 \frac{n_{i,j}^{\nu,\tau}}{\Delta w \Delta \mu} \right] , \quad (3.2.3\text{-}58)$$

with

$$S^\nu(w^\nu) = \int s^\nu(w^\nu) dw^\nu = \frac{2}{3} \left(\sqrt{w^\nu (1 + a_\nu w^\nu)} \right)^3 , \quad (3.2.3\text{-}59)$$

and $s^\nu(w^\nu)$ after (3.1.0-14). Note, due to the elastic nature of S^{ADP}, the terms for Pauli blocking vanish ((3.2.3-54) to (3.2.3-56)) and a *linear* collision term is obtained ($\overline{C^{\mathrm{ADP}}}$ in (3.2.3-57) and (3.2.3-58) is a linear function of $n_{i,j'}^{\nu,\tau}$ and $n_{i,j}^{\nu,\tau}$). As described in section 2.4.1, the result (3.2.3-58) is obtained regardless of whether the Pauli principle is neglected or being considered.

3.2.3.2 Intervalley optical phonon scattering

Based on section 2.3.1.2, the transition rate of intervalley optical phonon scattering (also referred as non-polar optical phonon scattering) reads in normalized form (cf. (3.1.0-31))

$$\overline{S^{\mathrm{NOP,out}}} = K^{\mathrm{NOP},\nu'} N_{\mathrm{ph}}^{\pm} \delta\left(\acute{w}^{\nu'} - w^\nu \pm \Delta w_{\mathrm{ph}} - \Delta w_0^{\nu-\nu'}\right) , \quad (3.2.3\text{-}60)$$

$$K^{\mathrm{NOP},\nu'} = \frac{1}{K_{\mathrm{N}}} \frac{Z^{\nu'} D_{ij}^2}{8\pi^2 \rho \omega_{ij}} , \quad (3.2.3\text{-}61)$$

$$N_{\mathrm{ph}}^{\pm} = n_{\mathrm{ph}} + \frac{1}{2} \pm \frac{1}{2} , \quad (3.2.3\text{-}62)$$

$$\Delta w_0^{\nu-\nu'} = W_0^\nu - W_0^{\nu'} , \quad (3.2.3\text{-}63)$$

with the phonon occupation number n_{ph} after (2.3.1-64), the normalized band edge energies W_0^ν and $W_0^{\nu'}$ of the current and final valley ν and ν', respectively, and the normalized phonon energy $\Delta w_{\mathrm{ph}} = \frac{\hbar \omega_{ij}}{E_{\mathrm{N}}}$. The upper (lower) sign in (3.2.3-60) and (3.2.3-62) corresponds to an emission (absorption) process. Evaluating (3.2.3-50) results in

$$(\sqrt{m_\nu^*})^3 s^\nu(w^\nu)\overline{C^{\mathrm{NOP}}} = (\sqrt{m_\nu^*})^3 s^\nu(w^\nu) N_{\mathrm{ph}}^{\pm} \sum_{\nu'} \left\{ K^{\mathrm{NOP},\nu'} \pi \right.$$

$$\left[\int_{-1}^{1} d\mu' \int_0^{w_{\max}} d\acute{w}^{\nu'} \left(\delta\left(\acute{w}^{\nu'} - w^\nu \mp \Delta w_{\mathrm{ph}} - \Delta w_0^{\nu-\nu'}\right) \Phi^{\nu'}(\acute{w}^{\nu'}, \mu') (1 - f^\nu) \right.\right.$$

$$\left.\left.\left. -\delta\left(\acute{w}^{\nu'} - w^\nu \pm \Delta w_{\mathrm{ph}} - \Delta w_0^{\nu-\nu'}\right) f^\nu \left((\sqrt{m_{\nu'}^*})^3 s^\nu(\acute{w}^{\nu'}) - \Phi^{\nu'}(\acute{w}^{\nu'}, \mu')\right) \right) \right] \right\} ,$$

$$(3.2.3\text{-}64)$$

which becomes after integrating over $\acute{w}^{\nu'}$ and employing (3.2.3-55)

$$(\sqrt{m_\nu^*})^3 s^\nu(w^\nu)\overline{C^{\mathrm{NOP}}} = (\sqrt{m_\nu^*})^3 s^\nu(w^\nu)\pi N_{\mathrm{ph}}^\pm \sum_{\nu'}\left\{K^{\mathrm{NOP},\nu'}\left[\int_{-1}^{1} d\mu'\right.\right.$$

$$\left(\Phi^{\nu'}(w^\nu \pm \Delta w_{\mathrm{ph}} + \Delta w_0^{\nu-\nu'},\mu')(1-f^\nu) - f^\nu\left(s^{\nu'}(w^\nu \mp \Delta w_{\mathrm{ph}} + \Delta w_0^{\nu-\nu'})\right.\right.$$

$$\left.\left.\left.\cdot(\sqrt{m_{\nu'}^*})^3 - \Phi^{\nu'}(w^\nu \mp \Delta w_{\mathrm{ph}} + \Delta w_0^{\nu-\nu'},\mu')\right)\right]\right\}. \tag{3.2.3-65}$$

For the integration over μ', the transition to the discretized solution variable $n^{\nu/\nu',\tau}$ is performed, like previously done for acoustic deformation potential scattering (cf. (3.2.3-56) to (3.2.3-57)). However, care has to be taken when performing this transition, because:

i.) the final energies $\acute{w}^{\nu'}$ for in- and out-scattering might be outside of the considered energy range ($\acute{w}^{\nu'} < 0$ or $\acute{w}^{\nu'} > w_{\mathrm{max}}$), and

ii.) the energy transfer between w^ν and $\acute{w}^{\nu'}$ is usually not a multiple of the step size Δw and thus the solution variables $n_{i',j'}^{\nu',\tau}$ at $\acute{w}^{\nu'}$ are displaced from the actual considered dirscretization.

In [19, 20] these issues are circumvented by choosing Δw in such a way that the energetic differences $\Delta w_0^{\nu-\nu'}$ and Δw_{ph} become a integer multiple of Δw, where $\Delta w_0^{\nu-\nu'}$ and Δw_{ph} are rounded in a suitable way. This procedure has the advantage of small implementation effort in a simulator and prevents displaced $n_{i',j'}^{\nu',\tau}$. However, these advantages are obtained at the cost of flexibility regarding the step size Δw and the accuracy to take multiple valleys and energy transfers due to phonon scattering accurately into account. In this work, a different approach is pursued which prevents these restrictions.

The employed method is inspired by the one used in [31] and basically focuses on the final integration over w^ν in (3.2.3-51). Due to the energetic difference between the initial and final state originating from $\Delta w_0^{\nu-\nu'}$ and Δw_{ph}, the integration interval $[w_{i-\frac{1}{2}}^\nu, w_{i+\frac{1}{2}}^\nu]$ in (3.2.3-51) overlaps at most with two adjacent control volumes (along the $\acute{w}^{\nu'}$-direction) in the final valley ν'. Therefore, the integration interval $[w_{i-\frac{1}{2}}^\nu, w_{i+\frac{1}{2}}^\nu]$ is split into two intervals

$[w_{i-\frac{1}{2}}^{\nu}, w_{i-\frac{1}{2}+r}^{\nu}]$ and $[w_{i-\frac{1}{2}+r}^{\nu}, w_{i+\frac{1}{2}}^{\nu}]$. The energy $w_{i-\frac{1}{2}+r}^{\nu}$ is chosen that the boundary of the two adjacent control volumes in the final valley ν' is obtained with $\Delta w_0^{\nu-\nu'}$ and Δw_{ph}. Thus, the integration interval $[w_{i-\frac{1}{2}}^{\nu}, w_{i+\frac{1}{2}}^{\nu}]$ is mapped to $[\acute{w}_{i'-\frac{1}{2}}^{\nu'} - \Delta w^-, \acute{w}_{i'-\frac{1}{2}}^{\nu'} + \Delta w^+]$ (with $\Delta w = \Delta w^- + \Delta w^+$, $r = \frac{\Delta w^-}{\Delta w}$) and the proportionate contributions of the two adjacent control volumes, counted by $\acute{r}$, are accurately considered for the integration over w^{ν}:

$$\acute{r} = 1 : \left[w_{i-\frac{1}{2}}^{\nu}, w_{i-\frac{1}{2}+r}^{\nu} \right] \xrightarrow{\Delta w_0^{\nu-\nu'}, \Delta w_{\mathrm{ph}}} \left[\acute{w}_{i'-\frac{1}{2}}^{\nu'} - \Delta w^-, \acute{w}_{i'-\frac{1}{2}}^{\nu'} \right] , \quad (3.2.3\text{-}66)$$

$$\acute{r} = 2 : \left[w_{i-\frac{1}{2}+r}^{\nu}, w_{i+\frac{1}{2}}^{\nu} \right] \xrightarrow{\Delta w_0^{\nu-\nu'}, \Delta w_{\mathrm{ph}}} \left[\acute{w}_{i'-\frac{1}{2}}^{\nu'}, \acute{w}_{i'-\frac{1}{2}}^{\nu'} + \Delta w^+ \right] . \quad (3.2.3\text{-}67)$$

Note, if the boundary $\acute{w}_{i'-\frac{1}{2}}^{\nu'}$ in the final valley ν' coincides with $\acute{w}_{i'-\frac{1}{2}}^{\nu'} = 0$ (bottom of the valley) or $\acute{w}_{i'-\frac{1}{2}}^{\nu'} = w_{\mathrm{max}}$ (maximum considered kinetic energy) the integration interval after (3.2.3-66) or after (3.2.3-67) is skipped, respectively. Thus, both the integration interval in the w^{ν} and $w^{\nu'}$ domain are adjusted to take only valid states into account by still preserving the energetic offset due to $\Delta w_0^{\nu-\nu'}$ and Δw_{ph}. The principle of the scheme is shown in figure 3.3. With the integration scheme described before, (3.2.3-65) becomes

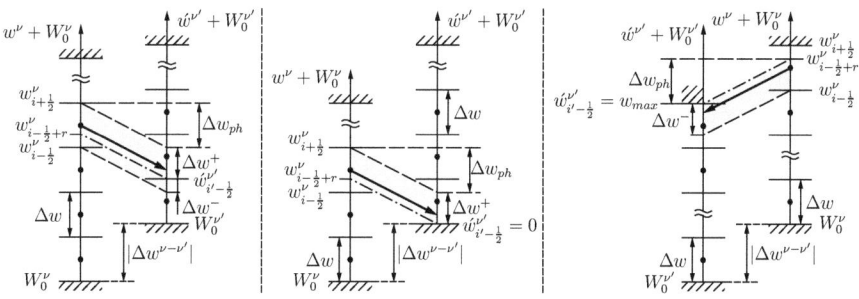

Figure 3.3: Integration scheme for inelastic phonon scattering. Here, two valleys (initial ν, final ν') are shown for three cases: $\acute{w}_{i'-\frac{1}{2}}^{\nu'}$ lies in the interior $\acute{w}^{\nu'}$ domain (left), $\acute{w}_{i'-\frac{1}{2}}^{\nu'} = 0$ (bottom of valley ν', center) and $\acute{w}_{i'-\frac{1}{2}}^{\nu'} = w_{\mathrm{max}}$ (upper end of valley ν', right). Here, only out-scattering from valley ν to ν' by emission of a phonon is shown.

$$(\sqrt{m_\nu^*})^3 s^\nu(w^\nu)\overline{C^{\text{NOP}}} = (\sqrt{m_\nu^*})^3 s^\nu(w^\nu)\pi N_{\text{ph}}^\pm \sum_{\nu'}\Big\{ K^{\text{NOP},\nu'}$$

$$\cdot\left[(1-f^\nu)\sum_{j'=1}^{N_\mu}\frac{n_{i'_{\text{in}}+\acute{r}-2,j'}^{\nu',\tau}}{\Delta w} - f^\nu\left(2(\sqrt{m_{\nu'}^*})^3 s^{\nu'}(w^\nu \mp \Delta w_{\text{ph}} + \Delta w_0^{\nu-\nu'})\right.\right.$$

$$\left.\left.-\sum_{j'=1}^{N_\mu}\frac{n_{i'_{\text{out}}+\acute{r}-2,j'}^{\nu',\tau}}{\Delta w}\right)\right]\Big\}, \tag{3.2.3-68}$$

with $\acute{r} = 1,2$ counting the two final control volumes in the valley ν' after (3.2.3-66) and (3.2.3-67) (depending on the actual w^ν). i'_{in} and i'_{out} are given by

$$i'_{\text{in}} = i \pm ip + iv_{\nu-\nu'}, \tag{3.2.3-69}$$

$$i'_{\text{out}} = i \mp ip + iv_{\nu-\nu'}, \tag{3.2.3-70}$$

and the integer offsets ip and iv are stemming from the phonon energy and valley offsets, respectively. ip and $iv_{\nu-\nu'}$ are determined in such a way that $i'_{\text{in}}/i'_{\text{out}}$ are pointing to the upper $\acute{w}^{\nu'}$ control volume ($\acute{r} = 2$). Using (3.1.0-16) ($\Phi^\nu(w^\nu,\mu)$) in combination with its average value approximation within $\Omega_{i,j}$ after (3.2.0-35) results for the final integration of (3.2.3-51) in

$$\sum_{\acute{r}=1}^{2}\int_{w_l^\nu(\acute{r})}^{w_u^\nu(\acute{r})}\int_{\mu_{j-\frac{1}{2}}}^{\mu_{j+\frac{1}{2}}}(\sqrt{m_\nu^*})^3 s^\nu(w^\nu)\overline{C^{\text{NOP}}}d\mu\,dw^\nu = \frac{\pi}{\Delta w}N_{\text{ph}}^\pm\sum_{\nu'}\Big\{K^{\text{NOP},\nu'}\sum_{\acute{r}=1}^{2}\Big[$$

$$\left(\Delta\mu(\sqrt{m_\nu^*})^3\,S^\nu(w^\nu)\big|_{w_l^\nu(\acute{r})}^{w_u^\nu(\acute{r})} - \frac{\overset{\text{IN}}{\Delta w(\acute{r})}}{\Delta w}n_{i,j}^{\nu,\tau}\right)\sum_{j'=1}^{N_\mu}n_{i'_{\text{in}}+\acute{r}-2,j'}^{\nu',\tau}$$

$$-n_{i,j}^{\nu,\tau}\left(2(\sqrt{m_{\nu'}^*})^3\,S^{\nu'}(\acute{w}^{\nu'})\big|_{\acute{w}_l^{\nu'}(\acute{r})}^{\acute{w}_u^{\nu'}(\acute{r})} - \frac{\overset{\text{OUT}}{\Delta w(\acute{r})}}{\Delta w}\sum_{j'=1}^{N_\mu}n_{i'_{\text{out}}+\acute{r}-2,j'}^{\nu',\tau}\right)\Big]\Big\} \tag{3.2.3-71}$$

with

$$\overset{\text{IN, OUT}}{\Delta w(\acute{r})} = \begin{cases} \overset{\text{IN, OUT}}{\Delta w^-}, & \text{for } \acute{r} = 1, \\ \overset{\text{IN, OUT}}{\Delta w^+}, & \text{for } \acute{r} = 2, \end{cases} \tag{3.2.3-72}$$

$$[w_1^\nu(\acute{r}), w_u^\nu(\acute{r})] = \begin{cases} \left[w_{i-\frac{1}{2}}^\nu, \overset{\text{IN}}{\overbrace{w_{i-\frac{1}{2}}^\nu + \Delta w^-}} \right], \text{ for } \acute{r} = 1 , \\ \left[\overset{\text{IN}}{\overbrace{w_{i-\frac{1}{2}}^\nu + \Delta w^-}}, w_{i+\frac{1}{2}}^\nu \right], \text{ for } \acute{r} = 2 , \end{cases} \qquad (3.2.3\text{-}73)$$

$$\left[\acute{w}_1^{\nu'}(\acute{r}), \acute{w}_u^{\nu'}(\acute{r}) \right] = \begin{cases} \left[\acute{w}_{i'_{\text{out}}-\frac{1}{2}}^{\nu'} - \overset{\text{OUT}}{\overbrace{\Delta w^-}}, \acute{w}_{i'_{\text{out}}-\frac{1}{2}}^{\nu'} \right], \text{ for } \acute{r} = 1 , \\ \left[\acute{w}_{i'_{\text{out}}-\frac{1}{2}}^{\nu'}, \overset{\text{OUT}}{\overbrace{\acute{w}_{i'_{\text{out}}-\frac{1}{2}}^{\nu'} + \Delta w^+}} \right], \text{ for } \acute{r} = 2 . \end{cases} \qquad (3.2.3\text{-}74)$$

$S^{\nu/\nu'}$ in (3.2.3-71) is obtained by (3.2.3-59) for the respective valley and energy. Note, in (3.2.3-71) it is assumed that both control volumes ($\acute{r} = 1, 2$) lie within the considered range of energy. If only one control volume of the valley ν' contributes to the in- or out-scattering part, the respective value of $\acute{r}$ is inserted and thus the summation over $\acute{r}$ is omitted for the respective scattering part. In addition, the final valleys ν' available for a scattering event are controlled via the valley degeneracy factor $Z^{\nu'}$ in (3.2.3-61). For non-contributing valleys ν' the valley degeneracy factor becomes $Z^{\nu'} = 0$.

3.2.3.3 Intravalley optical phonon scattering

The treatment of intravalley optical phonon scattering is similar to the one of intervalley optical phonon scattering described in the previous section 3.2.3.2. With (2.3.1-65) in section 2.3.1.3 the normalized transition rate reads

$$\overline{S^{\text{OP,out}}} = K^{\text{OP}} N_{\text{ph}}^{\pm} \delta \left(\acute{w}^{\nu'} - w^\nu \pm \Delta w_{\text{ph}} \right) \delta_{\nu,\nu'} , \qquad (3.2.3\text{-}75)$$

$$K^{\text{OP}} = \frac{1}{K_N} \frac{D_O^2}{8\pi^2 \rho \omega_O} , \qquad (3.2.3\text{-}76)$$

$$N_{\text{ph}}^{\pm} = n_{\text{ph}} + \frac{1}{2} \pm \frac{1}{2} , \qquad (3.2.3\text{-}77)$$

with the phonon occupation number n_{ph} after (2.3.1-64), the normalized phonon energy $\Delta w_{\text{ph}} = \frac{\hbar \omega_O}{E_N}$ and the normalization factor K_N (3.1.0-31). The upper (lower) sign in (3.2.3-75) and (3.2.3-77) corresponds to a phonon emission (absorption) process. Following the procedure of intervalley optical phonon scattering (3.2.3-60), (3.2.3-51) reads for intravalley optical phonon scattering:

$$\sum_{\acute{r}=1}^{2} \int_{w_{\mathrm{l}}^{\nu}(\acute{r})}^{w_{\mathrm{u}}^{\nu}(\acute{r})} \int_{\mu_{j-\frac{1}{2}}}^{\mu_{j+\frac{1}{2}}} (\sqrt{m_{\nu}^*})^3 s^{\nu}(w^{\nu}) \overline{C^{\mathrm{OP}}} d\mu \, dw^{\nu} = \frac{\pi}{\Delta w} N_{ph}^{\pm} K^{\mathrm{OP}}$$

$$\sum_{\acute{r}=1}^{2} \left[\left(\Delta\mu(\sqrt{m_{\nu}^*})^3 \, S^{\nu}(w^{\nu})|_{w_{\mathrm{l}}^{\nu}(\acute{r})}^{w_{\mathrm{u}}^{\nu}(\acute{r})} - \overset{\mathrm{IN}}{\frac{\Delta w(\acute{r})}{\Delta w}} n_{i,j}^{\nu,\tau} \right) \sum_{j'=1}^{N_{\mu}} n_{i'_{\mathrm{in}}+\acute{r}-2,j'}^{\nu,\tau} \right.$$

$$\left. - n_{i,j}^{\nu,\tau} \left(2(\sqrt{m_{\nu}^*})^3 \, S^{\nu}(\acute{w}^{\nu})|_{\acute{w}_{\mathrm{l}}^{\nu}(\acute{r})}^{\acute{w}_{\mathrm{u}}^{\nu}(\acute{r})} - \overset{\mathrm{OUT}}{\frac{\Delta w(\acute{r})}{\Delta w}} \sum_{j'=1}^{N_{\mu}} n_{i'_{\mathrm{out}}+\acute{r}-2,j'}^{\nu,\tau} \right) \right], \quad (3.2.3\text{-}78)$$

where i'_{in} and i'_{out} is obtained by (3.2.3-69) and (3.2.3-70) for $iv_{\nu-\nu'} = 0$ (intravalley process, $\nu = \nu'$), respectively. $\Delta w(\acute{r})$ is given by (3.2.3-72) and for the integration intervals $[w_{\mathrm{l}}^{\nu}(\acute{r}), w_{\mathrm{u}}^{\nu}(\acute{r})]$ and $[\acute{w}_{\mathrm{l}}^{\nu}(\acute{r}), \acute{w}_{\mathrm{u}}^{\nu}(\acute{r})]$ (including $\acute{r}$) the mapping after (3.2.3-66) and (3.2.3-67) (with $\Delta w_0^{\nu-\nu'} = 0$) holds, respectively.

3.2.3.4 Polar optical phonon scattering

As mentioned in section 2.3.1.4, polar optical phonon scattering is an anisotropic scattering process. Latter fact results in a more elaborate analytic preprocessing compared to isotropic scattering processes. A way to circumvent this burden is to use the isotropic approximation given in section 2.3.5.1. Both versions are considered here, but only for isotropic dispersion relations ($T^{\mathrm{HV},\nu} = I_3$, with I_3 as the identity matrix of size three).

Anisotropic version

In normalized form, the anisotropic version of polar optical phonon scattering after (2.3.1-66) reads:

$$\overline{S^{\mathrm{POP,out}}} = K^{\mathrm{POP}} \frac{1}{|\vec{k} - \vec{k'}|^2} N_{ph}^{\pm} \delta \left(\acute{w}^{\nu'} - w^{\nu} \pm \Delta w_{ph} \right) \delta_{\nu,\nu'}, \quad (3.2.3\text{-}79)$$

$$K^{\mathrm{POP}} = \frac{1}{K_N k_N^2} \frac{q^2 \omega_{\mathrm{p}}}{8\pi^2 \varepsilon_0} \left(\frac{1}{\varepsilon_{\mathrm{hf}}} - \frac{1}{\varepsilon_{\mathrm{st}}} \right), \quad (3.2.3\text{-}80)$$

with the involved natural constants and model parameters given in section 2.3.1.4 and $N_{\text{ph}}^{\pm}$ after (3.2.3-77) (with the phonon energy of POP scattering). The term $|\vec{k} - \vec{k'}|$, responsible for the anisotropic nature, is replaced by the expression (2.3.5-86) and becomes [1]

$$h_\nu(w) = w(1 + a_\nu w) \tag{3.2.3-81}$$

$$|\vec{k} - \vec{k'}|^2 = m_\nu^* \left[h_\nu(w^\nu) + h_\nu(\acute{w}^\nu) \right.$$

$$\left. - 2\sqrt{h_\nu(w^\nu)h_\nu(\acute{w}^\nu)} \left(\mu\mu' + \sqrt{1 - \mu^2}\sqrt{1 - \mu'^2}\cos(\varphi') \right) \right], \tag{3.2.3-82}$$

where the fact that polar optical phonon scattering is an intravalley scattering process ($\nu = \nu'$) has already been accounted for. Inserting (3.2.3-82) in (3.2.3-79) results for equation (3.2.3-50) of the procedural method in

$$(\sqrt{m_\nu^*})^3 s^\nu(w^\nu)\overline{C^{\text{POP}}} = \frac{K^{\text{POP}}\pi}{m_\nu^* \Delta w \Delta \mu} N_{\text{ph}}^{\pm}$$

$$\cdot \left\{ \begin{array}{l} \left((\sqrt{m_\nu^*})^3 s^\nu(w^\nu) - \Phi^\nu \right) \\ \cdot \sum_{j'=1}^{N_\mu} n_{i'_{\text{in}}+\acute{r}-2,j'}^{\nu,\tau} \left. G(w^\nu, w^\nu \pm \Delta w_{\text{ph}}, \mu, \mu') \right|_{\mu'=\mu_{j'}-\frac{1}{2}}^{\mu'=\mu_{j'}+\frac{1}{2}} \\[2ex] -\Phi^\nu \\ \cdot \sum_{j'=1}^{N_\mu} \left[\left((\sqrt{m_\nu^*})^3 s^\nu(w^\nu \mp \Delta w_{\text{ph}})\Delta w \Delta \mu - n_{i'_{\text{out}}+\acute{r}-2,j'}^{\nu,\tau} \right) \right. \\ \left. \cdot \left. G(w^\nu, w^\nu \mp \Delta w_{\text{ph}}, \mu, \mu') \right|_{\mu'=\mu_{j'}-\frac{1}{2}}^{\mu'=\mu_{j'}+\frac{1}{2}} \right] \end{array} \right\} . \tag{3.2.3-83}$$

---

[1] The expression (2.3.5-86) is used for isotropic dispersion relations without any loss of generality. φ' measures the polar angle of $\vec{k'}$ w.r.t. $\vec{k}$. Since the 1D transport in x-direction is considered, the rotational symmetry depicted in section 2.4.3 holds. Due to the assumption of an isotropic dispersion relation, $\vec{\vec{k}}$ equals $\vec{\vec{k}}$ and lies within the $\tilde{k}_x^\nu - \tilde{k}_y^\nu$-plane (see figure 2.21). Therefore, φ' coincides with the integration variable of (3.2.3-51).

The function G in (3.2.3-83) is defined as

$$G(w^\nu, \acute{w}^\nu, \mu, \mu^{'}) = \frac{\text{arcsinh}\left(\frac{2\sqrt{h_\nu(w^\nu)h_\nu(\acute{w}^\nu)}\mu^{'} - \mu(h_\nu(w^\nu) + h_\nu(\acute{w}^\nu))}{\sqrt{(h_\nu(w^\nu) - h_\nu(\acute{w}^\nu))^2}\sqrt{1-\mu^2}}\right)}{2\sqrt{h_\nu(w^\nu)h_\nu(\acute{w}^\nu)}}. \quad (3.2.3\text{-}84)$$

Focusing on the argument of the arcsinh-term in (3.2.3-84), a division by zero stemming from the terms h_ν is prevented due to the inelastic nature of POP-scattering ($h_\nu(w^\nu) \neq h_\nu(\acute{w}^\nu)$). In addition, the staggered grid of w^ν-μ simulation domain ensures $\mu^2 < 1$. However, the evaluation of the integral (3.2.3-51) in combination with (3.2.3-83) is, compared to isotropic scattering processes, a non-trivial task. The evaluation of the integral (3.2.3-51) associated with the procedural method for the collision term is problematic since G after (3.2.3-84) depends on both μ and w^ν. Focusing on the integration over μ, a single primitive of G covering the complete $\mu^{'}$-μ-domain can not be found. However, in the course of this work it has been found that by combining the analytic solutions for three distinct cases the complete $\mu^{'}$-μ-domain is covered. Figure 3.4 illustrates schematically the three cases for which analytic solutions exist. Note, $\mu^{'}$ and μ are interchangeable for all cases shown

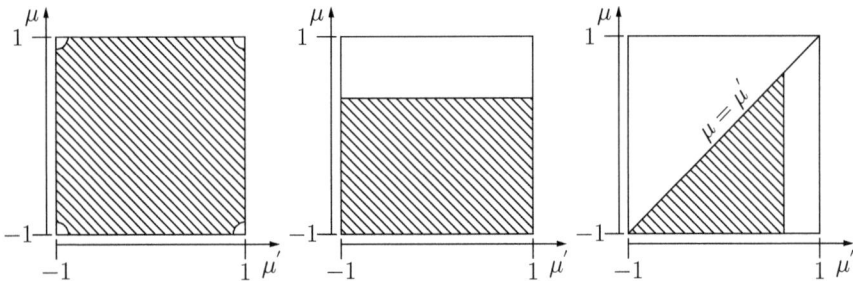

Figure 3.4: Schematic illustration of the three cases: interior of the $\mu^{'}$-μ-domain, except for the corners (left); region parallel to the $\mu^{'}$-axis ($-1 \leq \mu^{'} \leq 1$) from -1 to a given μ (center); region underneath the main-diagonal from -1 to a given $\mu^{'}$ (right).

in figure 3.4. It can be seen from figure 3.4 that the captured $\mu^{'}$-μ-domain on the left-hand side does not include the corners. However, if one or more of the four corners are involved in the evaluation of (3.2.3-51) in combination with (3.2.3-83), the μ'-μ-domain over which G is integrated is reassembled based on at most all three analytic primitives. All three analytic primitives are given in the appendix B.6. An example for the reassembly of the μ'-μ-domain is given in figure 3.5. The μ'-μ-region labeled by I_S is the one, over which G is

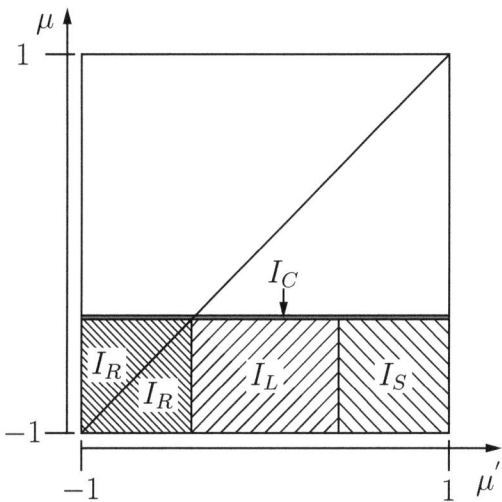

Figure 3.5: Example for the reassembly of the original domain based on the three domains shown in 3.4.

integrated. I_L, I_C and I_R are the three analytic solutions shown in figure 3.4 on the left-hand side, in the center and on right-hand side, respectively. In this case, the desired integral I_S (for G) becomes

$$I_S = I_C - 2I_R - I_L \ . \tag{3.2.3-85}$$

Proceeding analogously, any integral containing one or more corners of the μ'-μ-domain can analytically be obtained.

For the further treatment of (3.2.3-83), the result of the integral over the currently considered μ'-μ-region is label by $\left[H_{\mathrm{POP}}(w^\nu, \acute{w}^\nu)|_{\mu_{j'-\frac{1}{2}}}^{\mu_{j'+\frac{1}{2}}} \right]_{\mu_{j-\frac{1}{2}}}^{\mu_{j+\frac{1}{2}}}$.

Thus, in the case of POP-scattering one obtains for (3.2.3-51)

$$
\int_{w^\nu_{i-\frac{1}{2}}}^{w^\nu_{i+\frac{1}{2}}} \int_{\mu_{j-\frac{1}{2}}}^{\mu_{j+\frac{1}{2}}} (\sqrt{m^*_\nu})^3 s^\nu(w^\nu) \overline{C^{\mathrm{POP}}} d\mu\, dw^\nu = \frac{K^{\mathrm{POP}} \pi}{m^*_\nu \Delta w \Delta \mu} N^\pm_{\mathrm{ph}}
$$

$$
\cdot \int_{w^\nu_{i-\frac{1}{2}}}^{w^\nu_{i+\frac{1}{2}}} \left\{ \begin{aligned} & \left((\sqrt{m^*_\nu})^3 s^\nu(w^\nu) - \frac{n^{\nu,\tau}_{i,j}}{\Delta w \Delta \mu} \right) \\ & \cdot \sum_{j'=1}^{N_\mu} \left(\left[H_{\mathrm{POP}}(w^\nu, w^\nu \pm \Delta w_{\mathrm{ph}}) \big|_{\mu_{j'-\frac{1}{2}}}^{\mu_{j'+\frac{1}{2}}} \right]_{\mu_{j-\frac{1}{2}}}^{\mu_{j+\frac{1}{2}}} n^{\nu,\tau}_{i'_{\mathrm{in}}+\acute{r}-2,j'} \right) \\ & - \frac{n^{\nu,\tau}_{i,j}}{\Delta w \Delta \mu} \\ & \cdot \sum_{j'=1}^{N_\mu} \left(\left((\sqrt{m^*_\nu})^3 s^\nu(w^\nu \mp \Delta w_{\mathrm{ph}}) \Delta w \Delta \mu - n^{\nu,\tau}_{i'_{\mathrm{out}}+\acute{r}-2,j'} \right) \right. \\ & \left. \cdot \left[H_{\mathrm{POP}}(w^\nu, w^\nu \mp \Delta w_{\mathrm{ph}}) \big|_{\mu_{j'-\frac{1}{2}}}^{\mu_{j'+\frac{1}{2}}} \right]_{\mu_{j-\frac{1}{2}}}^{\mu_{j+\frac{1}{2}}} \right) \end{aligned} \right\} dw^\nu
$$

$$(3.2.3\text{-}86)$$

The last remaining integration over w^ν can not be performed analytically due to the procedure involved in obtaining $\left[H_{\mathrm{POP}}(w^\nu, \acute{w}^\nu) \big|_{\mu_{j'-\frac{1}{2}}}^{\mu_{j'+\frac{1}{2}}} \right]_{\mu_{j-\frac{1}{2}}}^{\mu_{j+\frac{1}{2}}}$ and the functions involved therein. In addition, since POP-scattering is an inelastic scattering process, care has also to be taken in conjunction with the discretized energy intervals, like for inter- and intravalley phonon scattering processes. Thus, an approximation is used instead. The most simple approximation is to employ the *mean value theorem of integration* [8]:

$$
\int_a^b f(x)\varphi(x)dx = f(\chi) \int_a^b \varphi(x)dx \ , \ (a < \chi < b) \ , \quad (3.2.3\text{-}87)
$$

$$
\approx f\left(\frac{a+b}{2} \right) \int_a^b \varphi(x)dx \ . \quad (3.2.3\text{-}88)
$$

With the approximation (3.2.3-88) the discretized collision term for POP-scattering reads

$$
\sum_{\acute{r}=1}^{2} \int_{w_1^\nu(\acute{r})}^{w_u^\nu(\acute{r})} \int_{\mu_{j-\frac{1}{2}}}^{\mu_{j+\frac{1}{2}}} (\sqrt{m_\nu^*})^3 s^\nu (w^\nu) \overline{C^{POP}} d\mu \, dw^\nu \approx \frac{K^{POP} \pi}{m_\nu^* \Delta w \Delta \mu} N_{ph}^{\pm}
$$

$$
\cdot \sum_{\acute{r}=1}^{2} \left\{
\begin{array}{l}
\left((\sqrt{m_\nu^*})^3 S^\nu (w^\nu) \Big|_{w_1^\nu(\acute{r})}^{w_u^\nu(\acute{r})} - \overset{\text{IN}}{\frac{\Delta w(\acute{r})}{\Delta w}} \frac{n_{i,j}^{\nu,\tau}}{\Delta \mu} \right) \\[4mm]
\cdot \sum_{j'=1}^{N_\mu} \left(n_{i_{in}'+\acute{r}-2,j'}^{\nu,\tau} \left[H_{POP}(w_m^\nu(\acute{r}), w_m^\nu(\acute{r}) \pm \Delta w_{ph}) \Big|_{\mu_{j'-\frac{1}{2}}}^{\mu_{j'+\frac{1}{2}}} \right]_{\mu_{j-\frac{1}{2}}}^{\mu_{j+\frac{1}{2}}} \right) \\[4mm]
-n_{i,j}^{\nu,\tau} \\[4mm]
\cdot \sum_{j'=1}^{N_\mu} \left[\left((\sqrt{m_\nu^*})^3 S^\nu (\acute{w}^\nu) \Big|_{\acute{w}_1^\nu(\acute{r})}^{\acute{w}_u^\nu(\acute{r})} - \overset{\text{OUT}}{\frac{\Delta w(\acute{r})}{\Delta w}} \frac{n_{i_{out}'+\acute{r}-2,j'}^{\nu,\tau}}{\Delta \mu} \right) \right. \\[4mm]
\left. \cdot \left[H_{POP}(w_m^\nu(\acute{r}), w_m^\nu(\acute{r}) \mp \Delta w_{ph}) \Big|_{\mu_{j'-\frac{1}{2}}}^{\mu_{j'+\frac{1}{2}}} \right]_{\mu_{j-\frac{1}{2}}}^{\mu_{j+\frac{1}{2}}} \right]
\end{array}
\right\},
$$

$$(3.2.3\text{-}89)$$

where $i_{in/out}'$ is given by (3.2.3-69)-(3.2.3-70) (with $iv_{\nu-\nu'} = 0$, since an intravalley scattering process is considered) and the energetic intervals after (3.2.3-66) and (3.2.3-67) with $\Delta w_0^{\nu-\nu'} = 0$. The energy $w_m^\nu(\acute{r})$ is the center of the intervals defined in (3.2.3-66) ($\acute{r} = 1$) and (3.2.3-67) ($\acute{r} = 2$) in accordance to the approximation (3.2.3-88). Although an approximation is employed for the last integration over w^ν, the anisotropic nature of POP-scattering is analytically captured by (3.2.3-89). Numerical experiments (BULK simulations) revealed that the *exact* mathematical treatment of the anisotropic nature (integration over μ' and μ) is crucial for the particle conservation and leads otherwise to an artificial and nonphysical generation or loss of particles.

Isotropic approximation

In normalized form, the isotropic approximation of the POP-transition rate after (2.3.5-95) (cf. section 2.3.5.1) reads

$$\overline{S^{\text{POP,ISO,out}}} = \frac{K^{\text{POP}} N_{\text{ph}}^{\pm}}{2m^*} \tilde{g}(w^{\nu}, \acute{w}^{\nu}) \delta(\acute{w}^{\nu} - w^{\nu} \pm \Delta w_{\text{ph}}) , \qquad (3.2.3\text{-}90)$$

$$\tilde{g}(w^{\nu}, \acute{w}^{\nu}) = \frac{1}{h_{\nu}(w^{\nu})} \left[1 + \frac{(1 - \tilde{a})(1 + \tilde{a})}{2\tilde{a}} \log \left(\left| \frac{1 + \tilde{a}}{1 - \tilde{a}} \right| \right) \right] , \qquad (3.2.3\text{-}91)$$

$$\tilde{a} = \sqrt{\frac{h_{\nu}(\acute{w}^{\nu})}{h_{\nu}(w^{\nu})}} , \qquad (3.2.3\text{-}92)$$

with the function h_{ν} after (3.2.3-81). Note, $\nu' = \nu$ has already been applied to (3.2.3-90) (intravalley scattering). With (3.2.3-90) one obtains for (3.2.3-50):

$$(\sqrt{m_{\nu}^*})^3 s^{\nu}(w^{\nu}) \overline{C^{\text{POP,ISO}}} = K^{\text{POP}} N_{\text{ph}}^{\pm} \frac{\pi}{2m_{\nu}^* \Delta w}$$

$$\cdot \left\{ \begin{array}{l} \tilde{g}(w^{\nu} \pm \Delta w_{\text{ph}}, w^{\nu}) \left((\sqrt{m_{\nu}^*})^3 s^{\nu}(w^{\nu}) - \frac{n_{i,j}^{\nu,\tau}}{\Delta w \Delta \mu} \right) \\ \cdot \sum_{j'=1}^{N_{\mu}} n_{i'_{\text{in}}+\acute{r}-2,j'}^{\nu,\tau} \\ -\tilde{g}(w^{\nu}, w^{\nu} \mp \Delta w_{\text{ph}}) \frac{n_{i,j}^{\nu,\tau}}{\Delta \mu} \\ \cdot \sum_{j'=1}^{N_{\mu}} \left((\sqrt{m_{\nu}^*})^3 s^{\nu}(w^{\nu} \mp \Delta w_{\text{ph}}) \Delta \mu - \frac{n_{i'_{\text{out}}+\acute{r}-2,j'}^{\nu,\tau}}{\Delta w} \right) \end{array} \right\} , \quad (3.2.3\text{-}93)$$

where (3.2.3-93) is already prepared for the integration scheme of inelastic scattering processes, sketched in section 3.2.3.2. Using, like for the anisotropic version of POP-scattering, the approximation (3.2.3-88) leads to

$$\sum_{\acute{r}=1}^{2} \int_{w_1^{\nu}(\acute{r})}^{w_u^{\nu}(\acute{r})} \int_{\mu_{j-\frac{1}{2}}}^{\mu_{j+\frac{1}{2}}} (\sqrt{m_{\nu}^*})^3 s^{\nu}(w^{\nu}) \overline{C^{\text{POP,ISO}}} d\mu \, dw^{\nu} \approx K^{\text{POP}} N_{\text{ph}}^{\pm} \frac{\pi}{2m_{\nu}^* \Delta w} \Delta \mu$$

$$\cdot \sum_{\acute{r}=1}^{2} \left\{ \begin{array}{l} \tilde{g}(w_{\text{m}}^{\nu}(\acute{r}) \pm \Delta w_{\text{ph}}, w_{\text{m}}^{\nu}(\acute{r})) \left((\sqrt{m_{\nu}^*})^3 S^{\nu}(w^{\nu})\Big|_{w_1^{\nu}(\acute{r})}^{w_u^{\nu}(\acute{r})} - \frac{\Delta w(\acute{r})}{\Delta w} \frac{n_{i,j}^{\nu,\tau}}{\Delta \mu} \right) \\ \cdot \sum_{j'=1}^{N_{\mu}} n_{i'_{\text{in}}+\acute{r}-2,j'}^{\nu,\tau} \\ -\tilde{g}(w_{\text{m}}^{\nu}(\acute{r}), w_{\text{m}}^{\nu}(\acute{r}) \mp \Delta w_{\text{ph}}) n_{i,j}^{\nu,\tau} \\ \cdot \sum_{j'=1}^{N_{\mu}} \left((\sqrt{m_{\nu}^*})^3 S^{\nu}(w^{\nu} \mp \Delta w_{\text{ph}})\Big|_{w_1^{\nu}(\acute{r})}^{w_u^{\nu}(\acute{r})} - \frac{\Delta w(\acute{r})}{\Delta w} \frac{n_{i'_{\text{out}}+\acute{r}-2,j'}^{\nu,\tau}}{\Delta \mu} \right) \end{array} \right\} .$$

$$(3.2.3\text{-}94)$$

3.2.3.5 Impurity scattering

Anisotropic version

With the normalization strategy given in section 3.1, the anisotropic transition rate of impurity scattering (section 2.3.2, eq. (2.3.2-69)) reads in normalized form:

$$\overline{S^{\text{IMP,out}}} = \frac{K^{\text{IMP}}}{\left(|\vec{\overline{k}} - \vec{\overline{k'}}|^2 + \overline{q_{\text{d}}}^2\right)^2} \delta\left(\acute{w}^\nu - w^\nu\right) , \qquad (3.2.3\text{-}95)$$

$$K^{\text{IMP}} = \frac{1}{D_{\text{N}} k_{\text{N}} K_{\text{N}}} \frac{N_{\text{I}}}{4\pi^2\hbar} \left(\frac{q^2}{\varepsilon_0\varepsilon_{\text{r}}}\right)^2 , \qquad (3.2.3\text{-}96)$$

where for (3.2.3-95) the Kronecker-Delta $\delta_{\nu,\nu'}$ (intravalley process) has already been evaluated. Like for the numerical treatment of POP-scattering (section 3.2.3.4), also for IMP-scattering the term $|\vec{\overline{k}} - \vec{\overline{k'}}|^2$ is replaced by (2.3.5-86) and leads to:

$$|\vec{\overline{k}} - \vec{\overline{k'}}|^2 = 2h_\nu(w^\nu)m_\nu^*\left[1 - \left(\mu\mu' + \sqrt{1-\mu^2}\sqrt{1-\mu'^2}\cos(\varphi')\right)\right] , \qquad (3.2.3\text{-}97)$$

with $h_\nu(w^\nu)$ after (3.2.3-81). Note, for (3.2.3-97) the elastic nature of IMP-scattering has already been taken into account and an isotropic dispersion relation is assumed. Using (3.2.3-97), the equation (3.2.3-50) of the procedural method evaluates to

$$(\sqrt{m_\nu^*})^3 s^\nu(w^\nu)\overline{C^{\text{IMP}}} = \frac{K^{\text{IMP}} s^\nu(w^\nu)}{8(h_\nu(w^\nu))^2\sqrt{m_\nu^*}} \int_{-1}^{1}\int_0^{2\pi}\left[\frac{\acute{\Phi}^\nu - \Phi^\nu}{(a + b\cos(\varphi'))^2}\right]d\varphi'\,d\mu'$$

$$= \frac{K^{\text{IMP}} s^\nu(w^\nu)}{8(h_\nu(w^\nu))^2\sqrt{m_\nu^*}} \int_{-1}^{1}\frac{2\pi a(\acute{\Phi}^\nu - \Phi^\nu)}{(a^2 - b^2)^{\frac{3}{2}}}d\mu' , \qquad (3.2.3\text{-}98)$$

with

$$a = 1 - \mu\mu' + \frac{\overline{q_{\text{d}}}^2}{2h_\nu(w^\nu)m_\nu^*} , \qquad (3.2.3\text{-}99)$$

101

and

$$b = -\sqrt{1 - \mu^2}\sqrt{1 - \mu'^2} \; . \tag{3.2.3-100}$$

As for ADP-scattering (section 3.2.3.1), also the terms related to the Pauli principle disappear in (3.2.3-98) due to the elastic nature of IMP-scattering. Since a (3.2.3-99) and b (3.2.3-100) are functions of μ', the following integral is needed:

$$\int \frac{a}{(a^2 - b^2)^{\frac{3}{2}}} d\mu' = \frac{\mu'\left(1 + \frac{\overline{q_{\mathrm{d}}}^2}{2h_\nu(w^\nu)m_\nu^*}\right) - \mu}{\frac{\overline{q_{\mathrm{d}}}^2}{h_\nu(w^\nu)m_\nu^*}\left(1 + \frac{\overline{q_{\mathrm{d}}}^2}{4h_\nu(w^\nu)m_\nu^*}\right)\sqrt{R}} = G(w^\nu, \mu, \mu') \; , \tag{3.2.3-101}$$

with the terms

$$R = \tilde{a} + \tilde{b}\mu' + \tilde{c}\mu'^2 \; , \tag{3.2.3-102}$$

$$\tilde{a} = \mu^2 + \frac{\overline{q_{\mathrm{d}}}^2}{h_\nu(w^\nu)m_\nu^*}\left(1 + \frac{\overline{q_{\mathrm{d}}}^2}{4h_\nu(w^\nu)m_\nu^*}\right) \; , \tag{3.2.3-103}$$

$$\tilde{b} = -2\mu\left(1 + \frac{\overline{q_{\mathrm{d}}}^2}{2h_\nu(w^\nu)m_\nu^*}\right) \; , \tag{3.2.3-104}$$

$$\tilde{c} = 1 \; . \tag{3.2.3-105}$$

Using (3.2.3-101) and (3.2.0-35) yields for (3.2.3-98)

$$(\sqrt{m_\nu^*})^3 s^\nu(w^\nu)\overline{C^{\mathrm{IMP}}} =$$

$$\frac{K^{\mathrm{IMP}}\pi}{4(h_\nu(w^\nu))^2\sqrt{m_\nu^*}}\frac{s^\nu(w^\nu)}{\Delta w \Delta \mu}\sum_{j'=1}^{N_\mu}\left(\left(n_{i,j'}^{\nu,\tau} - n_{i,j}^{\nu,\tau}\right)G(w^\nu,\mu,\mu')\Big|_{\mu'=\mu_{j'-\frac{1}{2}}}^{\mu'=\mu_{j'+\frac{1}{2}}}\right) \; . \tag{3.2.3-106}$$

Due to the considered anisotropy, the function G after (3.2.3-101) also depends on w^ν and μ, over which (3.2.3-106) has to be integrated in last stage (cf. eq. (3.2.3-51)). In order to ensure the particle conservation (like in the case of POP-scattering), at least the integration after μ has to be done analytically. In contrast to POP-scattering, here a closed-form analytical solution can be found:

$$\int G(w^\nu, \mu, \mu') d\mu = \frac{-\sqrt{R}}{\frac{\overline{q_d}^2}{h_\nu(w^\nu)m_\nu^*}\left(1 + \frac{\overline{q_d}^2}{4h_\nu(w^\nu)m_\nu^*}\right)} = H(w, \mu, \mu') , \quad (3.2.3\text{-}107)$$

where R in (3.2.3-107) equals (3.2.3-102). With (3.2.3-107) the last integration over w^ν (cf. eq. (3.2.3-51)) is again not analytically solvable. Thus, the approximation (3.2.3-88) is utilized and the collision term for IMP-scattering reads

$$\int_{w_{i-\frac{1}{2}}^\nu}^{w_{i+\frac{1}{2}}^\nu} \int_{\mu_{j-\frac{1}{2}}}^{\mu_{j+\frac{1}{2}}} (\sqrt{m_\nu^*})^3 s^\nu(w^\nu)\overline{C^{\mathrm{IMP}}} d\mu \, dw^\nu \approx \frac{K^{\mathrm{IMP}}\pi \, S^\nu(w^\nu)|_{w_{i-\frac{1}{2}}^\nu}^{w_{i+\frac{1}{2}}^\nu}}{4(h_\nu(w^\nu))^2\sqrt{m_\nu^*}\Delta w \Delta \mu}$$

$$\cdot \sum_{j'=1}^{N_\mu} \left(\left(n_{i,j'}^{\nu,\tau} - n_{i,j}^{\nu,\tau} \right) \left[H(w_i^\nu, \mu, \mu') \Big|_{\mu'=\mu_{j'-\frac{1}{2}}}^{\mu'=\mu_{j'+\frac{1}{2}}} \right]_{\mu=\mu_{j-\frac{1}{2}}}^{\mu=\mu_{j+\frac{1}{2}}} \right) , \quad (3.2.3\text{-}108)$$

where w_i^ν in (3.2.3-108) is the discretized kinetic energy after (3.2.0-33) which is therefore taken at the center of the currently considered control volume:

$$w_i^\nu = \frac{w_{i+\frac{1}{2}}^\nu + w_{i-\frac{1}{2}}^\nu}{2} . \quad (3.2.3\text{-}109)$$

Isotropic approximation

The approximation for the isotropic case of impurity scattering, given in section 2.3.5.2 by equation (2.3.5-101), reads in normalized form

$$\overline{S^{\mathrm{IMP,ISO,out}}} = \frac{K^{\mathrm{IMP}}}{8\left(m_\nu^*\right)^2 \left(h_\nu(w^\nu)\right)^2} \left(\log\left(\frac{\tilde{b}+2}{\tilde{b}}\right) - \frac{2}{\tilde{b}+2} \right) \delta\left(\acute{w}^\nu - w^\nu\right) ,$$

$$\hspace{11cm} (3.2.3\text{-}110)$$

$$\tilde{b} = \frac{\overline{q_d}^2}{2h_\nu(w^\nu)m_\nu^*} , \quad (3.2.3\text{-}111)$$

with K^{IMP} after (3.2.3-96) and $h_\nu(w^\nu)$ after (3.2.3-81). Similar to ADP-scattering (see section 3.2.3.1), which is also an isotropic and elastic scattering process, the intermediate step (3.2.3-50) of the procedural method reads:

$$
(\sqrt{m_\nu^*})^3 s^\nu(w^\nu)\overline{C^{\mathrm{IMP,ISO}}} = \frac{\pi K^{\mathrm{IMP}} s^\nu(w^\nu)}{8\sqrt{m_\nu^*}\,(h_\nu(w^\nu))^2\,\Delta w}\left(\log\left(\frac{\tilde{b}+2}{\tilde{b}}\right) - \frac{2}{\tilde{b}+2}\right)
$$
$$
\cdot \left[\sum_{j'=1}^{N_\mu}\left(n_{i,j'}^{\nu,\tau} - n_{i,j}^{\nu,\tau}\right)\right].
\tag{3.2.3-112}
$$

Again, the approximation (3.2.3-88) is used for the integration over w^ν involved in (3.2.3-51). Here, $\tilde{b}$ (3.2.3-111) and $h_\nu(w^\nu)$ (3.2.3-81) are functions of w^ν that eliminate an analytic solution for (3.2.3-51). Thus, the collision term is approximated by

$$
\int_{w_{i-\frac{1}{2}}^\nu}^{w_{i+\frac{1}{2}}^\nu}\int_{\mu_{j-\frac{1}{2}}}^{\mu_{j+\frac{1}{2}}}(\sqrt{m_\nu^*})^3 s^\nu(w^\nu)\overline{C^{\mathrm{IMP,ISO}}}\,d\mu\,dw^\nu \approx \frac{\pi K^{\mathrm{IMP}}\Delta\mu\,S^\nu(w^\nu)|_{w_{i-\frac{1}{2}}^\nu}^{w_{i+\frac{1}{2}}^\nu}}{8\sqrt{m_\nu^*}\Delta w}
$$
$$
\cdot\,\frac{\log\left(\frac{\tilde{b}+2}{\tilde{b}}\right) - \frac{2}{\tilde{b}+2}}{(h_\nu(w^\nu))^2}\Bigg|_{w_i^\nu}\left[\sum_{j'=1}^{N_\mu}\left(n_{i,j'}^{\nu,\tau} - n_{i,j}^{\nu,\tau}\right)\right].
\tag{3.2.3-113}
$$

3.2.3.6 Alloy scattering

As described in section 2.3.3, alloy (ALY) scattering is an elastic and isotropic scattering mechanism. Although this is the simplest case (like in the case of ADP-scattering, section 3.2.3.1) regarding the numerical treatment, the burden stems here from the fact that ALY is both an intra- as well as an intervalley scattering process. Thus, the potential energies of the valleys involved have to be considered to ensure the energy conservation. With (2.3.3-72) the normalized transition rate reads

$$
\overline{S^{\mathrm{ALY,out}}} = K^{\mathrm{ALY}}\delta\left(\acute{w}^{\nu'} - w^\nu\right),
$$
$$
K^{\mathrm{ALY}} = \frac{1}{K_{\mathrm{N}}}\frac{a_\perp a_\parallel^2 U_{\mathrm{ALY}}^2}{32\pi^2\hbar}(1 - x_{\mathrm{Ge}})\,x_{\mathrm{Ge}},
\tag{3.2.3-114}
$$

with the alloy scattering potential U_{ALY}, the material composition content x_{Ge} and quantities given by (2.3.3-73)-(2.3.3-76) in section 2.3.3. Due to the different valleys involved, the procedure used for intervalley phonon scattering (see section 3.2.3.2) is utilized, since it is inherently capable of considering energetic offsets of different valleys. The intermediate step (3.2.3-50) of the procedural method reads here

$$
(\sqrt{m_\nu^*})^3 s^\nu (w^\nu) \overline{C^{\text{ALY}}} = \frac{K^{\text{ALY}} \pi}{\Delta w}
$$

$$
\cdot \sum_{\nu'} Z^{\nu'} \left\{ (\sqrt{m_\nu^*})^3 s^\nu (w^\nu) \sum_{j'=1}^{N_\mu} n_{i',j'}^{\nu',\tau} - 2 \frac{n_{i,j}^{\nu,\tau}}{\Delta \mu} (\sqrt{m_{\nu'}^*})^3 s^{\nu'} (w^{\nu'}) \right\} \quad , (3.2.3\text{-}115)
$$

with $Z^{\nu'}$ being the valley degeneracy factor, and with $w^{\nu'} = w^\nu + \Delta w_0^{\nu-\nu'}$, with the energetic valley offset $\Delta w_0^{\nu-\nu'}$ after (3.2.3-63). i' in (3.2.3-115) is given by (3.2.3-69) and (3.2.3-70), with $ip = 0$ (elastic process). Following the integration scheme outlined in section 3.2.3.2, the discretized ALY-collision term develops to

$$
\sum_{\acute{r}=1}^{2} \int_{w_1^\nu(\acute{r})}^{w_u^\nu(\acute{r})} \int_{\mu_{j-\frac{1}{2}}}^{\mu_{j+\frac{1}{2}}} (\sqrt{m_\nu^*})^3 s^\nu(w^\nu) \overline{C^{\text{ALY}}} d\mu \, dw^\nu = \frac{K^{\text{ALY}} \pi \Delta \mu}{\Delta w} \sum_{\nu'} Z^{\nu'} \sum_{\acute{r}=1}^{2} \Bigg\{
$$

$$
(\sqrt{m_\nu^*})^3 S^\nu(w^\nu) \Big|_{w_1^\nu(\acute{r})}^{w_u^\nu(\acute{r})} \sum_{j'=1}^{N_\mu} n_{i'+\acute{r}-2,j'}^{\nu',\tau} - 2 \frac{n_{i,j}^{\nu,\tau}}{\Delta \mu} (\sqrt{m_{\nu'}^*})^3 S^{\nu'}(\acute{w}^{\nu'}) \Big|_{\acute{w}_1^{\nu'}(\acute{r})}^{\acute{w}_u^{\nu'}(\acute{r})} \Bigg\} ,
$$

$$
(3.2.3\text{-}116)
$$

where the integration limits are given by (3.2.3-73) and (3.2.3-74) ($\Delta w_{\text{ph}} = 0$).

3.2.3.7 Piezoelectric scattering

Anisotropic version

In normalized form, the transition rate of the as elastic approximated PZ scattering mechanism reads (cf. equation (2.3.4-79))

$$\overline{S^{\text{PZ,out}}} = K^{\text{PZ}} \frac{|\vec{\overline{k}} - \vec{\overline{k'}}|^2}{\left(|\vec{\overline{k}} - \vec{\overline{k'}}|^2 + q_{\text{d}}^2\right)^2} \delta(\acute{w}^{\nu'} - w^{\nu}) \delta_{\nu,\nu'}, \quad (3.2.3\text{-}117)$$

$$K^{\text{PZ}} = \frac{1}{K_{\text{N}} k_{\text{N}}^2} \frac{k_{\text{b}} T_{\text{L}}}{4\pi^2 \rho v_{\text{s,av}}^2} \left(\frac{q \varepsilon_{\text{PZ}}}{\varepsilon_0 \varepsilon_{\text{r}}}\right)^2, \quad (3.2.3\text{-}118)$$

where the constants involved in (3.2.3-117) and (3.2.3-118) are presented in section 2.3.4. Like for IMP-scattering, here also the substitution (3.2.3-97) for the term $|\vec{\overline{k}} - \vec{\overline{k'}}|^2$ is used, since (3.2.3-117) is an intravalley and as elastic approximated scattering process [2]. For the intermediate step (3.2.3-50) one obtains

$$(\sqrt{m_{\nu}^*})^3 s^{\nu}(w^{\nu}) \overline{C^{\text{PZ}}} =$$

$$\frac{K^{\text{PZ}} s^{\nu}(w^{\nu}) \sqrt{m_{\nu}^*}}{4 h_{\nu}(w^{\nu})} \int_{-1}^{1} \int_{0}^{2\pi} \left[\left(\acute{\Phi}^{\nu} - \Phi^{\nu}\right) \hat{f}(w^{\nu}, \mu, \mu', \varphi')\right] d\varphi' d\mu' , \quad (3.2.3\text{-}119)$$

with the dimensionless function $\hat{f}$ specifying the anisotropy of PZ-scattering and defined as

$$\hat{f}(w^{\nu}, \mu, \mu', \varphi') = \frac{a + b \cos(\varphi')}{(a + c + b \cos(\varphi'))^2} , \quad (3.2.3\text{-}120)$$

$$a = 1 - \mu\mu' , \quad (3.2.3\text{-}121)$$

$$b = -\sqrt{1 - \mu^2} \sqrt{1 - \mu'^2} , \quad (3.2.3\text{-}122)$$

$$c = \frac{\overline{q_{\text{d}}}^2}{2 h_{\nu}(w^{\nu}) m_{\nu}^*} . \quad (3.2.3\text{-}123)$$

---

[2]Note, (3.2.3-97) is only valid for isotropic dispersion relations.

To finalize (3.2.3-119), the function $\hat{f}$ (3.2.3-120) is integrated over φ' in the range of $[0, 2\pi]$:

$$\int_0^{2\pi} \hat{f}(w^\nu, \mu, \mu', \varphi')d\varphi' = 2\pi \frac{(\mu - \mu')^2 + (1 - \mu\mu')c}{[R]^{\frac{3}{2}}}$$

$$= 2\pi g(w^\nu, \mu, \mu') , \qquad (3.2.3\text{-}124)$$

$$R = (\mu - \mu')^2 + 2c(1 - \mu\mu') + c^2 \qquad (3.2.3\text{-}125)$$

and subsequently over μ', which leads to

$$\int g(w^\nu, \mu, \mu')d\mu' = \frac{\mu - \mu'(1+c)}{2\left(1+\frac{c}{2}\right)\sqrt{R}} + \operatorname{arcsinh}\left(\frac{\mu' - \mu(1+c)}{\sqrt{2c\left(1+\frac{c}{2}\right)}\sqrt{1-\mu^2}}\right)$$

$$= G_{\mathrm{PZ}}(w^\nu, \mu, \mu') . \qquad (3.2.3\text{-}126)$$

Thus, with the results (3.2.3-124) and (3.2.3-126) the intermediate step (3.2.3-50) for PZ-scattering becomes

$$(\sqrt{m_\nu^*})^3 s^\nu(w^\nu)\overline{C^{\mathrm{PZ}}} = \frac{\pi K^{\mathrm{PZ}} s^\nu(w^\nu)\sqrt{m_\nu^*}}{2h_\nu(w^\nu)\Delta w \Delta \mu} \cdot$$

$$\left\{ \sum_{j'=1}^{N_\mu} \left(n_{i,j'}^{\nu,\tau} \left. G_{\mathrm{PZ}}(w^\nu, \mu, \mu') \right|_{\mu'=\mu_{j'-\frac{1}{2}}}^{\mu'=\mu_{j'+\frac{1}{2}}} \right) - n_{i,j}^{\nu,\tau} \left. G_{\mathrm{PZ}}(w^\nu, \mu, \mu') \right|_{\mu'=-1}^{\mu'=1} \right\} .$$

$$(3.2.3\text{-}127)$$

As for POP- and IMP-scattering, here also the integration of (3.2.3-127) w.r.t. μ (for the discretized collision term after (3.2.3-51)) has to be done analytically in order to preserve the particle conservation. However, due to the arcsinh-term and the term $\sqrt{1-\mu^2}$ in the denominator of its argument in (3.2.3-126), the same elaborate strategy as for the treatment of anisotropic POP-scattering (see section 3.2.3.4) needs to be applied. Fortunately, the three required primitives for the domain composition strategy, introduced for anisotropic POP-scattering in section 3.2.3.4 and schematically shown in figure 3.4, are also found for PZ-scattering. These primitives can be found in appendix B.7 and the principle evaluation strategy for the integrals is identical to the one for POP-scattering in section 3.2.3.4. Thus, any integral of the kind

$$\int_{\mu=\mu_1}^{\mu=\mu_u} G_{PZ}(w^\nu, \mu, \mu')\Big|_{\mu'=\mu_1'}^{\mu'=\mu_u'}\, d\mu = \left[H_{PZ}(w^\nu, \mu, \mu')\Big|_{\mu'=\mu_1'}^{\mu'=\mu_u'}\right]_{\mu=\mu_1}^{\mu=\mu_u} \quad (3.2.3\text{-}128)$$

can analytically be solved. Again, for the last integration over w^ν involved in the procedural method (3.2.3-51) the approximation (3.2.3-88) is used due to the elaborate w^ν-dependence of (3.2.3-128) (see appendix B.7). Therefore, (3.2.3-51) for PZ-scattering evaluates to

$$\int_{w_{i-\frac{1}{2}}^\nu}^{w_{i+\frac{1}{2}}^\nu} \int_{\mu_{j-\frac{1}{2}}}^{\mu_{j+\frac{1}{2}}} (\sqrt{m_\nu^*})^3 s^\nu(w^\nu)\overline{C^{PZ}} \approx \frac{\pi K^{PZ}\, S^\nu(w^\nu)\Big|_{w_{i-\frac{1}{2}}^\nu}^{w_{i+\frac{1}{2}}^\nu}\sqrt{m_\nu^*}}{2h_\nu(w_i^\nu)\Delta w}$$

$$\cdot \left\{ \sum_{j'=1}^{N_\mu} \left(n_{i,j'}^{\nu,\tau} \left[H_{PZ}(w_i^\nu, \mu, \mu')\Big|_{\mu'=\mu_{j'-\frac{1}{2}}'}^{\mu'=\mu_{j'+\frac{1}{2}}'} \right]_{\mu=\mu_{j-\frac{1}{2}}}^{\mu=\mu_{j+\frac{1}{2}}} \right) \right.$$

$$\left. - n_{i,j}^{\nu,\tau} \left[H_{PZ}(w_i^\nu, \mu, \mu')\Big|_{\mu'=-1}^{\mu'=1} \right]_{\mu=\mu_{j-\frac{1}{2}}}^{\mu=\mu_{j+\frac{1}{2}}} \right\} . \quad (3.2.3\text{-}129)$$

Isotropic approximation

After normalization, the transition rate of the as isotropic approximated PZ scattering mechanism (2.3.5-106) (section 2.3.5.3) reads

$$\overline{S^{PZ,ISO,out}} = \frac{K^{PZ}}{2h_\nu(w^\nu)m_\nu^*} \left[c\log\left(\frac{c}{c+2}\right) + 2\frac{c+1}{c+2} \right] \delta(\hat{w}^{\nu'} - w^\nu)\delta_{\nu,\nu'}, \quad (3.2.3\text{-}130)$$

with K^{PZ} after (3.2.3-118) and c after (3.2.3-123). Similar to ADP- and IMP-scattering (isotropic approximations), the intermediate step after (3.2.3-50) leads to

$$(\sqrt{m_\nu^*})^3 s^\nu(w^\nu)\overline{C^{\text{PZ,ISO}}} = \frac{\pi K^{\text{IMP}}\sqrt{m_\nu^*}s^\nu(w^\nu)}{2h_\nu(w^\nu)\Delta w}\left[c\log\left(\frac{c}{c+2}\right)+2\frac{c+1}{c+2}\right]$$

$$\cdot\left[\sum_{j'=1}^{N_\mu}n_{i,j'}^{\nu,\tau}-\frac{2}{\Delta\mu}n_{i,j}^{\nu,\tau}\right] \quad . \tag{3.2.3-131}$$

Since c after (3.2.3-123) is a function of w^ν (due to $h_\nu(w^\nu)$), the integration w.r.t. w^ν for (3.2.3-51) is non-trivial, hence, the approximation (3.2.3-88) is employed again. Here, the same procedure as for IMP-scattering (see section 3.2.3.5) is pursued. Thus, (3.2.3-51) for the isotropic approximation of PZ-scattering results in

$$\int_{w_{i-\frac{1}{2}}^\nu}^{w_{i+\frac{1}{2}}^\nu}\int_{\mu_{j-\frac{1}{2}}}^{\mu_{j+\frac{1}{2}}}(\sqrt{m_\nu^*})^3 s^\nu(w^\nu)\overline{C^{\text{PZ,ISO}}} \approx \frac{\pi K^{\text{PZ}}\left.S^\nu(w^\nu)\right|_{w_{i-\frac{1}{2}}^\nu}^{w_{i+\frac{1}{2}}^\nu}\sqrt{m_\nu^*}\Delta\mu}{2h_\nu(w_i^\nu)\Delta w}$$

$$\cdot\left.\left[c\log\left(\frac{c}{c+2}\right)+2\frac{c+1}{c+2}\right]\right|_{w_i^\nu}\left[\sum_{j'=1}^{N_\mu}n_{i,j'}^{\nu,\tau}-\frac{2}{\Delta\mu}n_{i,j}^{\nu,\tau}\right] \quad . \tag{3.2.3-132}$$

3.2.3.8 Conclusions and remarks

With the employed discretization scheme, it is shown that isotropic scattering processes, both elastic and inelastic, are accurately transformed to their discretized pendants without any simplification. In terms of inelastic scattering processes, the splitting technique introduced in section 3.2.3.2 offers the possibility to capture the portions of the discretized energy intervals and thus of the solution variables n therein accurately. In contrast to the method used in [19, 20], the employed method allows an arbitrary energy step-size Δw, arbitrary phonon energies and energetic valley separations. Here, phonon energies and energetic valley separations are not restricted to be an integer multiple of Δw, enabling a more accurate consideration of inelastic scattering mechanisms and band structures.

Regarding anisotropic scattering mechanisms, discretized formulations are given which capture the anisotropy of the scattering mechanisms analytically. However, some simplifications were assumed and employed:

i.) isotropic dispersion relation: this simplification stems from the term $|\vec{k} - \vec{k'}|$ involved in all considered anisotropic scattering mechanisms. For anisotropic dispersion relations, the discretization method of section 3.2.3 in combination with the term $|\vec{k} - \vec{k'}|$ results in expressions hardly manageable for the further analytic pre-processing. The Herring-Vogt transformation cannot solve the problem, since the transformation maps $\vec{k}$ and $\vec{k'}$ to isotropic $\vec{\tilde{k}}$ and $\vec{\tilde{k}'}$ individually. Therefore, expressing $|\vec{k} - \vec{k'}|$ in terms of $\vec{\tilde{k}}$ and $\vec{\tilde{k}'}$ does not resolve the original complexity. As a work-around, for anisotropic dispersion relations the isotropic approximations can be used instead. With a possible re-adjustment of the scattering parameters involved, the accuracy for strong anisotropic dispersion relations can be increased.

ii.) *Mean value theorem of integration*: the most important part for the treatment of the anisotropic scattering processes is the $\mu - \mu'$-dependence. As described in the respective sections, the accurate treatment of $\mu - \mu'$-dependence is crucial for the particle conservation. The *mean value theorem of integration* after (3.2.3-88) and used for the integration over the kinetic energy w^ν (see equation (3.2.3-51)) is detached from the particle conservation. Due to the Dirac-functions involved, the final energies are uniquely associated with the initial energy. Thus, $|\vec{k} - \vec{k'}|$ for a given initial energy (and thus with the final energies) becomes solely dependent on μ and μ', which consequently specifies the anisotropy. This anisotropy ($\mu - \mu'$-dependence) is analytically captured and processed. The remaining integration w.r.t. w^ν (cf. eq. (3.2.3-51)) can be viewed as a summation of the analytically captured anisotropy ($\mu - \mu'$-dependence) within the energetic range of the considered control volume. Here, the *mean value theorem of integration* after (3.2.3-88) is employed for the sake of simplicity. However, more sophisticated numerical integration schemes can be used instead.

Regarding IMP-scattering, in [31, 37] the *strength* of the isotropic approximation of impurity scattering (K^{IMP} in section 3.2.3.5) is re-adjusted as function of the doping concentration. Latter has been done in order to capture experimental data of the carrier mobility. Based on the given graphs, analytical functions whose shape resemble those given in [31, 37] are presented in the appendix B.5.

3.2.4 Boundary conditions in the reciprocal space

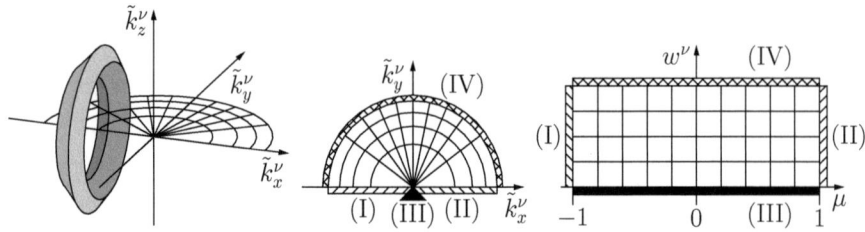

Figure 3.6: Schematic illustration of the simulation domain (valley ν in the reciprocal space) for a 1D BULK/real space simulation where the x-direction is assumed to be the transport direction. Here, the control volumes are sketched in all plots. Due to the rotational symmetry in the 1D case (around $\tilde{k}^\nu_x$, cf. section 2.4.3, figure 2.21), a ring (φ-dependence) with a constant density $n^{\nu,\tau}$ is considered in the original 3D k-space (left). Therefore, it is sufficient to examine the semicircle (center) with the corresponding domain boundaries labeled by (I-IV). The right picture illustrates the mapping of the domain boundaries by the coordinate transformation described in section 2.4.2.3.

In the 1D case, the homogeneous *Neumann* boundary condition is applied to each boundary of the w^ν-μ domain. The fluxes of the drift-term (v^{w^ν} and v^μ) perpendicular to the four boundaries (figure 3.6, right) become

I) $v^\mu|_{\mu=-1} = 0$, due to $\langle a^{\nu,\tau}_{i,1-\frac{1}{2}} \rangle = 0$ ($\mu^2_{1-\frac{1}{2}} = 1$) in (3.2.2-45). The suppression of the flux is caused by the flux coefficient $\langle a^{\nu,\tau}_{i,1-\frac{1}{2}} \rangle$. For $\mu = -1$ and $+1$ the negative and positive $\tilde{k}^\nu_x$-axis, respectively, is specified regardless of w^ν and φ (cf. figure 3.6, center and right). From the physical point of view, this boundary condition prevents an artificial flow of particles through the infinitesimal thin $\tilde{k}^\nu_x$-axis into the opposite lying region of the k-space (e.g. from the northern to the southern hemisphere).

II) $v^\mu|_{\mu=1} = 0$, due to $\langle a^{\nu,\tau}_{i,N_\mu+\frac{1}{2}} \rangle = 0$ ($\mu^2_{N_\mu+\frac{1}{2}} = 1$) in (3.2.2-45), where the same explanation as for (I) holds.

III) $v^{w^\nu}|_{w^\nu=0} = 0$, since that boundary represents the bottom of the currently considered valley ν and thus the respective band edge; below (energetically) the band edge the forbidden region for electrons is located and an electron flux to the forbidden region is suppressed. This condition is inherently fulfilled by $\langle a^{\nu,\tau}_{1-\frac{1}{2},j}\rangle = 0$ in (3.2.2-44) ($w^\nu_{1-\frac{1}{2}} = 0$).

IV) $v^{w^\nu}\big|_{w^\nu = w^\nu_{N_w+\frac{1}{2}}} = 0$ is motivated by the conservation of the carrier density within the total simulation domain and thus suppresses a carrier flux through that boundary.

So, the boundary conditions (I-III) are already given by the flux coefficients, where for (IV) the flux v^{w^ν} has to be set to zero. Nevertheless, the boundary conditions are provided by equations to the BULK system of equations. Depending on the wind-direction, the information of *Neumann* boundary condition has to be provided to the Upwind-MinMod interpolation scheme (see section 3.2.2) for consistency. Two distinct cases are affected (ξ is a placeholder for w^ν and μ):

1) at $\xi_{1+\frac{1}{2}}$ for $\langle a^{\nu,\tau}_{1+\frac{1}{2}}\rangle > 0$ (wind in positive ξ-direction) and

2) at $\xi_{N_\xi-\frac{1}{2}}$ for $\langle a^{\nu,\tau}_{N_\xi-\frac{1}{2}}\rangle < 0$ (wind in negative ξ-direction).

In these cases, the values $n^{\nu,\tau}_{\iota=0}$ for 1) and $n^{\nu,\tau}_{\iota=N_\xi+1}$ for 2) have to be provided to the Upwind-MinMod interpolation scheme (3.2.2-47) and (3.2.2-48), respectively. For both cases, ghost points at $\xi_{\iota=0}$ and $\xi_{\iota=N_\xi+1}$ are used and the corresponding densities are set to $n^{\nu,\tau}_0 = -n^{\nu,\tau}_1$ for 1) and $n^{\nu,\tau}_{N_\xi+1} = -n^{\nu,\tau}_{N_\xi}$ for 2). Thus, a linear interpolation for $n^{\nu,\tau}_{\frac{1}{2}}$ or $n^{\nu,\tau}_{N_\xi+\frac{1}{2}}$ (densities at the domain boundaries) will give a density of zero and thus a flux of zero after (3.2.2-44) or (3.2.2-45) (regardless of $\langle a^{\nu,\tau}_{\frac{1}{2}}\rangle$ or $\langle a^{\nu,\tau}_{N_\xi+1}\rangle$). Thus, inserting the values at the ghost points in the corresponding equations ((3.2.2-47) and (3.2.2-48)) gives

1) $\langle a^{\nu,\tau}_{1+\frac{1}{2}}\rangle > 0$:

$$n^{\nu,\tau}_{1+\frac{1}{2}} = n^{\nu,\tau}_1 + \frac{\mathrm{MM}(\overbrace{n^{\nu,\tau}_2 - n^{\nu,\tau}_1}^{m_2}, \overbrace{2n^{\nu,\tau}_1}^{m_1})}{2}$$

$$= \begin{cases} \frac{1}{2}(n^{\nu,\tau}_2 + n^{\nu,\tau}_1) \,, \, |m_1| > |m_2| \,, \, m_1 m_2 > 0 \text{ ($\circ$ in fig. 3.7(l))}, \\ 2n^{\nu,\tau}_1 \,, \, |m_1| < |m_2| \,, \, m_1 m_2 > 0 \text{ ($\diamond$ in fig. 3.7(l))}, \\ n^{\nu,\tau}_1 \,, \text{ else ($\square$ in fig. 3.7(l))}. \end{cases} \quad (3.2.4\text{-}133)$$

2) $\langle a_{N_\xi - \frac{1}{2}} \rangle < 0$:

$$n^{\nu,\tau}_{N_\xi - 1 + \frac{1}{2}} = n^{\nu,\tau}_{N_\xi} - \frac{MM(\overbrace{-2n^{\nu,\tau}_{N_\xi}}^{m_2}, \overbrace{n^{\nu,\tau}_{N_\xi} - n^{\nu,\tau}_{N_\xi - 1}}^{m_1})}{2}$$

$$= \begin{cases} 2n^{\nu,\tau}_{N_\xi} , |m_1| > |m_2| , m_1 m_2 > 0 \; (\diamond \text{ in fig. 3.7(r)}), \\ \frac{1}{2}(n^{\nu,\tau}_{N_\xi} + n^{\nu,\tau}_{N_\xi - 1}) , |m_1| < |m_2| , m_1 m_2 > 0 \; (\circ \text{ in fig. 3.7(r)}), \\ n^{\nu,\tau}_{N_\xi} , \text{else } (\square \text{ in fig. 3.7(r)}). \end{cases} \quad (3.2.4\text{-}134)$$

Figure 3.7 shows schematically the results of (3.2.4-133) and (3.2.4-134), respectively.

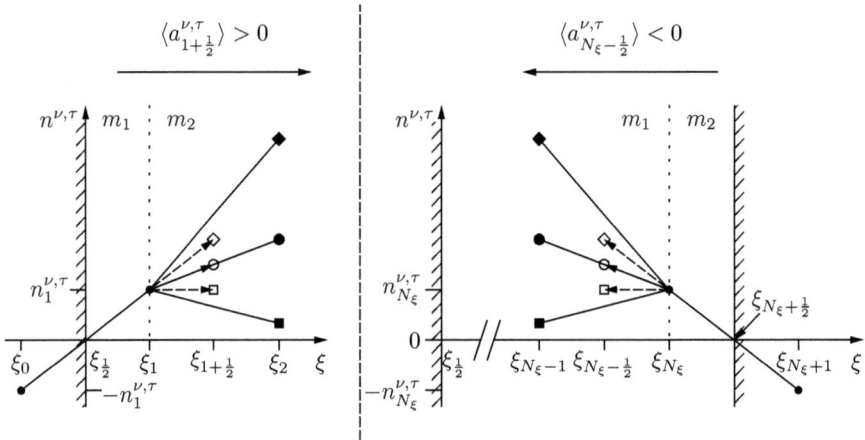

Figure 3.7: Schematic illustration of the impact of the *Neumann* boundary condition for the fluxes $v^\xi_{\frac{1}{2}}$ and $v^\xi_{N_\xi + \frac{1}{2}}$ on the interpolation scheme for $n^{\nu,\tau}_{1 + \frac{1}{2}}$ and $n^{\nu,\tau}_{N_\xi - \frac{1}{2}}$, respectively. Note, that due to the ghost points (ξ_0 and $\xi_{N_\xi + 1}$, domain boundaries) the densities $n^{\nu,\tau}_{\frac{1}{2}}$ and $n^{\nu,\tau}_{N_\xi + \frac{1}{2}}$ are interpolated to be zero and thus the respective fluxes (see section 3.2.2, equation (3.2.2-44) and (3.2.2-45)). Again, ξ represents a placeholder for w^ν and μ.

3.2.5 Method for the stationary solution

The method described so far is only capable of solving the BULK BTE for transient simulations. However, stationary (or time independent) solutions are often of interest, since these results allow to adjust the physical models, e.g. carrier mobility model, involved in the DD and HD transport model.

In order to obtain the stationary solution directly, an additional equation has to be supplied to the inherently transient BULK BTE system. Here, the demand of a charge neutral BULK region

$$\rho = q \left(p - n + N_{\mathrm{D}}^{+} - N_{\mathrm{A}}^{-} \right) = 0 \qquad (3.2.5\text{-}135)$$

is applied to the BULK BTE system to exchange its temporal dependence. Equation (3.2.5-135) gives the space charge with the elementary charge q, the donor (acceptor) doping concentration N_{D}^{+} (N_{A}^{-}) and the electron (hole) density n (p), respectively. The hole density can be obtained by employing for example Boltzmann statistics in combination with an effective density of states for the valence band (N_{V}), like usually done for DD and HD simulations. The electron density is numerically obtained by the BULK BTE (see (4.1.0-6), section 4.1):

$$n = \frac{D_{\mathrm{N}}}{4\pi^2} \sum_{\nu} Z^{\nu} \sum_{j=1}^{N_{\mu}} \sum_{i=1}^{N_w} n_{i,j}^{\nu} \ , \qquad (3.2.5\text{-}136)$$

where D_{N} is the normalization factor for doping and carrier densities after (3.1.0-12), Z^{ν} is the valley degeneracy factor and the solution variable of the BTE $n_{i,j}^{\nu}$. The normalized version of (3.2.5-135) reads

$$\left(\frac{p}{D_{\mathrm{N}}} - \frac{1}{4\pi^2} \sum_{\nu} Z^{\nu} \sum_{j=1}^{N_{\mu}} \sum_{i=1}^{N_w} n_{i,j}^{\nu} + \frac{N_{\mathrm{D}}^{+}}{D_{\mathrm{N}}} - \frac{N_{\mathrm{A}}^{-}}{D_{\mathrm{N}}} \right) = 0 \ . \qquad (3.2.5\text{-}137)$$

Note, (3.2.5-137) is applied for both stationary and transient simulations. In the case of transient simulations, (3.2.5-137) guarantees accurate results without the transient effects known from Monte-Carlo simulations.

3.3 Extension for the 1D Device BTE

For the Device BTE, the discretized terms of the BULK BTE introduced in section 3.2 and its subsections can be employed almost unchanged here. However, care has to be taken regarding a position dependent dispersion relation. For the drift-term, the effective driving force after (2.4.0-108) has to be used in contrast to the position independent (and user-supplied) field $\overline{E_{\text{eff.}}^{\nu,x}}$ of the BULK case. In (2.4.0-108) the spatial derivative of the kinetic energy ϵ^ν and thus a position dependence of dispersion relation is included. Likewise, for the diffusion-term the parts stemming from the total differential (section 2.4.2.1, equation (2.4.2-125)) and associated to the spatial dependence of the dispersion relation need to be considered. These terms are given in normalized form in section 3.1, equation (3.1.0-23), (3.1.0-25) and (3.1.0-26). However, instead of a straightforward discretization of all these terms, it is more advantageous to combine first the drift-term and the driving force after (2.4.0-108) analytically. In normalized form and for the 1D case, (2.4.0-108) reads

$$\overline{E_{\text{eff.}}^{\nu,x}} \;=\; -\frac{\partial \overline{\psi}}{\partial \overline{x}} + \frac{\partial \overline{V_{C,0}^{\nu}}}{\partial \overline{x}} + \frac{\partial w^{\nu}}{\partial \overline{x}} = \overline{E_{\text{eff., pot.}}^{\nu,x}} + \frac{\partial w^{\nu}}{\partial \overline{x}} \;,\qquad (3.3.0\text{-}138)$$

where $\overline{\psi}$ is the normalized electrostatic potential obtained by the *Poisson* equation and $\overline{V_{C,0}^{\nu}}$ is the operating point independent conduction band potential of the valley ν. $\overline{\psi}$ and $\overline{V_{C,0}^{\nu}}$ are forming the potential carrier energy and therefore their real-space derivative is abbreviated by $\overline{E_{\text{eff., pot.}}^{\nu,x}}$. Inserting the driving force after (3.3.0-138) in the normalized drift-term after (3.1.0-27) and adding the parts of the diffusion-term after (3.1.0-23), (3.1.0-25) and (3.1.0-26) leads to

$$T_{\text{diff.}} + T_{\text{drift}} = \frac{\partial}{\partial \overline{x}_T}\left(\frac{\Phi^\nu}{\sqrt{m_\nu^*}} \frac{2\sqrt{w^\nu\left(1 + a_\nu w^\nu\right)}}{1 + 2a_\nu w^\nu} T_{1,1}^{\text{HV},\nu}\mu \right)$$

$$-\left\{ \frac{\partial}{\partial w^\nu}\left(\frac{2\Phi^\nu\sqrt{w^\nu(1 + a_\nu w^\nu)}}{\sqrt{m_\nu^*}(1 + 2a_\nu w^\nu)}\mu T_{1,1}^{\text{HV},\nu}\overline{E_{\text{eff., pot.}}^{\nu,x}} \right) + \frac{\partial}{\partial\mu}\left(\frac{(1 - \mu^2)T_{1,1}^{\text{HV},\nu}\overline{E_{\text{eff.}}^{\nu,x}}\Phi^\nu}{\sqrt{m_\nu^*}\sqrt{w^\nu(1 + a_\nu w^\nu)}} \right) \right.$$

$$\left. -\frac{\Phi^\nu}{\sqrt{m_\nu^*}}\frac{2\sqrt{w^\nu\left(1 + a_\nu w^\nu\right)}}{1 + 2a_\nu w^\nu}T_{1,1}^{\text{HV},\nu}\mu\frac{\partial\mu}{\partial\overline{x}} \right) \right\}\;.$$

$$(3.3.0\text{-}139)$$

Due to the effective field (3.3.0-138) almost all terms responsible for a position dependent dispersion relation and associated to (3.1.0-23), (3.1.0-25) and (3.1.0-26) are vanishing in (3.3.0-139). For a more detailed derivation the reader is referred to appendix B.3.

A direct comparison of the drift-term for the Device BTE (partial derivatives $\frac{\partial}{\partial w^\nu}$ and $\frac{\partial}{\partial \mu}$) with the one of the BULK BTE (equation (3.2.0-32) and (3.2.2-40)- (3.2.2-41)) reveals that the flux description in w^ν-direction is not altered. However, instead of the full formulation for the effective field $\overline{E_{\text{eff.}}^{\nu,x}}$ after (3.3.0-138) only the portions associated with potential energy $\overline{E_{\text{eff., pot.}}^{\nu,x}}$ needs to be considered.

In the case of the flux in μ-direction, the description for the BULK BTE after (3.2.2-45) needs to be extended by the term associated to $\frac{\partial \mu}{\partial \overline{x}}$ in (3.3.0-139), which is a remnant of the position dependent dispersion relation. Due to h_x, φ-dependent formulations for $\frac{\partial w^\nu}{\partial \overline{x}}$ (normalized $\frac{\partial \epsilon^\nu}{\partial x}$ in (3.3.0-138) and (3.3.0-139)) and $\frac{\partial \mu}{\partial \overline{x}}$ (normalized $\frac{\partial \mu}{\partial x}$) are obtained (cf. sect. 2.4.2.2, (2.4.2-140)-(2.4.2-142)). Utilizing the rotational symmetry described in section 2.4.4 for the 1D case, h_x and subsequent the derivatives $\frac{\partial w^\nu}{\partial \overline{x}}$ and $\frac{\partial \mu}{\partial \overline{x}}$ become φ-independent (see appendix B.3, (B.3.0-28)-(B.3.0-30)). Following the procedure described in section 3.2.2 results for (3.2.2-45) in

$$
\int_{w_{i-\frac{1}{2}}^\nu}^{w_{i+\frac{1}{2}}^\nu} v_{i,j\pm\frac{1}{2}}^\mu \, dw^\nu = -\frac{T_{1,1}^{\text{HV},\nu}}{\sqrt{m_\nu^*}} \frac{n_{i,j\pm\frac{1}{2}}^{\nu,\tau}}{\Delta w \Delta \mu} \frac{(1-\mu_{j\pm\frac{1}{2}}^2)}{\sqrt{a_\nu}} \left[\log\left(2\sqrt{a_\nu}t_1 + t_2\right)\overline{E_{\text{eff., pot.}}^{\nu,x}} \right.
$$

$$
-\frac{1}{2\sqrt{a_\nu}}\left\{ \frac{1}{a_\nu}\frac{\partial a_\nu}{\partial \overline{x}}\left(t_1 - t_3 - \frac{1}{\sqrt{a_\nu}}\log\left(\frac{t_1}{\sqrt{a_\nu}} + \frac{t_2}{2a_\nu}\right)\right)\right.
$$

$$
\left.\left. - (t_1 + t_3)\left(\frac{\partial \log\left(T_{2,2}^{\text{HV},\nu}\right)}{\partial \overline{x}} + \frac{\partial \log\left(T_{3,3}^{\text{HV},\nu}\right)}{\partial \overline{x}} - \frac{\partial m_\nu^*}{\partial \overline{x}}\right)\right\}\right]\Bigg|_{w_{i-\frac{1}{2}}^\nu}^{w_{i+\frac{1}{2}}^\nu}
$$

$$
= \langle a_{i,j\pm\frac{1}{2}}^{\nu,\tau}\rangle n_{i,j\pm\frac{1}{2}}^{\nu,\tau} \, , \tag{3.3.0-140}
$$

with the abbreviations $t_1 = \sqrt{w^\nu(1 + a_\nu w^\nu)}$, $t_2 = (1 + 2a_\nu w^\nu)$ and $t_3 = \frac{1}{2\sqrt{a_\nu}}\arcsin\left(\frac{1}{t_2}\right)$. The term associated with $\overline{E_{\text{eff., pot.}}^{\nu,x}}$ in (3.3.0-140) agrees with the one obtained for the BULK BTE, like for the flux in w^ν-direction.

The remaining lines are due to the position dependence of the dispersion relation. It should be noted here that for a position independent dispersion relation the fluxes in w^ν- and μ-direction are identical with those of the BULK BTE after (3.2.2-44) and (3.2.2-45), respectively.

3.3.1 Diffusion term

Like for the BULK BTE, also for the Device BTE each term is integrated over the discretized control volume. Here, the diffusion-term is the only term left to be integrated. Starting point here is its formulation involved in (3.3.0-139):

$$
T_{\text{diff.}} = \frac{\partial}{\partial \overline{x_T}} \left(\frac{\Phi^\nu}{\sqrt{m_\nu^*}} \frac{2\sqrt{w^\nu \left(1 + a_\nu w^\nu\right)}}{1 + 2a_\nu w^\nu} T_{1,1}^{\text{HV},\nu} \mu \right) . \qquad (3.3.1\text{-}141)
$$

The term in parenthesis in (3.3.1-141) corresponds to a particle flux in the real-space direction, here x. The partial derivative $\left(\frac{\partial}{\partial \overline{x_T}}\right)$ itself stems from the real-space divergence. For consistency with the numerical treatment of fluxes by BIM, *Gauss's* theorem after (3.2.0-36) is also employed here:

$$
\oint_{dV_{i,j,k}} \vec{v}^{\,\tilde{x}} \cdot \vec{n} \, dS = \int_{w_{i-\frac{1}{2}}^\nu}^{w_{i+\frac{1}{2}}^\nu} \int_{\mu_{j-\frac{1}{2}}}^{\mu_{j+\frac{1}{2}}} v_{i,j,k+\frac{1}{2}}^x \, d\mu \, dw^\nu
$$
$$
- \int_{w_{i-\frac{1}{2}}^\nu}^{w_{i+\frac{1}{2}}^\nu} \int_{\mu_{j-\frac{1}{2}}}^{\mu_{j+\frac{1}{2}}} v_{i,j,k-\frac{1}{2}}^x \, d\mu \, dw^\nu . \quad (3.3.1\text{-}142)
$$

However, in order to apply (3.3.1-142) directly, the remaining terms of the BTE, which are forming the BULK BTE, also need to be integrated over x, since the considered space spans now over w^ν, μ and x. Due to the assumptions made for BIM, the BULK BTE is assumed to be spatially independent within the control volume in x-direction. Thus, the missing integration results in a simple multiplication of the BULK BTE terms by an average distance $\overline{\Delta x_k}$:

$$
\overline{\Delta x_k} = \frac{x_{k+1} - x_{k-1}}{2} . \qquad (3.3.1\text{-}143)
$$

In order to avoid an additional scaling of the BTE, the multiplication by $\overline{\Delta x_k}$ is omitted and the diffusion-term is divided by $\overline{\Delta x_k}$:

$$T_{\text{diff., disc.}} = \frac{\langle a^{\nu,\tau}_{i,j,k+\frac{1}{2}} \rangle n^{\nu,\tau}_{i,j,k+\frac{1}{2}} - \langle a^{\nu,\tau}_{i,j,k-\frac{1}{2}} \rangle n^{\nu,\tau}_{i,j,k-\frac{1}{2}}}{\Delta x_k} \, , \qquad (3.3.1\text{-}144)$$

where for (3.3.1-144) the relation

$$\langle a^{\nu,\tau}_{i,j,k\pm\frac{1}{2}} \rangle n^{\nu,\tau}_{i,j,k\pm\frac{1}{2}} \approx \int_{w^\nu_{i-\frac{1}{2}}}^{w^\nu_{i+\frac{1}{2}}} \int_{\mu_{j-\frac{1}{2}}}^{\mu_{j+\frac{1}{2}}} v^x_{i,j,k\pm\frac{1}{2}} \, d\mu \, dw^\nu \qquad (3.3.1\text{-}145)$$

is used. Equation (3.3.1-145) represents the approximated flux of the diffusion term integrated over the bounding surface between two adjacent real-space points. The fluxes and the considered bounding surfaces are shown in figure 3.8 schematically. For the numerical treatment of the fluxes after (3.3.1-145),

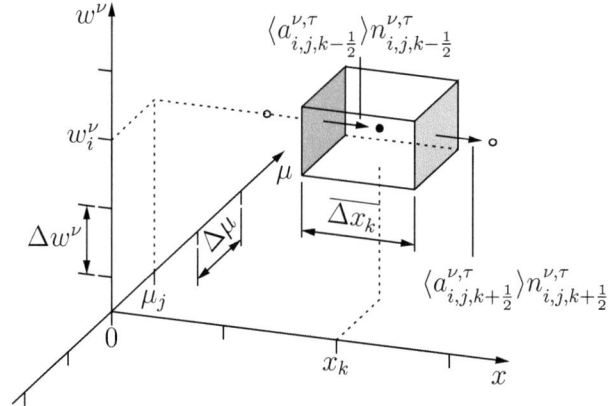

Figure 3.8: Schematic illustration of fluxes considered for the diffusion term and the corresponding bounding surfaces (grey shaded rectangles).

the coefficients $\langle a^{\nu,\tau}_{i,j,k\pm\frac{1}{2}} \rangle$ are obtained in an analogous manner as for the drift-term (see section 3.2.2). Based on term in parenthesis in (3.3.1-141) and by replacing Φ^ν with its numerical counterpart (see (3.2.0-35)), $\langle a^{\nu,\tau}_{i,j,k\pm\frac{1}{2}}\rangle$ reads

$$\langle a^{\nu,\tau}_{i,j,k\pm\frac{1}{2}}\rangle =$$

$$\int_{w^\nu_{i-\frac{1}{2}}}^{w^\nu_{i+\frac{1}{2}}} \int_{\mu_{j-\frac{1}{2}}}^{\mu_{j+\frac{1}{2}}} \frac{\mu}{\sqrt{m^*_{\nu,k\pm\frac{1}{2}}}} \frac{2\sqrt{w^\nu\left(1+a_{\nu,k\pm\frac{1}{2}}w^\nu\right)}}{1+2a_{\nu,k\pm\frac{1}{2}}w^\nu} \frac{T^{\mathrm{HV},\nu}_{1,1}\Big|_{x_{k\pm\frac{1}{2}}}}{\Delta w \Delta\mu} d\mu \, dw^\nu$$

$$= \left[\mu^2 \left(\frac{\arcsin\left(\frac{1}{1+2a_{\nu,k\pm\frac{1}{2}}w^\nu}\right)}{2\sqrt{a_{\nu,k\pm\frac{1}{2}}}} + \sqrt{w^\nu\left(1+a_{\nu,k\pm\frac{1}{2}}w^\nu\right)} \right) \right]\Bigg|_{w^\nu_{i-\frac{1}{2}}}^{w^\nu_{i+\frac{1}{2}}}\Bigg|_{\mu_{j-\frac{1}{2}}}^{\mu_{j+\frac{1}{2}}}$$

$$\cdot \frac{T^{\mathrm{HV},\nu}_{1,1}\Big|_{x_{k\pm\frac{1}{2}}}}{2\sqrt{m^*_{\nu,k\pm\frac{1}{2}}}\, a_{\nu,k\pm\frac{1}{2}}\Delta w\Delta\mu}, \tag{3.3.1-146}$$

where the Herring-Vogt matrix element $T^{\mathrm{HV},\nu}_{1,1}\Big|_{x_{k\pm\frac{1}{2}}}$, the effective mass $m^*_{\nu,k\pm\frac{1}{2}}$ and real space flux coefficient $a_{\nu,k\pm\frac{1}{2}}$ are evaluated at the bounding surfaces of the control volume (cf. figure 3.8) to account for the position dependence of the dispersion relation. To complete the diffusion (real space) flux description, the values of the solution variables $n^{\nu,\tau}_{i,j,k\pm\frac{1}{2}}$ at the bounding surfaces need to be approximated (see equation (3.3.1-144)). In this work, the approximation of $n^{\nu,\tau}_{i,j,k\pm\frac{1}{2}}$ is performed by the WENO53-method (see section 3.3.2). Analogous to the flux calculation involved in the drift term (see section 3.2.2), the coefficients $\langle a^{\nu,\tau}_{i,j,k\pm\frac{1}{2}}\rangle$ are used for determining the wind-direction needed by the interpolation scheme. With the wind-direction it is guaranteed that the flux through the bounding surface at $x_{k-\frac{1}{2}}$ is not affected by the control variable k of the discretized real space points or control volumes:

$$\langle a^{\nu,\tau}_{i,j,k-1+\frac{1}{2}}\rangle n^{\nu,\tau}_{i,j,k-1+\frac{1}{2}} = \langle a^{\nu,\tau}_{i,j,k-\frac{1}{2}}\rangle n^{\nu,\tau}_{i,j,k-\frac{1}{2}}. \tag{3.3.1-147}$$

The employed scheme is identical to the one given in [19, 20]. The coefficient $\langle a^{\nu,\tau}_{i,j,k\pm\frac{1}{2}}\rangle$ after (3.3.1-146) becomes negative for $\mu_{j-\frac{1}{2}}$ and $\mu_{j+\frac{1}{2}} < 0$, which corresponds to particles with negative k_x-component that move in negative x-direction. Likewise, a positive $\langle a_{i,j,k\pm\frac{1}{2}}^{\nu,\tau}\rangle$ ($\mu_{j-\frac{1}{2}}$ and $\mu_{j+\frac{1}{2}} > 0$) represents particles moving in positive x-direction (see figure 2.21 in section 2.4.4). The transport direction and thus the wind-direction of the interpolation scheme is included by $\langle a_{i,j,k\pm\frac{1}{2}}^{\nu,\tau}\rangle$, for example

$$\langle a_{i,j,k+\frac{1}{2}}^{\nu,\tau}\rangle > 0 \rightarrow n_{i,j}^{\nu,\tau}(x_{k+\frac{1}{2}}) = n_{i,j,k+\frac{1}{2}}^{\nu,\tau} , \qquad (3.3.1\text{-}148)$$

$$\langle a_{i,j,k+\frac{1}{2}}^{\nu,\tau}\rangle < 0 \rightarrow n_{i,j}^{\nu,\tau}(x_{k+\frac{1}{2}}) = n_{i,j,k+1-\frac{1}{2}}^{\nu,\tau} . \qquad (3.3.1\text{-}149)$$

For the bounding surface at $x_{k+\frac{1}{2}}$, $n_{i,j}^{\nu,\tau}(x_{k+\frac{1}{2}})$ is interpolated from the currently considered control volume at x_k in positive x-direction ($\langle a_{i,j,k+\frac{1}{2}}^{\nu,\tau}\rangle > 0$, (3.3.1-148)) and otherwise $n_{i,j}^{\nu,\tau}(x_{k+\frac{1}{2}})$ is interpolated from the adjacent control volume at x_{k+1} in negative x-direction (3.3.1-149). Figure 3.9 visualizes the direction of interpolation schematically. Note, in order to preserve a well defined wind-direction – excluding flux coefficients $\langle a_{i,j,k\pm\frac{1}{2}}^{\nu,\tau}\rangle = 0$ – the number of intervals in μ-direction (N_μ) should be even.

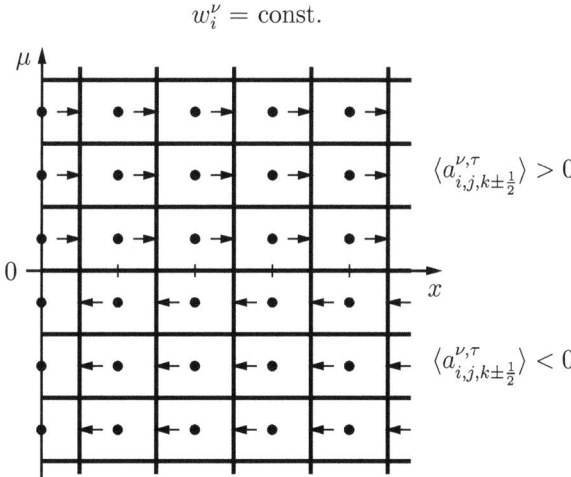

$$w_i^\nu = \text{const.}$$

Figure 3.9: Schematic illustration of diffusion flux interpolation at the bounding surfaces of the control volumes.

3.3.2 A brief survey of the WENO53 method

The WENO (weighted essentially non-oscillatory) scheme was introduced by Liu, Osher and Chan in a third order finite-volume version [45] first and extended by Jiang and Shu [36] for a fifth order finite difference scheme. The key idea behind the WENO scheme is to reconstruct the primitive of the solution variable, based on the discretized values of the solution variable. Thus, the first-derivative of the reconstructed primitive interpolates the solution variable. For the considered WENO53 method, five stencils are needed for the reconstruction of the primitive: $(x_{k-2}, n^{\nu,\tau}_{i,j,k-2})$, $(x_{k-1}, n^{\nu,\tau}_{i,j,k-1})$, $(x_k, n^{\nu,\tau}_{i,j,k})$, $(x_{k+1}, n^{\nu,\tau}_{i,j,k+1})$, $(x_{k+2}, n^{\nu,\tau}_{i,j,k+2})$. Using a fifth-order polynomial

$$
\begin{aligned}
p_5(x) &= a_5(x - x_k)^5 + b_5(x - x_k)^4 + c_5(x - x_k)^3 \\
&\quad + d_5(x - x_k)^2 + e_5(x - x_k) + f_5
\end{aligned}
\tag{3.3.2-150}
$$

and the solution variables $n^{\nu,\tau}_{i,j,k-2}$, $n^{\nu,\tau}_{i,j,k-1}$, $n^{\nu,\tau}_{i,j,k}$, $n^{\nu,\tau}_{i,j,k+1}$, $n^{\nu,\tau}_{i,j,k+2}$, the coefficients a_5-e_5 can be uniquely obtained by

$$
\int_{x_{k-\frac{5}{2}}}^{x_{k-\frac{3}{2}}} n^{\nu,\tau}_{i,j}(x)dx \approx p_5(x_{k-\frac{3}{2}}) - p_5(x_{k-\frac{5}{2}}) = n^{\nu,\tau}_{i,j,k-2}\overline{\Delta x_{k-2}} \; ,
$$

$$
\int_{x_{k-\frac{5}{2}}}^{x_{k-\frac{1}{2}}} n^{\nu,\tau}_{i,j}(x)dx \approx p_5(x_{k-\frac{1}{2}}) - p_5(x_{k-\frac{5}{2}}) = n^{\nu,\tau}_{i,j,k-2}\overline{\Delta x_{k-2}} + n^{\nu,\tau}_{i,j,k-1}\overline{\Delta x_{k-1}} \; ,
$$

$$
\vdots
$$

$$
\int_{x_{k-\frac{5}{2}}}^{x_{k+\frac{5}{2}}} n^{\nu,\tau}_{i,j}(x)dx \approx p_5(x_{k+\frac{5}{2}}) - p_5(x_{k-\frac{5}{2}}) = n^{\nu,\tau}_{i,j,k-2}\overline{\Delta x_{k-2}} + n^{\nu,\tau}_{i,j,k-1}\overline{\Delta x_{k-1}}
$$

$$
+ n^{\nu,\tau}_{i,j,k}\overline{\Delta x_k} + n^{\nu,\tau}_{i,j,k+1}\overline{\Delta x_{k+1}}
$$

$$
+ n^{\nu,\tau}_{i,j,k+2}\overline{\Delta x_{k+2}} \; ,
\tag{3.3.2-151}
$$

where (3.3.2-151) forms a linear system of equations and the coefficients a_5-e_5 are the unknowns to be solved for. Note, f_5 in (3.3.2-150) does not play any role, since definite integrals are evaluated in (3.3.2-151) and thus f_5 cancels out. Another advantage can also be seen from the right-hand side of (3.3.2-151). Here, the solution variables are integrated over the respective control volumes. Since it is assumed that the solution variable is spatially independent within its control volume (assumptions for BIM), the coefficients a_5-e_5 can exactly be calculated within the framework of the BIM. Once a_5-e_5 are determined, the actual polynomial used for the interpolation reads

$$n_5(x) = \frac{dp_5(x)}{dx} = 5a_5(x-x_k)^4 + 4b_5(x-x_k)^3 + 3c_5(x-x_k)^2$$
$$+2d_5(x-x_k) + e_5 \ . \tag{3.3.2-152}$$

Using (3.3.2-152), $n^{\nu,\tau}_{i,j,k\pm\frac{1}{2}}$ at the bounding surfaces for (3.3.1-144) can be approximated by

$$n^{\nu,\tau}_{i,j,k\pm\frac{1}{2}} \cong n_5(x_{k\pm\frac{1}{2}}) \ . \tag{3.3.2-153}$$

However, the interpolation of $n^{\nu,\tau}_{i,j,k\pm\frac{1}{2}}$ might result in negative values and thus in oscillations near discontinuities (abrupt changes in $n^{\nu,\tau}$). To circumvent this issue, lower order polynomials are used near discontinuities and the smoothest candidate is selected. However, in smooth regions (3.3.2-152) is used. If the initial interpolation is based on an lth (here $l = 5$) degree polynomial, the lower order (rth order) polynomials have to fulfill the relation $l = 2r - 1$ [36, 45]. Therefore, third order polynomials are used and thus the WENO method considered here is usually denoted in literature by WENO53. For the rth order polynomials (here $r = 3$), the initial set of stencils is split. Three regions have to be considered:

$x_{k-2}, x_{k-1}, x_k \rightarrow$

$$p_{3,1}(x) = a_{3,1}(x-x_{k-1})^3 + b_{3,1}(x-x_{k-1})^2 + c_{3,1}(x-x_{k-1}) + d_{3,1} \ ,$$
$$n_{3,1}(x) = 3a_{3,1}(x-x_{k-1})^2 + 2b_{3,1}(x-x_{k-1}) + c_{3,1} \ , \tag{3.3.2-154}$$

$x_{k-1}, x_k, x_{k+1} \rightarrow$

$$p_{3,2}(x) = a_{3,2}(x-x_k)^3 + b_{3,2}(x-x_k)^2 + c_{3,2}(x-x_k) + d_{3,2} \ ,$$
$$n_{3,2}(x) = 3a_{3,2}(x-x_k)^2 + 2b_{3,2}(x-x_k) + c_{3,2} \ , \tag{3.3.2-155}$$

$x_k, x_{k+1}, x_{k+2} \rightarrow$

$$p_{3,3}(x) = a_{3,3}(x-x_{k+1})^3 + b_{3,3}(x-x_{k+1})^2 + c_{3,3}(x-x_{k+1}) + d_{3,3} \ ,$$
$$n_{3,3}(x) = 3a_{3,3}(x-x_{k+1})^2 + 2b_{3,3}(x-x_{k+1}) + c_{3,3} \ , \tag{3.3.2-156}$$

where the coefficients $a_{3,1-3}$, $b_{3,1-3}$ and $c_{3,1-3}$ are obtained in the same manner as for $p_5(x)$ in (3.3.2-151). The classical ENO method [27] uses one of the three polynomials near discontinuities for interpolating $n_{i,j,k\pm\frac{1}{2}}^{\nu,\tau}$, depending on their smoothness. Thus, third order accuracy is guaranteed near discontinuities and oscillations are prevented, if at least one of (3.3.2-154)-(3.3.2-156) is smooth in its interval.However, linear weights $\gamma_\iota^\pm$ can be determined such that

$$n_5(x_{k\pm\frac{1}{2}}) = \sum_{\iota=1}^{3} \gamma_\iota^\pm n_{3,\iota}(x_{k\pm\frac{1}{2}}) \qquad (3.3.2\text{-}157)$$

is fulfilled, where

$$\sum_{\iota=1}^{3} \gamma_\iota^\pm = 1 . \qquad (3.3.2\text{-}158)$$

Note, the superscript $\pm$ at the linear weights $\gamma_\iota^\pm$ indicates an interpolation at $x_{k+\frac{1}{2}}$ or $x_{k-\frac{1}{2}}$, respectively. Based on the linear weights $\gamma_\iota^\pm$ nonlinear weights $\omega_\iota^\pm$ are calculated, which suppress $n_{3,\iota}$ in non-smooth regions ($\omega_\iota^\pm \approx 0$) or assign $n_{3,\iota}$ its linear weight in smooth regions ($\omega_\iota^\pm \approx \gamma_\iota^\pm$). Like for its linear counterpart, also the nonlinear weights have to fulfill

$$\sum_{\iota=1}^{3} \omega_\iota^\pm = 1 . \qquad (3.3.2\text{-}159)$$

The nonlinear weights are obtained by

$$\omega_\iota^\pm = \frac{\alpha_\iota^\pm}{\alpha_1^\pm + \alpha_2^\pm + \alpha_3^\pm} \text{ with} \qquad (3.3.2\text{-}160)$$

$$\alpha_\iota^\pm = \frac{\gamma_\iota^\pm}{(\epsilon + \beta_\iota)^2} , \qquad (3.3.2\text{-}161)$$

where β_ι is the smoothness-indicator and ϵ is a small number (usually $\epsilon = 10^{-6}$ in literature) preventing a division by zero. The smoothness-indicator is calculated by

$$\beta_\iota = \sum_{\nu=1}^{r-1=2} \overline{\Delta x_k}^{-2\nu-1} \int_{x_{k-\frac{1}{2}}}^{x_{k-\frac{1}{2}}} \left(\frac{d^\nu}{dx^\nu} n_{3,\iota}(x)\right)^2 dx , \qquad (3.3.2\text{-}162)$$

where $\overline{\Delta x_k}^{-2\nu-1}$ is responsible for canceling out the unit of the length (in conjunction with the coefficients $a_{3,1-3}-c_{3,1-3}$). The integral over the squared derivatives measures the strength of the deviation of $n_{3,\iota}(x)$ from an ideally smooth and thus constant $n_{3,\text{ideal}}(x) = \text{const.}$ within the currently considered control volume $[x_{k-\frac{1}{2}}, x_{k+\frac{1}{2}}]$. Finally, the desired value of $n^{\nu,\tau}$ at the box boundary is obtained by

$$n^{\nu,\tau}_{i,j,k\pm\frac{1}{2}} = \omega_1^{\pm} n_{3,1}(x_{k\pm\frac{1}{2}}) + \omega_2^{\pm} n_{3,2}(x_{k\pm\frac{1}{2}}) + \omega_3^{\pm} n_{3,3}(x_{k\pm\frac{1}{2}}) . (3.3.2\text{-}163)$$

Figure 3.10 illustrates the interpolation scheme. The black lined boxes repre-

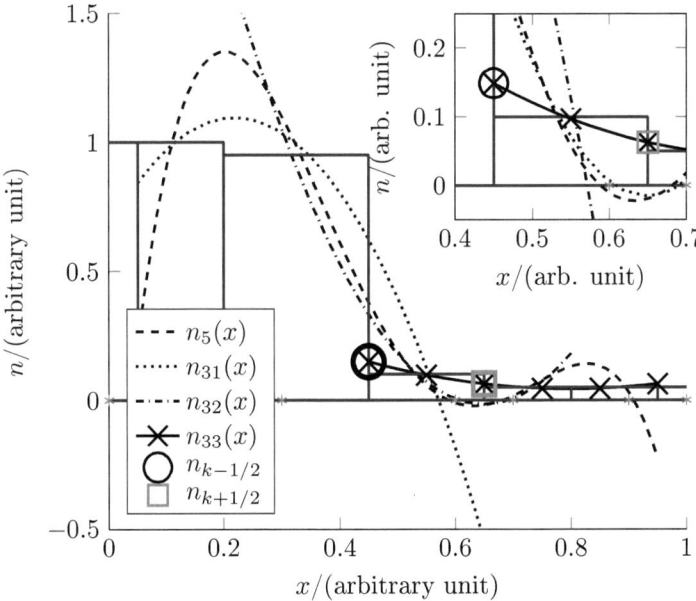

Figure 3.10: Polynomials involved in the WENO53 method and inter-polated values.

sent the solution variable (n) on an arbitrary non-equidistant grid. In addition, the polynomials (3.3.2-152) and (3.3.2-154)-(3.3.2-156) are shown. Due to the almost abrupt change in n (at $x = 0.45$), the 5th order polynomial after (3.3.2-152) shows oscillations and interpolates negative values of n in the region of interest $(0.45 \leq x \leq 0.65)$. The parabola $n_{31}(x)$ (3.3.2-154) interpolates a negative (non-practicable) value for $n_{k+\frac{1}{2}}$. Both $n_{31}(x)$ and $n_{32}(x)$ are the least smooth polynomials in contrast to $n_{33}(x)$. Therefore, $n_{33}(x)$ is weighted by a factor of $\omega_3^{\pm} \approx 1$, which results in $n_{k\pm\frac{1}{2}} \approx n_{33}(x_{k\pm\frac{1}{2}})$. In most literature [36, 45] the WENO53 method is given for equidistant grids. However, the method is also applicable for non-equidistant grids [63, 71], which is shown in figure 3.10 and also considered in this work.

3.3.3 Boundary conditions in the real space

For the 1D-BTE, boundary conditions in the real space have to be applied at x_1 and x_{N_x} (first and last real space discretization point). In this work, ideal ohmic contacts are considered. First, the ohmic contacts are assumed to be charge neutral and in thermodynamic equilibrium:

$$\left(p_{\text{eq.}} - n_{\text{eq.}} + N_D^+ - N_A^-\right) = 0 . \qquad (3.3.3\text{-}164)$$

In addition, inflow and outflow like boundary conditions are applied to the ohmic contacts [21]. The inflow boundary condition ensures that particles entering the contact from the device are ideally absorbed by the contact. Depending on the amount of particles absorbed by the ohmic contact, particles are ideally injected into the device. The amount of injected particles is determined such that the charge neutrality is ensured (outflow boundary condition). Thus, for an ohmic contact at x_1 (left-hand side of the device) the densities $n_{i,j,2}^{\nu,\tau}$ for $\mu_j < 0$ (negative k_x-component) are taken over to the corresponding control volumes at the contact point. The densities to be injected into the device (for $\mu_j > 0$, positive k_x-component) are obtained by the equilibrium distribution function (Boltzmann when Pauli principle is neglected, Fermi-Dirac with Pauli principle) scaled by the factor $a_{\text{scal.}}$ ensuring the charge neutrality (3.3.3-164):

$$a_{\text{scal.}} = 2 \left(1 - \frac{n_{\text{abs.}}}{n_{\text{eq.}}} \right) , \qquad (3.3.3\text{-}165)$$

with $n_{\text{abs.}}$ as the total electron density absorbed by the contact. Analogously, for the ohmic contact at x_{N_x}, $n_{\text{abs.}}$ is obtained from the control volumes at x_{N_x-1} for $\mu_j > 0$ (positive k_x-component). For $\mu_j < 0$ (negative k_x-component), the densities are injected by the equilibrium distribution function and $a_{\text{scal.}}$ after (3.3.3-165). The basic ideas behind the ohmic contact model are also used in [31], but implemented as a generation rate. The contact boundary conditions are schematically shown in figure 3.11.

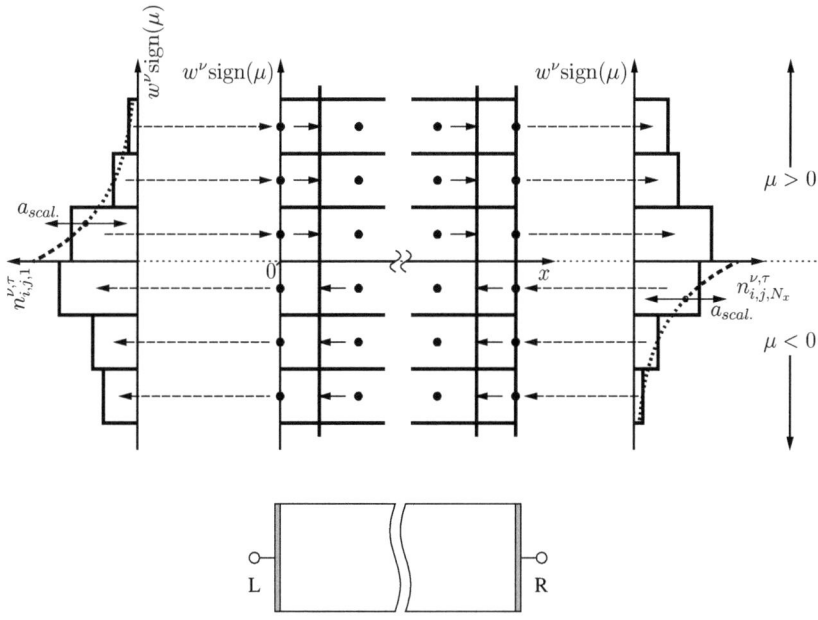

Figure 3.11: Schematic illustration of contact boundary conditions.

3.3.4 Method for the stationary solution

In contrast to the BULK BTE (see section 3.2.5), no additional equation is required for calculating the stationary solution of the Device BTE directly. The stationary solution can simply be obtained by setting the transient term of the Device BTE to zero. From the more engineering point of view, the negligence of the transient term is justified due to the contact boundary condition described in section 3.3.3. With the BTE after (2.4.0-107) (section 2.4) and considering, for convenience, only a 1D real- and 1D k-space, one ends up with

$$
\vec{v}_g^\nu(x, k_x^\nu) \frac{\partial f^\nu(x, k_x^\nu)}{\partial x} - \frac{q}{\hbar} \vec{E}_{\text{eff.}}^\nu(x, k_x^\nu) \frac{\partial f^\nu(x, k_x^\nu)}{\partial k_x^\nu}
$$
$$
= C(x, f^\nu(x, k_x^{\nu'}), f^{\nu'}(x, k_x^{\nu'})) \ . \tag{3.3.4-166}
$$

Furthermore, for a fixed $k_x^\nu > 0$, the simplified BTE after (3.3.4-166) becomes

$$
\left. \frac{\partial f^\nu(x, k_x^\nu)}{\partial x} \right|_{k_x^\nu = \text{const.}} = a(x) \ , \tag{3.3.4-167}
$$

where $a(x)$ contains both the drift- and collision-term divided by the group velocity $\vec{v}_g^\nu(x, k_x^\nu)$ for a fixed k_x^ν. Assuming that $a(x)$ is an integrable function, the solution for $f^\nu(x, k_x^\nu = \text{const.})$ is simply obtained by

$$
f^\nu(x, k_x^\nu = \text{const.}) = \int_0^x a(\tilde{x})d\tilde{x} = A(x) - A(0) \ , \tag{3.3.4-168}
$$

where $A(x)$ is the primitive of $a(x)$. $A(0)$ in (3.3.4-168) is the initial value of f^ν at the left contact (at $x = 0$) for $k_x^\nu > 0$. Therefore, $A(0)$ is given by the injected particles due to the contact boundary condition. Analogous considerations hold for $k_x^\nu < 0$. Thus, the applied contact boundary conditions determine the integration constants in (3.3.4-168) (here $A(0)$), which in turn are determined by the absorbed particles in conjunction with the demand of charge neutrality at the contacts. With the boundary conditions for the BULK BTE (see section 3.2.4, for the interior of the device), no degree of freedom is left, which might only be resolved by using the transient term.

3.4 Solution method

For BULK simulations, the discretized BULK BTE (see section 3.2) in conjunction with equation (3.2.5-137) (section 3.2.5) results in a system of $N_k + 1$ equations, where the $+1$ stems from equation (3.2.5-137) (demand for charge neutrality) and the number of discretization points used for the reciprocal space N_k:

$$N_k = N_w N_\mu N_{\text{val.}} \, . \tag{3.4.0-169}$$

N_w and N_μ are originating from grid definition after (3.2.0-33) and (3.2.0-34), respectively. The number of considered valleys is given by $N_{\text{val.}}$. Thus, for both transient and stationary simulations a $(N_k + 1) \times (N_k + 1)$ matrix is obtained. In the case of the 1D Device BTE, the complete system of equations to be solved consists of three discretized partial differential equations:

1) the BTE for electrons ($N_x N_k$ equations),

2) the *Poisson* equation for obtaining the electrostatic potential ψ (N_x equations),

3) and the continuity equation of holes, where its solution variable is the Quasi-Fermipotential of holes φ_p (N_x equations),

where N_x is the number of discretized points in the real space. For more information about the discretized versions of the *Poisson* and the hole continuity equation, the reader is referred to [56, 58, 72]. Therefore, for the 1D Device-BTE, a sparse matrix with the size of $N_x(N_k + 2) \times N_x(N_k + 2)$ is obtained, if a simultaneous solution method is employed. The computational burden is insignificantly decreased for a successive solution of the system, since the matrix associated with the Device BTE has already the size $N_x N_k \times N_x N_k$ and N_k is much larger than two. In addition, a successive solution sequence usually suffers from a bad convergence behavior [56]. Thus, the simultaneous solution method is pursued in this work. The total number of points and thus the size of the resulting matrix can be decreased by considering a spatially dependent N_k. In this case, the number of discretization points for the kinetic energy N_w is set for each real space point such that a user defined maximum total energy is not exceeded. On the one hand, this strategy offers the possibility to save up to 50% of the total points, but on the other hand a spatial and also operating point dependent N_w – due to the electrostatic potential ψ – is introduced. The operating point dependence might result in numerical issues due to the coupling of the Poisson equation (ψ) to the 1D Device-BTE/continuity equation of holes.

For BULK simulations, the considered equations are linear functions of the solution variable $n_{i,j}^{\nu,\tau}$, if the Pauli principle is neglected. A non-linearity is only introduced to the system by considering the Pauli principle in the collision term. In this case, the non-linear system of equations is numerically solved by the *Newton*-method [8]. Nevertheless, in both cases the matrix pattern is identical. The position of the matrix entries, associated with the collision term, is affected by the energy exchange involved in the scattering mechanism. Therefore, the underlying discretization of the reciprocal space specifies the matrix pattern. Figure 3.12 illustrates the matrix pattern. Here, a natural ordering is employed starting from the μ-direction, followed by the w^ν-direction and finally the considered valleys. In figure 3.12, the pattern is shown for $N_\mu = 10$, $N_w = 100$ and $N_{\text{val.}} = 2$. The matrix exhibits a block structure, consisting of $N_\mu \times N_\mu$ block-matrices. The block-matrices on the main-diagonal consist of entries stemming from elastic intra-valley scattering and – due to ordering – of entries associated with the drift term in μ-direction. The first upper and lower secondary diagonal block-matrices contain entries due to the drift term in w^ν-direction (terms associated with $w_{i\pm1}^\nu$). Only the main diagonals of those block-matrices are occupied due to the employed ordering. The block-matrices on the consecutive diagonals are associated with the collision term (intravalley inelastic phonon scattering). Their actual position depends on the phonon energy w_{ph} and the step-size Δw ($\frac{w_{\text{ph}}}{\Delta w}$). Adjacent block-matrices might occur due to the integration and discretization scheme employed for inelastic scattering mechanisms (see section 3.2.3.2). The first matrix of the size $N_\mu N_w \times N_\mu N_w$ represents the first valley. The adjacent matrix of the right-hand side (see figure 3.12, bottom) is originating from (in-)elastic intervalley scattering mechanisms (from valley #1 to valley #2).

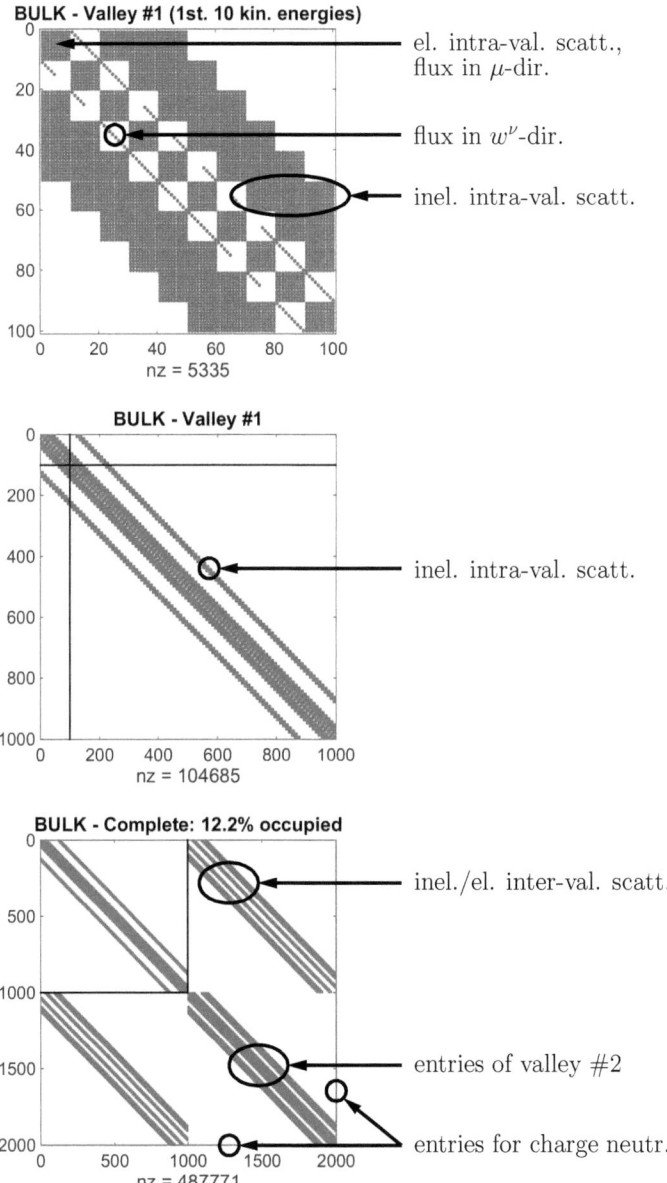

Figure 3.12: Illustration of matrix pattern (BULK case): top: the first 10 discretized kinetic energies; middle: matrix of the first valley; bottom: complete matrix of a two-valley simulation.

Likewise, the $N_\mu N_w \times N_\mu N_w$ matrix on the lower right part (3.12, bottom) contains entries associated with the drift term and intravalley scattering within valley #2. The adjacent matrix on the left-hand side contains entries due to intervalley scattering mechanisms (from valley #2 to valley #1). The entries in the row and column $N_k + 1$ are associated with equation (3.2.5-137) (section 3.2.5). The row and column entries specify and couple back the demand for charge neutrality, respectively.

In figure 3.13, the matrix pattern associated with a device simulation is shown. The same ordering like for the BULK case is employed. The number of discretization points are $N_\mu = 10$, $N_w = 61$, $N_{val.} = 2$ and $N_x = 161$. On top of figure 3.13, the pattern of the first 10 discretized kinetic energies at $x = 0$ is shown. This corresponds to the 1st real space point (left ohmic contact, see figure 3.11), where the boundary condition described in 3.3.3 is applied. For entries within $((r - 1)N_\mu + t)$ [3] $(-1 < \mu_j < 0$ for each kinetic energy), the main diagonal and the upper diagonal with the distance of $N_k + 1$ is set only (see figure 3.13, top & center). These entries are linked to the absorption of particles entering the contact from the interior of the device (particles from x_2 entering $x_1 = 0$). For elements within $\left(rN_\mu + \frac{N_\mu}{2} + t\right)$, the main diagonal is populated and $N_\mu/2 \times N_\mu/2$ block-matrices are located to its left- and right-hand side, which are coupling the main diagonal entries for $\mu_j > 0$ to ones for $-1 < \mu_j < 0$. These entries are stemming from the injection of particles for $\mu_j > 0$ (see figure 3.11) and the charge neutrality at the contact. In the center of figure 3.13, the matrix entries of the first five real space points are sketched. The entries associated with both the absorption and injection of particles at the contact, located at $x_1 = 0$, are marked. In addition, the $N_k \times N_k$ block diagonal matrices (for $x > x_1$) are highlighted, representing the k-space entries (drift- and collision-term), and are denoted by *Bulk-matrices*. These *Bulk-matrices* resemble the one shown in figure 3.12 (bottom). However, the *Bulk-matrices* in figure 3.13 are not completely identical with the one in figure 3.12, since the treatment of charges is captured by the Poisson equation (instead of equation (3.2.5-137)). Thus, the horizontal and vertical entries at $N_k + 1$ are missing in contrast to the BULK case. Furthermore, the

---

[3] $r \in \left[1, \frac{N_k}{N_\mu}\right]$, $t \in \left[1, \frac{N_\mu}{2}\right]$

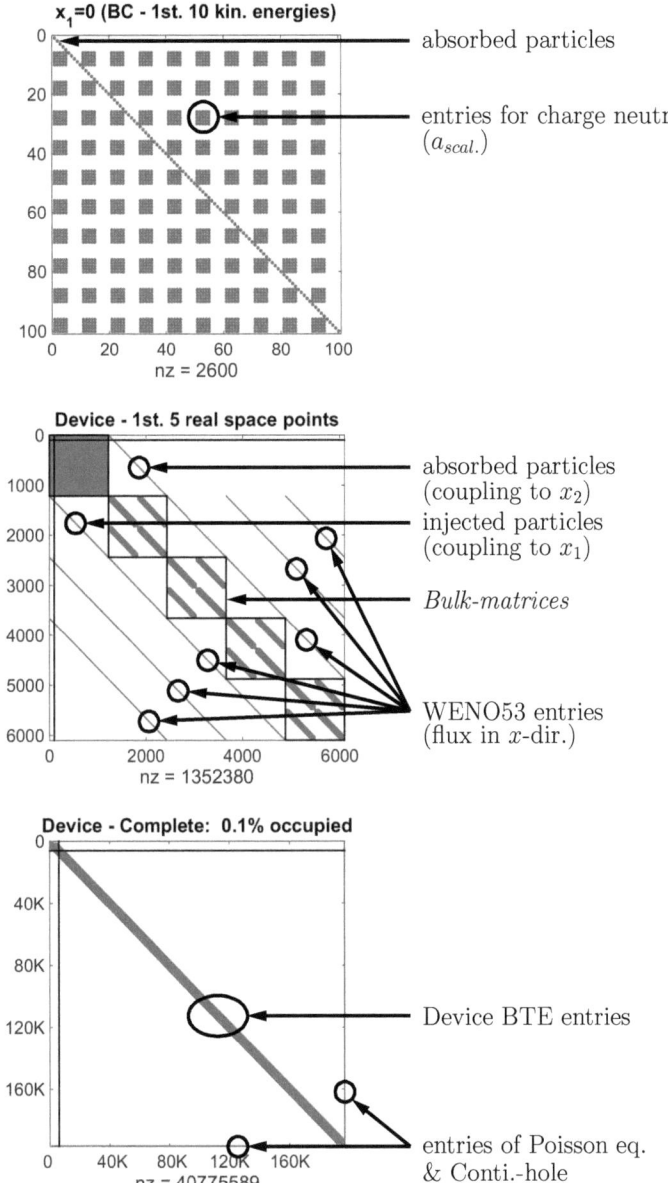

Figure 3.13: Illustration of matrix pattern (device case): top: the first 10 discretized kinetic energies at the first real space point (BC); center: matrix of the first 5 real space points; bottom: complete matrix of a two-valley device simulation.

WENO53 entries positioned with an integer multiple of N_k off the main diagonal are shown. On the bottom of figure 3.13 the complete device matrix is illustrated. Besides the entries associated with the BTE, also the entries belonging to the Poisson/hole-continuity equation are indicated. As mentioned before, the matrix size is insignificantly increased for a simultaneous solution sequence.

In contrast to the matrix structures for DD/HD simulations [56,58,72], the matrices for the BTE do not show the characteristic and a prori known shape. Moreover, the location and number of entries depend here mainly on the energy discretization and the considered scattering mechanisms. Therefore, most sparse-matrix solvers either need an excessive amount of memory/time to solve the problem or even completely fail in solving it. In this work, the package ILUPACK [6] based on the work [5] is employed. Compared to other solvers, latter offers both a fast and memory efficient solution of the BTE-systems (BULK and device case). This is of major importance since the complete matrix needs to be built up and solved for each Newton-step until the desired accuracy is achieved.

CHAPTER 4

Results

4.1 Calculation of the most important quantities

For the comparison with DD/HD simulations and measured data, macroscopic quantities are needed. In this section, the analytical formulas are given first. Based on them, the transition to the numerical equivalent is given, where the numerical solution variable of the BTE ($n_{i,j}^{\nu}$ after (3.2.0-35) in section 3.2) is involved. Starting point here is the electron density.

In its basic version, the total number of electrons N within an infinitely large and homogeneous device (the volume is denoted by Ω) is obtained by summing up the occupancy of all available states within the first Brillouin zone [46]:

$$N = \sum_{\vec{k}} f(\vec{k}) = \sum_{\nu=1}^{N_{\text{val.}}} \sum_{r=1}^{Z^{\nu}} \sum_{\vec{k}^{\nu}, \uparrow \downarrow} f_r^{\nu}(\vec{k}^{\nu}) \, , \tag{4.1.0-1}$$

where $N_{\text{val.}}$ is the number of considered valleys and Z^{ν} is the valley degeneracy factor for each considered valley counted by ν. In (4.1.0-1), the sum over $\vec{k}$ (measured from the origin of the 1st Brillouin zone, Γ-point) is split into a summation over the valleys and the k-vectors $\vec{k}^\nu$ measured from the position of the valley minimum within the 1st Brillouin zone. By taking the spin-degeneracy with a factor of two (spin $\uparrow\downarrow$) into account and assuming that each considered valley is equally occupied by $f_r^\nu(\vec{k}^\nu) = f^\nu(\vec{k}^\nu)$ leads to

$$N = 2 \sum_{\nu=1}^{N_{\text{val.}}} Z^\nu \sum_{\vec{k}^\nu} f^\nu(\vec{k}^\nu) \ . \tag{4.1.0-2}$$

In the next stage, the sum over $\vec{k}^\nu$ in (4.1.0-2) is converted into an integral (assuming a continuous set of $\vec{k}^\nu$-vectors), similar to the treatment of the collision term in section 2.4.1, equation (2.4.1-114). Therefore, a spherical and isotropic dispersion relation is needed or at least a description, which can be converted with the Herring-Vogt transformation into a spherical dispersion relation (see section 2.2 or appendix A.1). With k-vectors $\vec{k}^\nu$ associated with the spherical dispersion relation, (4.1.0-2) becomes

$$\begin{aligned} N &= 2 \sum_{\nu=1}^{N_{\text{val.}}} Z^\nu \frac{\Omega}{8\pi^3} \int_{\text{BZ}} f^\nu(\vec{k}^\nu) d\vec{k}^\nu \ , \\ &= \frac{\Omega}{4\pi^3} \sum_{\nu=1}^{N_{\text{val.}}} Z^\nu \int_{\text{BZ}} f^\nu(\vec{k}^\nu) d\vec{k}^\nu \ . \end{aligned} \tag{4.1.0-3}$$

Due to the employed coordinate transformation of the reciprocal space (see section 2.4.2.3), the infinitesimal volume element $d\vec{k}^\nu$ has to be transformed accordingly. Latter can be found in section 2.4.2.3, equation (2.4.2-155)-(2.4.2-156) and leads together with (4.1.0-3) to

$$N = \frac{\Omega}{4\pi^3} \sum_{\nu=1}^{N_{\text{val.}}} Z^\nu \int_0^{2\pi} d\varphi \int_{-1}^1 d\mu \int_0^{\epsilon_{\text{max.}}^\nu} d\epsilon^\nu f^\nu(\epsilon^\nu, \mu, \varphi) \, \text{dos}^\nu(\epsilon^\nu) \ . \tag{4.1.0-4}$$

With the non-parabolic dispersion relation after (2.2.0-43) – used throughout this work – the density-of-states dos^ν turns into the formulation after (3.0.0-8). In combination with the normalization factors listed in section 3.1, table 3.1

and 3.2, (4.1.0-4) becomes

$$
\begin{aligned}
N &= \frac{D_\text{N}}{2} \frac{\Omega}{4\pi^3} \sum_{\nu=1}^{N_\text{val.}} Z^\nu \int_0^{2\pi} d\varphi \int_{-1}^1 d\mu \int_0^{w_\text{max.}^\nu} dw^\nu f^\nu(w^\nu, \mu, \varphi) \left(\sqrt{m_\nu^*}\right)^3 s^\nu(w^\nu) \\
&= D_\text{N} \frac{\Omega}{8\pi^3} \sum_{\nu=1}^{N_\text{val.}} Z^\nu \int_0^{2\pi} d\varphi \int_{-1}^1 d\mu \int_0^{w_\text{max.}^\nu} dw^\nu \Phi^\nu(w^\nu, \mu, \varphi) ,
\end{aligned}
\tag{4.1.0-5}
$$

with the analytical counterpart Φ^ν (see (3.1.0-16) in section 3.1) of the numerical solution variable $n_{i,j}^\nu$ (3.2.0-35). Assuming a 1D simulation for both BULK and device simulations, the integration over φ in (4.1.0-5) simplifies to a multiplication with 2π (due to the rotational symmetry depicted in section 2.4.3). Using (3.2.0-35) and thus performing the transition from the analytical to the numerical description, one obtains for (4.1.0-5)

$$
\begin{aligned}
N = n\Omega &= D_\text{N} \frac{\Omega}{4\pi^2} \sum_{\nu=1}^{N_\text{val.}} Z^\nu \sum_{j=1}^{N_\mu} \Delta\mu \sum_{i=1}^{N_w} \Delta w \frac{n_{i,j}^\nu}{\Delta w \Delta \mu} , \\
\rightarrow n &= \frac{D_\text{N}}{4\pi^2} \sum_{\nu=1}^{N_\text{val.}} Z^\nu \sum_{j=1}^{N_\mu} \sum_{i=1}^{N_w} n_{i,j}^\nu ,
\end{aligned}
\tag{4.1.0-6}
$$

with the desired electron density n. $N_\mu(N_w)$ is the number of discretization points in $\mu-(w^\nu-)$ direction. Note, that for obtaining the electron density in (4.1.0-6) a division by the infinitely large sample volume Ω is performed. The occurrence of Ω is a remnant of the conversion of the sum in (4.1.0-2) into the integral of (4.1.0-3) by assuming a continuous set of $\vec{k}^\nu$-vectors enabling the integration in (4.1.0-3). Therefore, no special attention should be payed to Ω. Based on (4.1.0-6), the electron density for each valley can easily be deduced:

$$
n^\nu = \frac{D_\text{N}}{4\pi^2} Z^\nu \sum_{j=1}^{N_\mu} \sum_{i=1}^{N_w} n_{i,j}^\nu ,
\tag{4.1.0-7}
$$

where the valley degeneracy is included.

Next, the electron current density for the BULK case in the spatial direction $d \in \{x, y, z\}$ is considered:

$$
\begin{aligned}
J_n^d &= \frac{q}{4\pi^3} \sum_{\nu=1}^{N_{\text{val.}}} Z^\nu \int f^\nu(\vec{k}^\nu) v_{\text{g}}^{\nu,d}(\vec{k}^\nu) d\vec{k}^\nu \\
&= \frac{q}{4\pi^3} \sum_{\nu=1}^{N_{\text{val.}}} Z^\nu \int f^\nu(\vec{k}^\nu) \left\{ \{T^{\text{HV},\nu}\} \cdot \vec{\tilde{v}}_{\text{g}}^\nu(\vec{k}^\nu) \right\}^d d\vec{k}^\nu ,
\end{aligned}
\qquad (4.1.0\text{-}8)
$$

with the elementary charge q, the spatial direction $d \in \{x, y, z\}$, the non-transformed group velocity $v_{\text{g}}^{\nu,d}$ (after (2.4.0-109), section 2.4) and the transformed one $\tilde{v}_{\text{g}}^{\nu,d}$ (see (2.4.2-138) and (2.4.2-139) in section 2.4.2.2), respectively. By using the transformed group velocity $\tilde{v}_{\text{g}}^{\nu,d}$ in combination with the Herring-Vogt transformation matrix $T^{\text{HV},\nu}$, both the coordinate transformation of the reciprocal space and anisotropic dispersion relations are regarded. However, since only 1D BULK or device simulations are considered in this work, only for $d = x$, a current density is obtained. Proceeding with the integration analogously to the previously considered electron density, results with (2.4.2-139) for $d = x$ in

$$
J_n^x = \frac{q D_{\text{N}} v_{\text{N}}}{4\pi^2}
$$

$$
\sum_{\nu=1}^{N_{\text{val.}}} Z^\nu T_{1,1}^{\text{HV},\nu} \sum_{j=1}^{N_\mu} \sum_{i=1}^{N_w} \frac{n_{i,j}^\nu}{\Delta\mu\Delta w} \left[h(w_{i+\frac{1}{2}}^\nu, \mu) - h(w_{i-\frac{1}{2}}^\nu, \mu) \right]_{\mu_{j-\frac{1}{2}}}^{\mu_{j+\frac{1}{2}}} ,
\qquad (4.1.0\text{-}9)
$$

where the auxiliary function $h(w^\nu, \mu)$ is defined as

$$
h(w^\nu, \mu) = \frac{\mu^2}{2a_\nu\sqrt{m_\nu^*}} \left\{ \frac{1}{2\sqrt{a_\nu}} \arcsin\left(\frac{1}{1 + 2a_\nu w^\nu} \right) + \sqrt{w^\nu(1 + a_\nu w^\nu)} \right\} ,
$$

$$
(4.1.0\text{-}10)
$$

which stems from the integrals of the group-velocity (x-component) of (2.4.2-139) over w^ν and μ, in conjunction with the assumed non-parabolic dispersion relation (2.2.0-43). The normalization factors D_{N}, v_{N} and the normalized non-parabolicity factor a_ν (cf. (3.1.0-15)) are defined in section 3.1.

Based on the electron density and electron current density, the average electron velocity in each spatial direction can be obtained by

$$
\begin{aligned}
\langle v_n^d \rangle &= \frac{J_n^d}{qn} \\
&= \frac{\sum_{\nu=1}^{N_{\text{val.}}} Z^\nu \int f^\nu(\vec{k}^\nu) v_g^{\nu,d}(\vec{k}^\nu) d\vec{k}^\nu}{\sum_{\nu=1}^{N_{\text{val.}}} Z^\nu \int f^\nu(\vec{k}^\nu) d\vec{k}^\nu} \,,
\end{aligned}
\tag{4.1.0-11}
$$

with $d \in \{x, y, z\}$. Considering the x-direction only, (4.1.0-11) evaluates to

$$
\langle v_n^x \rangle = v_{\text{N}} \frac{\sum_{\nu=1}^{N_{\text{val.}}} Z^\nu T_{1,1}^{\text{HV},\nu} \sum_{j=1}^{N_\mu} \sum_{i=1}^{N_w} \frac{n_{i,j}^\nu}{\Delta\mu\Delta w} \left[h(w_{i+\frac{1}{2}}^\nu, \mu) - h(w_{i-\frac{1}{2}}^\nu, \mu) \right]_{\mu_{j-\frac{1}{2}}}^{\mu_{j+\frac{1}{2}}}}{\sum_{\nu=1}^{N_{\text{val.}}} Z^\nu \sum_{j=1}^{N_\mu} \sum_{i=1}^{N_w} n_{i,j}^\nu} \,,
\tag{4.1.0-12}
$$

by using the electron density and electron current density after (4.1.0-6) and (4.1.0-9), respectively. Note, analogously to the electron density per valley after (4.1.0-7), also the electron current density after (4.1.0-9) and the electron velocity after (4.1.0-12) can be partitioned by skipping the summation over ν and considering each valley ν separately. For BULK simulations, where spatial variations are suppressed, equation (4.1.0-9) and (4.1.0-12) can be implemented directly.

However, for device simulations, where the diffusion term of the BTE is non-zero (e.g. regions with strong changes in the doping profile), those equations tend to give inaccurate results. This behavior originates from the numerical treatment and discretization of the diffusion flux. Within the BIM-framework, fluxes are assumed to be constant (spatially independent) between two adjacent discretization points. In the 1D case (x-direction only), the two fluxes at an arbitrary coordinate x_k within the device read

1) $J_{n,i,j,k-\frac{1}{2}}^x =$ const., $x \in (x_{k-1}, x_k)$,

2) $J_{n,i,j,k+\frac{1}{2}}^x =$ const., $x \in (x_k, x_{k+1})$.

Both fluxes are obtained by the scheme depicted in section 3.3.1 in conjunction with the WENO53-method (see section 3.3.2). Based on the fluxes $J_{n,i,j,k-\frac{1}{2}}^x$

and $J^x_{n,i,j,k+\frac{1}{2}}$, the flux at grid point x_k has to be *interpolated*, where, due to the BIM assumptions, a jump from $J^x_{n,i,j,k-\frac{1}{2}}$ to $J^x_{n,i,j,k+\frac{1}{2}}$ is present. One might be tempted to use, for example, an interpolation method based on the distances between the discretization points, e.g. linear interpolation:

$$J^x_{n,i,j}(x_k) = J^x_{n,i,j,k-\frac{1}{2}} + \frac{J^x_{n,i,j,k+\frac{1}{2}} - J^x_{n,i,j,k-\frac{1}{2}}}{x_{k+\frac{1}{2}} - x_{k-\frac{1}{2}}} \left(x_k - x_{k-\frac{1}{2}} \right) . \quad (4.1.0\text{-}13)$$

However, such kind of interpolation techniques would violate the basic assumptions of the BIM framework (constant flux between the discretization points) and thus, result in inaccuracies for non-equidistant real space grids. The most intuitive method for interpolating $J^x_{n,i,j,k}$ at x_k is to take the average value (without considering the spatial discretization):

$$J^x_{n,i,j,k} = \frac{1}{2} \left(J^x_{n,i,j,k-\frac{1}{2}} + J^x_{n,i,j,k+\frac{1}{2}} \right) . \quad (4.1.0\text{-}14)$$

Therefore, (4.1.0-14) should be used for device simulations and the electron current densities at each discretized real space point become

$$J^x_{n,k} = \frac{qD_N v_N}{4\pi^2}$$

$$\sum_{\nu=1}^{N_{\text{val.}}} Z^\nu \sum_{j=1}^{N_\mu} \sum_{i=1}^{N_w} \frac{1}{2} \left(\langle a^\nu_{i,j,k-\frac{1}{2}} \rangle n^\nu_{i,j,k-\frac{1}{2}} + \langle a^\nu_{i,j,k+\frac{1}{2}} \rangle n^\nu_{i,j,k+\frac{1}{2}} \right) , \quad (4.1.0\text{-}15)$$

with $\langle a^\nu_{i,j,k\pm\frac{1}{2}} \rangle$ after (3.3.1-146) and $n^\nu_{i,j,k\pm\frac{1}{2}}$ based on the WENO53 method. Proceeding analogously for the average electron velocity after (4.1.0-12) for the BULK case (or for device simulations after (4.1.0-15)), $\langle v^x_{n,k} \rangle$ becomes

$$\langle v^x_{n,k} \rangle = v_N \frac{\sum_{\nu=1}^{N_{\text{val.}}} Z^\nu \sum_{j=1}^{N_\mu} \sum_{i=1}^{N_w} \frac{1}{2} \left(\langle a^\nu_{i,j,k-\frac{1}{2}} \rangle n^\nu_{i,j,k-\frac{1}{2}} + \langle a^\nu_{i,j,k+\frac{1}{2}} \rangle n^\nu_{i,j,k+\frac{1}{2}} \right)}{\sum_{\nu=1}^{N_{\text{val.}}} Z^\nu \sum_{j=1}^{N_\mu} \sum_{i=1}^{N_w} n^\nu_{i,j}} .$$

$$(4.1.0\text{-}16)$$

Note, in contrast to the BULK case, the Herring-Vogt matrix elements (here $T^{\text{HV},\nu}_{1,1}$) are already included in $\langle a^\nu_{i,j,k\pm\frac{1}{2}} \rangle$ and therefore do not directly appear in the equations (4.1.0-15)-(4.1.0-16).

4.2 Si/SiGe

First, results for Si/SiGe structures are presented. For all simulations the Pauli principle and an anisotropic dispersion relation is considered, unless otherwise specified. The material and the physical model parameters used for the BTE are summarized in table 4.1. In order to emphasize the influence of the Pauli principle, the scattering parameters were kept, contrary to the recommendation given in [31]. For the simulations, a two-valley model is employed, which is necessary for covering the anisotropy of the six X-valleys in conjunction with the 1D transport direction. Due to the assumed 1D transport in x-direction, four of the six ellipsoidal X-valleys (lying in the $k_y - k_z$−plane) are equivalent.Therefore, these valleys are modeled as a single valley with the valley degeneracy $Z^\nu = 4$. The remaining two X-valleys (lying on the k_x−axis) are also equivalent resulting in a two-fold valley. The situation is schematically shown in figure 4.1.

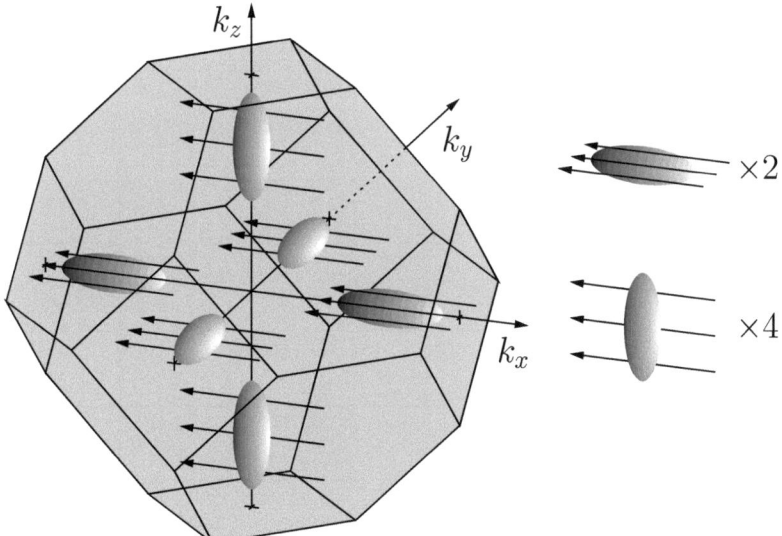

Figure 4.1: Schematic illustration of valley degeneracy in Si/SiGe considered for the 1D transport in x-direction. The arrows indicate an effective field with negative x-component only.

Quantity	Symbol	Value	Unit
General material parameters			
Permittivity	ε_r	$11.7 + 3.13\, x_{\mathrm{GE}} + 1.17\, x_{\mathrm{GE}}^2$	-
Sound velocity	v_{s}	$9040 - 3640\, x_{\mathrm{GE}}$	m/s
Mass density	ρ	$2330 + 2990\, x_{\mathrm{GE}}$	kg/m^3
Band parameters			
Electron affinity	χ_{r} (SI)	4.05	eV
Electron affinity change	$\Delta\chi$	$0.196\, x_{\mathrm{GE}} - 0.396\, x_{\mathrm{GE}}^2$	eV
Energetic valley offset	ΔE_0^{ν}	$X_2 : 0.63\, x_{\mathrm{GE}}$ $X_4 : 0$	eV
Valley degeneracy factor	Z^{ν}	$X_2 : 2$ $X_4 : 4$	-
Non-parabolicity factor	α_{ν}	$X_{2/4} : 0.5$	eV^{-1}
Anisotropic masses	m_x^{ν} m_y^{ν} m_z^{ν}	$\begin{array}{cc} X_2 & X_4 \\ 0.916 & 0.190 \\ 0.190 & 0.190 \text{ or } 0.916 \\ 0.190 & 0.916 \text{ or } 0.190 \end{array}$	-
Scattering parameters			
Acoustic deform. pot.	D_{A}	SI : 9.0 GE : 9.0	eV
Intervalley phonon scat.		SI GE	
TA mode	D_{ij}	$0.5 \cdot 10^{10}$ $0.49 \cdot 10^{10}$	eV/m
g-type	$\hbar\omega_{\mathrm{ij}}$	12.1 5.6	meV
TA mode	D_{ij}	$0.3 \cdot 10^{10}$ $0.29 \cdot 10^{10}$	eV/m
f-type	$\hbar\omega_{\mathrm{ij}}$	19.0 9.9	meV
LA mode	D_{ij}	$0.8 \cdot 10^{10}$ $0.79 \cdot 10^{10}$	eV/m
g-type	$\hbar\omega_{\mathrm{ij}}$	18.5 8.6	meV
LA mode	D_{ij}	$2.0 \cdot 10^{10}$ $1.99 \cdot 10^{10}$	eV/m
f-type	$\hbar\omega_{\mathrm{ij}}$	47.4 28.0	meV
LO mode	D_{ij}	$11.0 \cdot 10^{10}$ $9.5 \cdot 10^{10}$	eV/m
g-type	$\hbar\omega_{\mathrm{ij}}$	62.0 37.0	meV
TO mode	D_{ij}	$2.0 \cdot 10^{10}$ $1.73 \cdot 10^{10}$	eV/m
f-type	$\hbar\omega_{\mathrm{ij}}$	58.6 32.7	meV

Table 4.1: SI/SIGE parameters ($x_{\mathrm{GE}} \in [0, 0.3]$) [31, 37].

In table 4.1, a model for the change in the electron affinity as function of the composition x_{GE} is given. This model is responsible of the composition dependent shift of the conduction band edge $E_{\mathrm{C},0}$:

$$E_{\mathrm{C},0} = E_0 - \chi_{\mathrm{r}} - \Delta\chi - q\psi - \Delta E_{\mathrm{C},0,\mathrm{dop.}} \;, \qquad (4.2.0\text{-}17)$$

with a vacuum energy E_0, the electron affinity χ_{r} of the reference material (here Si), the electrostatic potential ψ obtained by the *Poisson* equation and the doping dependent shift of the conduction band edge $\Delta E_{\mathrm{C},0,\mathrm{dop.}} \geq 0$ [62]

$$\Delta E_{\mathrm{C},0,\mathrm{dop.}} = \gamma_{\mathrm{hd}} E_{\mathrm{hd}} \left(\ln\left(\frac{N_{\mathrm{A}} + |N_{\mathrm{D}}|}{N_{\mathrm{Ref}}} \right) + \sqrt{\left(\ln\left(\frac{N_{\mathrm{A}} + |N_{\mathrm{D}}|}{N_{\mathrm{Ref}}} \right) \right)^2 + C_{\mathrm{hd}}} \right) \;,$$

$$(4.2.0\text{-}18)$$

with $\gamma_{\mathrm{hd}}{=}0.5$, $E_{\mathrm{hd}}{=}6.92\,\mathrm{meV}$, $N_{\mathrm{Ref}}{=}1.3 \times 10^{17}\,\mathrm{cm}^{-3}$ and $C_{\mathrm{hd}}{=}0.5$ [37]. Based on (4.2.0-17), the potential valley energy $E_{\mathrm{C},0}^{\nu}$ becomes

$$E_{\mathrm{C},0}^{\nu} = E_{\mathrm{C},0} + \Delta E_0^{\nu} \;, \qquad (4.2.0\text{-}19)$$

where ΔE_0^{ν} is responsible of the energetic valley offsets w.r.t. the conduction band-edge $E_{\mathrm{C},0}$ (see table 4.1).

The scattering parameters listed in table 4.1 are related to pure Si and Ge. For SiGe, the total collision term consists of Siand Ge portions weighted with the respective mole fractions. In addition to acoustic deformation potential and intervalley phonon scattering, listed in table 4.1, also impurity and alloy scattering are considered. In the case of alloy scattering, the models and parameters given in section 2.3.3 are used. For impurity scattering the isotropic approximation given in section 2.3.5.2 is used in combination with a fit-function (see appendix B.5). The main reason for choosing the isotropic approximation of impurity scattering, besides comparison purposes, stems from the anisotropic dispersion relation used for Siand SiGe. Because of the anisotropic dispersion relation, the term $|\vec{k} - \vec{k'}|$ involved in the anisotropic transition rate (see section 2.3.2, equation (2.3.2-69)) results in additional complications for the analytic preparation of its discretized version. Therefore, the isotropic approximation and the fit-function in appendix B.5 is employed.

Despite of the fact that hetero-structures (SiGe) are considered and contrary to the composition dependence of most physical quantities, the dispersion relation parameters given in table 4.1 are not altered. In terms of the anisotropic masses, the results shown in [52] imply a negligible composition dependence and following [11], also the non-parabolicity factor α_ν is treated as composition independent.

In terms of SiGe device simulations (section 4.2.2), the simulation results obtained by the method in this work are compared with those obtained by SHE [31]. These comparisons assist both validating the correctness of the method used in this work and to identify possible issues.

4.2.1 BULK simulations

BULK simulations are mainly of interest for verifying the employed band structure and scattering models (comparison with experimental results) and for adjusting physical models involved in the DD/HD transport models towards BTE results. Here, the main focus is put on the electron mobility and drift velocities with neglected Pauli principle. Later on, the influence of the Pauli principle on the electron low-field mobility in conjunction with the employed scattering models is emphasized. In addition, impact of the Pauli principle on band gap used for DD/HD transport models is briefly discussed.

Figure 4.2 compares the simulated electron low-field mobility (majority case) as function of the donor doping concentration with experimental results. The agreement between the simulated and experimentally determined

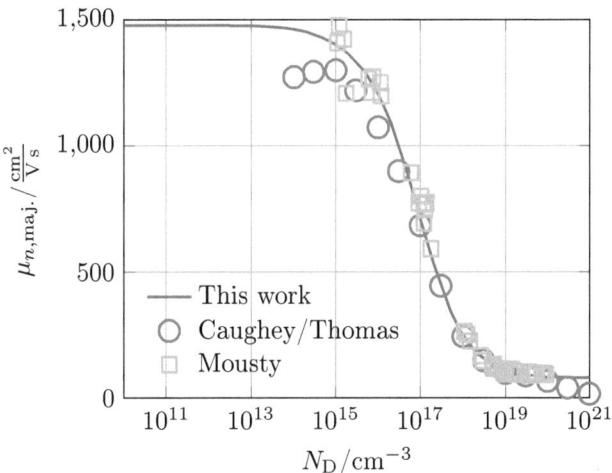

Figure 4.2: Calculated (no Pauli principle) and experimentally determined electron majority low-field mobility at $T_{\mathrm{L}}{=}300\,\mathrm{K}$ [49, 68].

data is good over a wide range of doping concentrations. For low and very high doping concentrations ($N_{\mathrm{D}} < 1 \times 10^{16}\,\mathrm{cm}^{-3}$ & $N_{\mathrm{D}} > 1 \times 10^{20}\,\mathrm{cm}^{-3}$), deviations are visible, both between the simulated/experimental data and among the experimental data themselves. Since the experimental results in figure 4.2

were determined decades ago, the exactness of the impurity concentration, especially for low doping concentrations, is questionable. However, considering the simulation results as averaging curve, the deviations among the experimental data at low and very high doping concentrations become less severe. Since impurity scattering is the only scattering mechanism directly influenced by the doping concentration (see section 2.3.2), the applied isotropic approximation in conjunction with the corresponding fit-function (see appendix B.5) seems to be suitable for resembling the experimental results.

Figure 4.3 compares the simulated and experimentally determined electron drift velocities as function of the electric field for three different lattice temperatures. The considered doping concentration is $N_D = 5 \times 10^{12}\,\mathrm{cm}^{-3}$ in accordance with [13]. In contrast to figure 4.2, where the influence of impurity

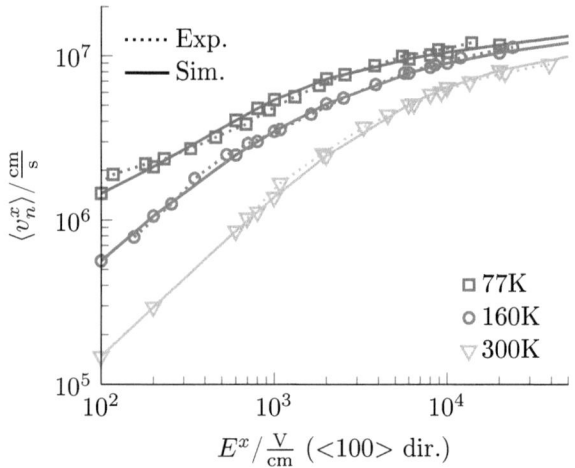

Figure 4.3: Calculated and experimentally determined (data taken from [13]) electron drift velocities for different lattice temperatures ($N_D < 1 \times 10^{13}\,\mathrm{cm}^{-3}$).

scattering is visible, from figure 4.3 (high-field case) the accuracy of both the dispersion relation and the phonon-scattering mechanisms can be observed. With increasing electric fields, the carrier distribution function is displaced from its equilibrium position in the reciprocal space. First, due to the displacement the average electron velocity after (4.1.0-11) starts to increase, which is related to the dispersion relation by the group-velocity. Second, due to the displacement, more high-energetic carriers are available (higher than the phonon energy $\hbar\omega_p$) enabling the emission of phonons[1] of the related phonon scattering mechanism. Thus, with increasing field strength carriers are more likely scattered to lower energies, reducing the slope of the average electron velocity. Since not all considered scattering mechanisms exhibit the same lattice temperature dependence, results for different lattice temperatures are shown in figure 4.3. For low electric fields ($E^x < 1 \times 10^3$ V/cm), the slopes of the illustrated velocities correspond to the low-field electron mobility. According to figure 4.2 ($N_D < 1 \times 10^{13}$ cm^{-3}), the impact of impurity scattering is negligible for the considered doping concentration ($N_D = 5 \times 10^{12}$ cm^{-3}). Thus, the electron velocities shown in figure 4.3 are dominated by phonon scattering. Especially for low temperatures (T_L roughly below 100 K), the carrier velocities for low electric fields are controlled by acoustic deformation potential scattering [69]. For high electric fields, the saturation of the electron drift velocity is seen. Although absorption and emission of phonons are involved, for rough estimations it is sufficient to consider the emission of phonons to be dominant [33] and thus, leading to a saturation of the carrier velocities at high electric fields. Both the qualitative and quantitative good agreement suggests that the dispersion relation and the considered phonon-scattering mechanisms are modeled with sufficient accuracy.

Having verified both the scattering mechanisms and the dispersion relation for Si, the composition dependence is inspected next. However, due to the scarcely available experimental data in literature, published Monte-Carlo results are used instead. In figure 4.4 and 4.5 the electron low-field mobility for the majority and minority case is shown, respectively. The Monte-Carlo data are taken from [11,12]. In both figures, results in growth-direction of the SiGe layer are illustrated. In literature, the mobility component **in growth-direction** is often labeled **out-of-plane** mobility. The mobility components **parallel** to the Si/SiGe-interface are referred to as **in-plane** mobility.

---

[1]Note, due to the phonon occupation number $n_{ph} \in (0, 1)$ after (2.3.1-64) (section 2.3.1.2) and the weighting involved in phonon scattering transition rates with n_{ph} and $(n_{ph} + 1)$ for absorption and emission processes, respectively, emission processes are dominating.

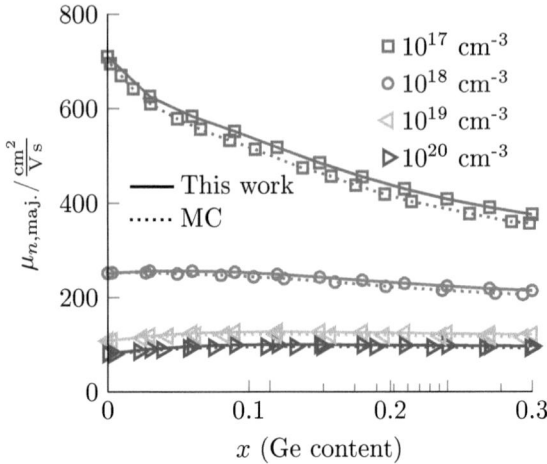

Figure 4.4: Germanium dependent electron majority low-field mobility for different donor (N_D) doping concentrations at T_L=300 K. The Monte-Carlo (MC) data are taken from [11, 12].

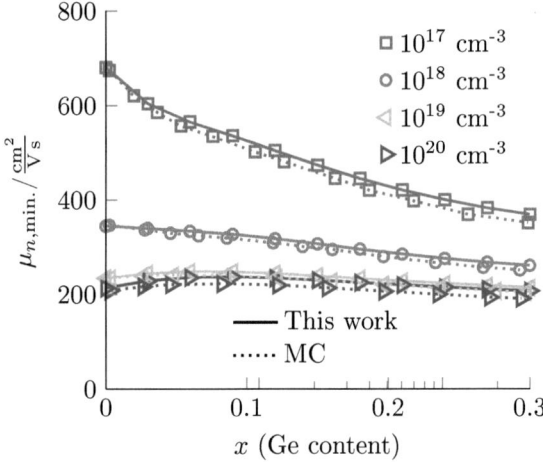

Figure 4.5: Germanium dependent electron minority low-field mobility for different acceptor (N_A) doping concentrations at T_L=300 K. The Monte-Carlo (MC) data are taken from [11, 12].

Note, the in- and out-of-plane mobility components exhibit different GE dependencies. Therefore, care has to be taken when comparing published results with those obtained by simulation. According to figure 4.4 and 4.5, good agreement is obtained between the MC data and the simulation results obtained by the deterministic method used in this work. The overall trend of the low-field electron mobilities is a decrease with increasing GE-content, since alloy-scattering becomes more pronounced. However, for high doping concentrations ($\geq 1 \times 10^{19}\,\mathrm{cm}^{-3}$) a slight increase in the low-field mobility is observed for both the majority and the minority case. With increasing GE-content, the X_2-valleys are shifted to higher energies (cf. section 2.2, figure 2.11). Thus, four instead of six valleys are available at the conduction band edge and the influence of intervalley scattering is weakened. Since the valleys at the conduction band edge are carrying most of the carriers, the distribution function of the X_4-valleys (f^{X_4}) is shifted to higher values in order to fulfill the charge neutrality. The shift of the distribution function f^{X_4} is sketched in figure 4.6 for the majority case with $N_\mathrm{D} = 1 \times 10^{19}\,\mathrm{cm}^{-3}$. Higher values

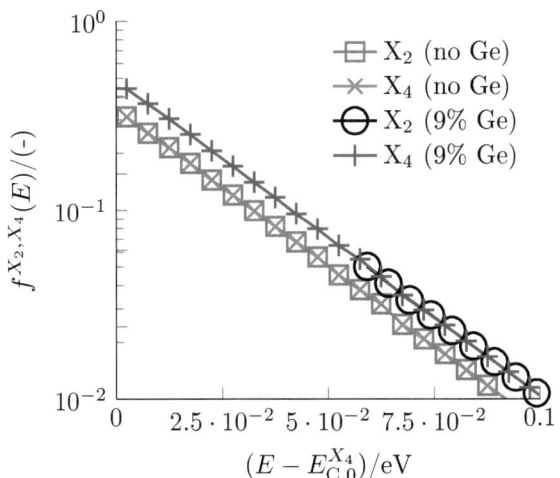

Figure 4.6: Electron distribution function over energy for $N_\mathrm{D} = 1 \times 10^{19}\,\mathrm{cm}^{-3}$. Here, the shift of the X_2-valleys is indicated for 9% GE-content (marker: o).

of f^{X_4} translate into an increased average electron velocity directly, since the group velocity is weighted by f^{X_4} much stronger. A further increase in the GE-content leads to a shift of the X_2-valleys to even higher energies and the lower X_4-valleys in combination with alloy-scattering are dominating the mobility characteristics. For low doping concentrations, the upward shift of f^{X_4} within the X_4-valleys is also present, but its impact on the average electron velocity is already for low GE-contents being compensated for alloy-scattering. Figure 4.7 and 4.8 show the impact of alloy-scattering on the low-field electron mobility for low ($1 \times 10^{17}\,\mathrm{cm^{-3}}$) and high ($1 \times 10^{19}\,\mathrm{cm^{-3}}$) doping concentrations. Neglecting alloy-scattering for N_D or $N_\mathrm{A} = 1 \times 10^{17}\,\mathrm{cm^{-3}}$, results in

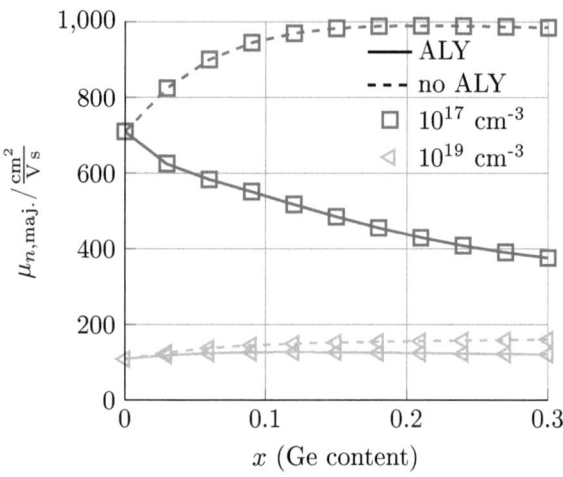

Figure 4.7: Germanium dependent electron majority low-field mobility for different donor (N_D) doping concentrations at T_L=300 K. Comparison between activated and deactivated alloy scattering (ALY).

an increased low-field mobility, as observed in figure 4.4 and 4.5 for N_D and $N_\mathrm{A} = 1 \times 10^{19}\,\mathrm{cm^{-3}}$, respectively. For N_D or $N_\mathrm{A} = 1 \times 10^{17}\,\mathrm{cm^{-3}}$, the effect of the upward shift of f^{X_4} is more than compensated for by activated alloy-scattering. Contrary to N_D or $N_\mathrm{A} = 1 \times 10^{17}\,\mathrm{cm^{-3}}$ and according to figure 4.7 and 4.8, a remnant of the higher f^{X_4} (higher mobility) is still visible with activated alloy-scattering for N_D or $N_\mathrm{A} = 1 \times 10^{19}\,\mathrm{cm^{-3}}$. Here, alloy-scattering is not strong enough to counteract the upward shift of f^{X_4}

and the accompanied increase in the carrier velocities.

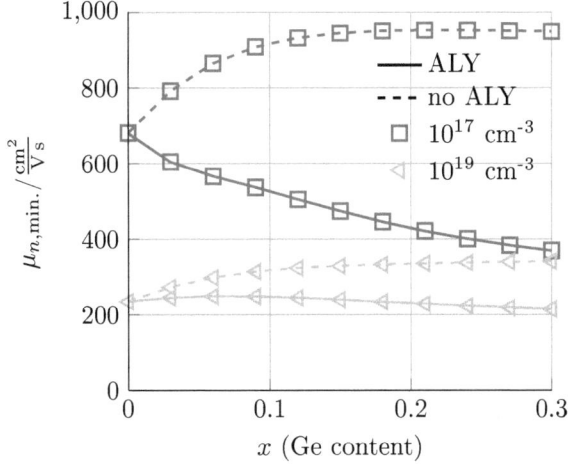

Figure 4.8: Germanium dependent electron minority low-field mobility for different acceptor (N_A) doping concentrations at T_L=300 K. Comparison between activated and deactivated alloy-scattering (ALY).

Next, the impact of the Pauli principle on the electron mobilities is shown. Figure 4.9 and 4.10 illustrate the doping and GE dependent low-field mobilities without and with considered Pauli principle, respectively. In both figures, the minority and majority case is shown. Good agreement between activated and deactivated Pauli principle is obtained for the majority carrier case over a wide range of doping concentrations. Deviations are observed at high doping concentrations ($G > 1 \times 10^{19}\,\mathrm{cm}^{-3}$), only. Due to the isotropic approximation of impurity scattering and the fit-functions involved, these deviations can be diminished by re-adjusting the fit-functions for the majority case with activated Pauli principle. The fit-functions in conjunction with activated Pauli principle also cause the identical shape of the minority and majority case ($G \le 1 \times 10^{19}\,\mathrm{cm}^{-3}$ in figure 4.10). Based on the data in appendix B.5, figure B.1, no separation is made for the minority and majority case. Thus, the electron mobilities resemble each other. In summary, the Pauli principle impacts the GE-dependent electron majority mobility marginally due to the involved fit-functions for $G \le 1 \times 10^{19}\,\mathrm{cm}^{-3}$. However, a revision of the fit-functions in the minority case is needed for $G > 1 \times 10^{19}\,\mathrm{cm}^{-3}$.

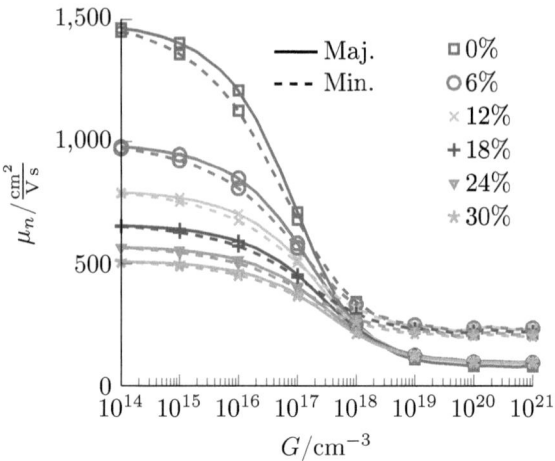

Figure 4.9: Electron minority and majority low-field mobility **without** Pauli principle as function of the total doping concentration $G = N_\mathrm{A}+|N_\mathrm{D}|$ at $T_\mathrm{L}=300\,\mathrm{K}$ for different GE-contents.

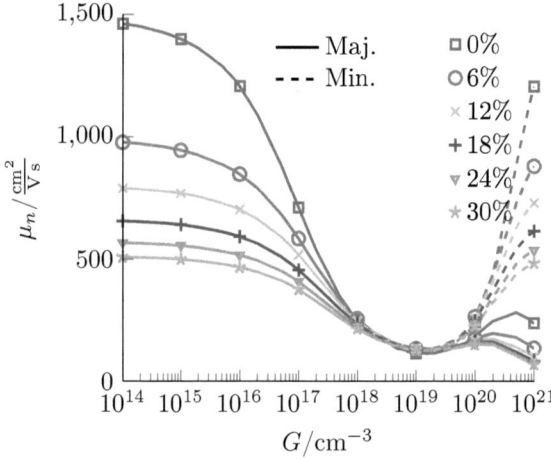

Figure 4.10: Electron minority and majority low-field mobility **with** Pauli principle as function of the total doping concentration $G = N_\mathrm{A}+|N_\mathrm{D}|$ at $T_\mathrm{L}=300\,\mathrm{K}$ for different GE-contents.

Besides its impact on the low-field electron mobility, the Pauli principle also influences the actual values of the band gap E_G used in the simulation. Imposing the effective intrinsic carrier density $n_{i,\text{eff}}$ on the BTE simulation[2], the doping dependent band gap shown in figure 4.11 is obtained for the electron majority case. The band gap $E_{G,\text{BTE}}^{\text{BD}}$ and $E_{G,\text{BTE}}^{\text{FD}}$ is obtained

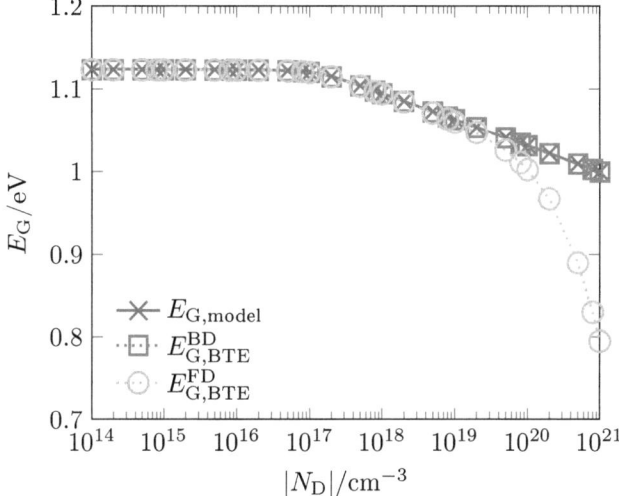

Figure 4.11: Doping dependent band gap of Si at $T_L = 300\,\text{K}$. Here, the impact of the Pauli principle on the band gap is illustrated for electrons as majorities.

for deactivated and activated Pauli principle, respectively. The model data $E_{G,\text{model}}$ shown in figure 4.11 are calculated with the well-known Slotboom band gap narrowing model (4.2.0-18) [62] with slightly changed model parameters according to [60]. The band gap predicted by the model and $E_{G,\text{BTE}}^{\text{BD}}$ and $E_{G,\text{BTE}}^{\text{FD}}$ agree with each other for $|N_D| < 1 \times 10^{19}\,\text{cm}^{-3}$, where for higher doping concentrations $E_{G,\text{BTE}}^{\text{FD}}$ (with Pauli principle) starts to deviate. However, for the electron minority case, shown in figure 4.12, no difference in the doping dependent band gap between the model $E_{G,\text{model}}$ and $E_{G,\text{BTE}}^{\text{FD}}$ or $E_{G,\text{BTE}}^{\text{BD}}$ (BTE simulations) is observed. Both the deviations seen for the ma-

---

[2]$n_{i,\text{eff}}$ can experimentally be determined from the saturation current density expression for BJTs and HBTs.

jority (figure 4.11) and the agreement for the minority case (figure 4.12) are associated with the Pauli principle and the assumed hole distribution function. The band gap extracted from BTE is obtained by BULK simulations

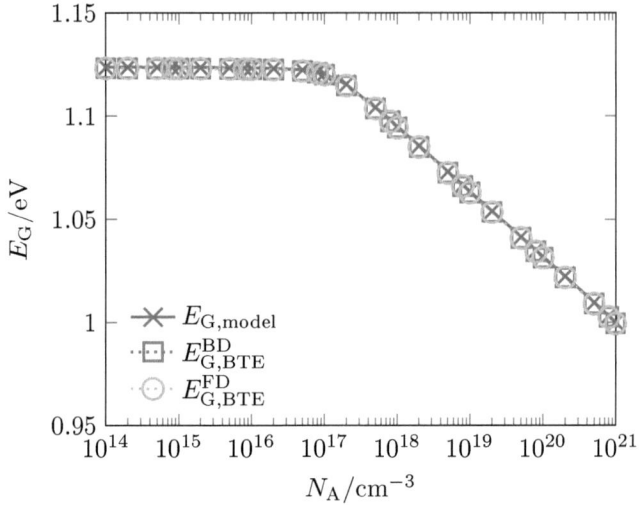

Figure 4.12: Doping dependent band gap of Si at $T_L = 300\,\mathrm{K}$. The impact of the Pauli principle on the band gap is illustrated for electrons as minorities.

under thermodynamic equilibrium. Thus, both the *law of mass action* and the charge neutrality are fulfilled. For the electron majority case, the Fermi level is shifted closer to the conduction band edge with increasing donor doping concentration $|N_D|$ and even above the conduction band edge for very high doping concentrations. Neglecting the Pauli principle for electrons, the distribution being dependent on the energy – obtained by solving the BTE – satisfies the Boltzmann distribution (BD), where the inclusion of the Pauli principle results in a Fermi-Dirac distribution (FD):

$$f_{\mathrm{FD},n}(E) = \frac{1}{1 + \exp\left(\frac{E - E_F}{k_b T_L}\right)} \quad (4.2.1\text{-}20)$$

$$\overset{(E - E_F) \gg k_b T_L}{\approx} \exp\left(-\frac{E - E_F}{k_b T_L}\right) = f_{\mathrm{BD},n}(E)\,, \quad (4.2.1\text{-}21)$$

with the Fermi level E_F, the total energy E and $f_{FD,n}$ and $f_{BD,n}$, respectively, as the Fermi-Dirac and Boltzmann electron distribution function. It can be taken from figure 4.13 that for a moderate doping concentration ($n \approx |N_D| = 1 \times 10^{18}\,\mathrm{cm}^{-3}$) both distributions coincide with each other and so do their Fermi levels E_F ($E_F < E_C$, non-degenerate case). However, for

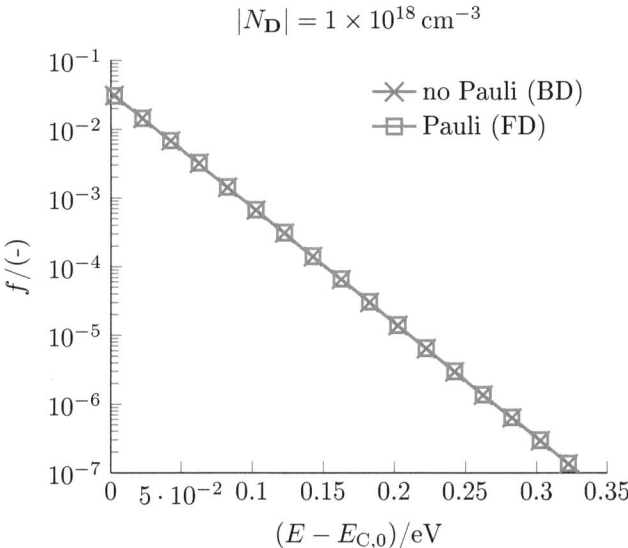

Figure 4.13: Electron distribution functions over energy obtained by BTE BULK-simulations for Si, $|N_D| = 1 \times 10^{18}\,\mathrm{cm}^{-3}$, $T_L = 300\,\mathrm{K}$ with/without Pauli principle. $E_{C,0}$ is the lowest conduction band edge obtained by (4.2.0-17).

high doping concentrations, the different shapes of the distribution functions – in the vicinity of the Fermi level – are influencing the Fermi level. Figure 4.14 illustrates the two distribution functions (BD and FD) and the position of the Fermi levels for $n \approx |N_D| = 2 \times 10^{20}\,\mathrm{cm}^{-3}$ ($E_F > E_C$, degenerate case). Due to the energetic positions of the Fermi levels $E_F^{FD} > E_F^{BD}$, the valence band edge has to be shifted to higher energies with considered Pauli principle (FD) in order to *generate* enough holes for fulfilling the *law of mass action*. Thus, a more pronounced band gap narrowing is seen in figure 4.11 with activated Pauli principle.

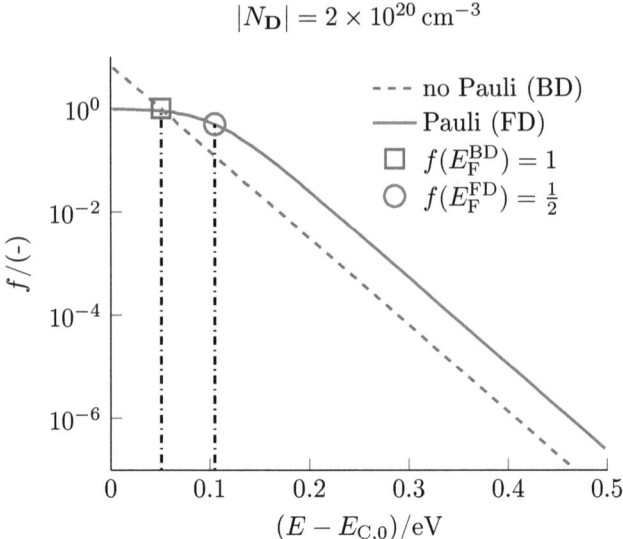

Figure 4.14: Electron distribution functions over energy obtained by BTE BULK-simulations for S_I, $|N_D| = 2 \times 10^{20}\,\mathrm{cm}^{-3}$, T_L=300 K with/without Pauli principle. $E_{C,0}$ is the lowest conduction band edge obtained by (4.2.0-17).

Since a Boltzmann distribution is assumed for the holes, a comparable effect on the band gap is not seen in figure 4.12 for the electron minority (hole majority) case. In this case, the Fermi level is dominated by the hole (Boltzmann) distribution function and is located far below the conduction band edge. Thus, BD and FD resemble each other for electrons, similar to the case of moderate donor doping concentrations shown in figure 4.13. Thus, the agreement between the BD and FD for electrons results in an identical band gap, regardless of the Pauli principle.

In summary, the Pauli principle results in a stronger band gap narrowing at high doping concentrations in the electron majority case. Similar investigations can be found in [60] for DD/HD simulations in combination with BD (apparent band gap) and FD (true band gap). As noted in [60], the designation *true* is purely related to DD/HD simulations in order to obtain agreement between FD/BD implemented in most commercial DD/HD simulators. Thus, the physical band gap reduction is not considered in [60], due to the simplifications usually made for DD/HD simulations, e.g. parabolic and isotropic dispersion relation and effective density of states. Depending on the employed dispersion relation and parameters, BTE BULK simulations deliver more physical insight into the band gap narrowing. For the sake of completeness, the impact of GE on the doping dependent band gap is illustrated in figure 4.15, where theresults of BTE BULK simulations with activated Pauli principle for electrons as majorities are shown. Based on the imposed effective intrinsic carrier density, the results include both the shift of the Fermi level due to FD and the GE-dependent shift of the X_2-valleys. The upward shift of the X_2-valleys forces an upward shift of the Fermi level (preservation of $n \approx |N_\mathrm{D}|$). Thus, a higher band gap narrowing is needed in order to fulfill the charge neutrality in conjunction with the Boltzmann distribution function for holes.

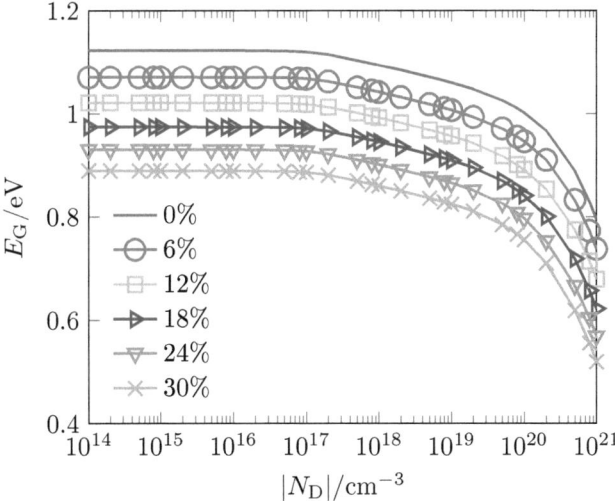

Figure 4.15: Doping dependent band gap of SiGe at $T_\mathrm{L}=300\,\mathrm{K}$ for different GE contents. The electron majority case (with Pauli principle) is shown.

4.2.2 Device simulations

IHP SiGe HBT

First, the SiGe HBT published in [40] and fabricated by IHP is investigated. The doping and Ge-profile is shown in figure 4.16. The simulated transfer characteristic and normalized transconductance $\overline{g_m} = \frac{V_T g_m}{J_C}$ are shown in figure 4.17, and the transit frequency (f_T) is shown in figure 4.18. The results obtained by both the simulator developed in this work (with/without Pauli principle) and the SHE code [31] (without Pauli principle) are shown. Accord-

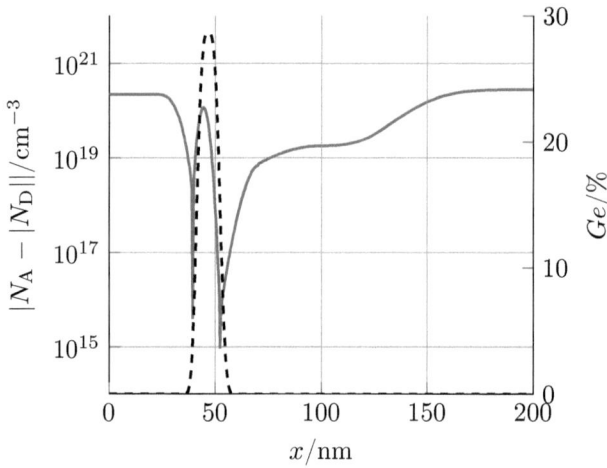

Figure 4.16: Doping and Ge-profile of the IHP SiGe HBT [40].

ing to figure 4.17 and 4.18, good agreement between SHE and the developed simulator is obtained over a wide range of applied biases. Considering the Pauli principle, only marginal deviations in the transfer characteristic and the transit frequencies are observed. Compared to SHE, slightly higher collector current densities J_C are obtained at moderate base-emitter voltages $(V_{BE} = 0.7\ldots0.8\,\mathrm{V})$ only (cf. figure 4.17, w/ and w/o Pauli). These deviations translate into deviations in the normalized transconductance directly, which are observed in the particular V_{BE}-region, too.

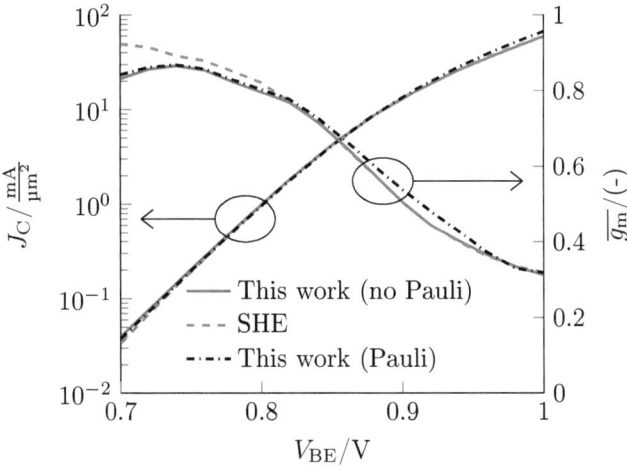

Figure 4.17: Transfer characteristic and normalized transconductance of the IHP SiGe HBT [40] at $V_{BC}=0\,$V. BTE-results obtained by both the solver developed in this work and the SHE code [31] are shown.

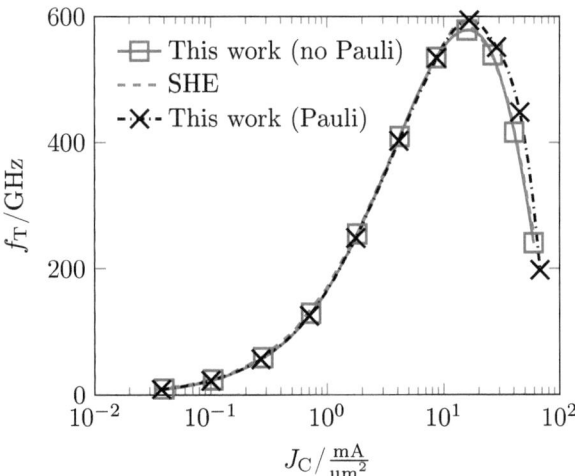

Figure 4.18: Transit frequency of the IHP SiGe HBT [40] at $V_{BC}=0\,$V. BTE-results obtained by both the solver developed in this work and the SHE code [31] are shown.

Before the onset of high current injection, the transfer current density and transconductance are dominated by the base-emitter (BE) barrier in the conduction band. To identify the cause of the deviations in J_C, the conduction band edges obtained from SHE and from the method used in this work (w/o Pauli principle) are compared in figure 4.19. It can be seen that the SHE

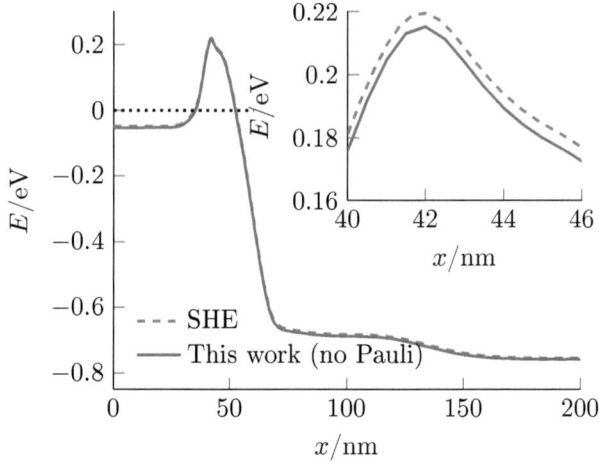

Figure 4.19: Conduction band edge of the IHP SiGe HBT [40] at $V_{BE}=0.7$ V, $V_{BC}=0$ V. For the curves shown above, the Pauli principle is neglected. The black dotted line indicates the Fermi level (at $E = 0$ eV).

BE-barrier height (see inset) is slightly higher than the one obtained from the method used in this work. Due to a higher BE-barrier, less electrons are able to overcome the barrier. Consequently, the current density obtained by SHE is smaller than the one obtained by the developed method. With the conduction band edge after equation (4.2.0-17), the material models for χ_r, $\Delta\chi$, $\Delta E_{C,0,dop.}$ (including their parameters) or the electrostatic potential ψ might cause the deviations seen in figure 4.19. However, it has been ensured that both the material models and their parameters are identical. Therefore, the electrostatic potentials, obtained by both SHE and the developed method, are drawn in figure 4.20 for the considered operating point ($V_{BE}=0.7$ V, $V_{BC}=0$ V). The inset of figure 4.20 focuses on the emitter contact region for emphasizing the deviations in ψ. However, deviations in ψ are also observed over the whole device length. Focusing on the contact point at

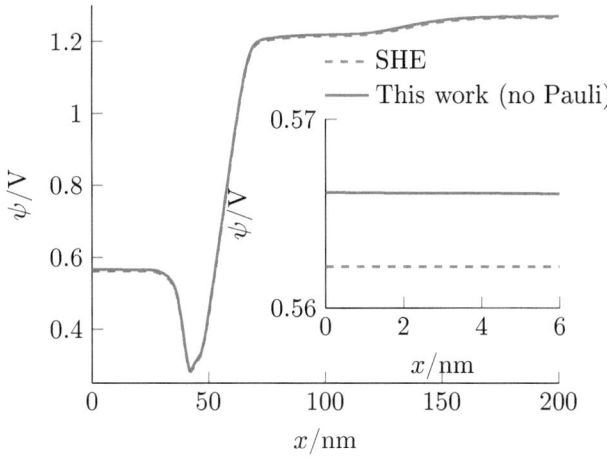

Figure 4.20: Electrostatic potential ψ of the IHP SiGe HBT [40] at V_{BE}=0.7 V, V_{BC}=0 V. For the curves shown above, the Pauli principle is neglected.

x=0 nm, the deviation in the electrostatic potential implies a disagreement of the calculated build-in potentials $\psi_{equ.}$ involved in the ohmic contact boundary conditions for the *Poisson* equation:

$$\psi_{cont.} = \psi_{equ.} + V_A \,, \tag{4.2.2-22}$$

$$\psi_{equ.} = \frac{E_{F,equ.} - E_{F,ir}}{q} \,, \tag{4.2.2-23}$$

with the applied contact voltage V_A, the Fermi level in equilibrium $E_{F,equ.}$ and the intrinsic Fermi level $E_{F,ir}$ of the reference material (here Si). Usually, an analytic formula for $\psi_{equ.}$ as function of doping and composition is used. However, this formula relies on the assumption of Boltzmann distribution function for both electrons and holes. In addition, band structure effects like the Ge induced energetic shift of the X_2-valleys are not directly captured. Therefore, the simulator developed in this work determines $\psi_{equ.}$ numerically based on the defined band structure and dispersion relation and the doping and composition concentration. The deviations shown in figure

161

4.20 imply that in SHE $\psi_{equ.}$ is not calculated numerically, but by use of a different approach, e.g. by the aforementioned analytical formula. However, for consistency with the defined band structure and dispersion relation, it is recommended to determine $\psi_{equ.}$ numerically. Albeit $\psi_{equ.}$, the positive slope in the normalized transconductance $\overline{g_m}$ seen at V_{BE}=0.7 V - 0.73 V (cf. figure 4.17), obtained by the method in this work only, is nonphysical. Compared to SHE, a coarser real space discretization has been employed to save simulation time. By using the same real space discretization as for the SHE simulations, the nonphysical behavior diminishes. In figure 4.21, the initially obtained $\overline{g_m}$s (cf. figure 4.17) are compared with the $\overline{g_m}$ obtained by using the SHE real space discretization for V_{BE}=0.7 V, V_{BC}=0.0 V (w/ Pauli Principle). An improvement to a more physical behavior is seen. However, the collector current density is still higher than the one of SHE, since the re-discretization does not influence the $\psi_{equ.}$ deviations. With figure 4.21, the sensitivity of

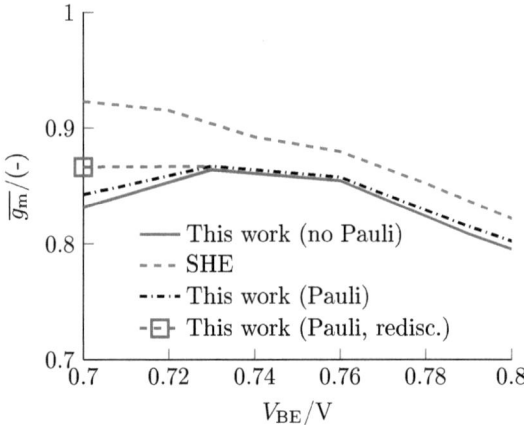

Figure 4.21: Normalized transconductance of the IHP SiGe HBT [40] at V_{BC}=0 V. BTE-results obtained by both the solver developed in this work and the SHE code [31] are shown.

the simulation results on the real space discretization for low V_{BE} voltages is emphasized. For a complete analysis, the impact of both the real space as well as the energy discretization on J_C and $\overline{g_m}$ is needed. Additionally, lower V_{BE} voltages need to be considered. These investigations are very time consuming and could not be performed within the time frame of this work. Thus, they are subject of further investigations. Despite of the deviations in $\psi_{\text{equ.}}$ and

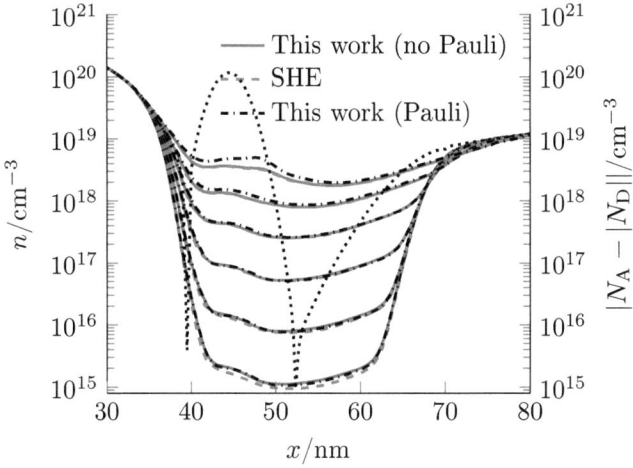

Figure 4.22: Electron densities of IHP SIGE HBT for $V_{\text{BE}}=0.70\,\text{V}$, $0.76\,\text{V}$, $0.82\,\text{V}$, $0.88\,\text{V}$, $0.94\,\text{V}$ and $1.00\,\text{V}$ at $V_{\text{BC}}=0\,\text{V}$. Here, the results of the BTE-solver developed in this work (with/without Pauli principle) and the one of SHE code [31] (without Pauli principle) are shown.

$\overline{g_{\text{m}}}$, the electron densities shown in figure 4.22 agree well over the range of considered operating points. Except for $V_{\text{BE}}=0.70\,\text{V}$ (due to $\psi_{\text{equ.}}$, BE-barrier), deviations in the electron densities between SHE and the method of this work are hardly observable (w/o Pauli principle). Considering the Pauli principle, only for $V_{\text{BE}}=0.94\,\text{V}$ and $1.00\,\text{V}$ small deviations are observed, which are insignificantly influencing the terminal characteristics shown in figure 4.17. Also the small-signal quantities like f_{T} (see figure 4.18) and the normalized transconductance (figure 4.17) agree with the SHE results. The figures 4.23 and 4.24 are showing the electron distribution function over energy and position within the X_2-valley for $V_{\text{BE}}=0.70\,\text{V}$. Due to the GE-induced valley shift and its degeneracy, the X_2-valley is the least occupied valley within the base region, and changes in f^{X_2} over orders of magnitude are well captured. Thus, the developed method is capable of resolving the minority carriers (electrons in the base region) accurately and smoothly over orders of magnitude.

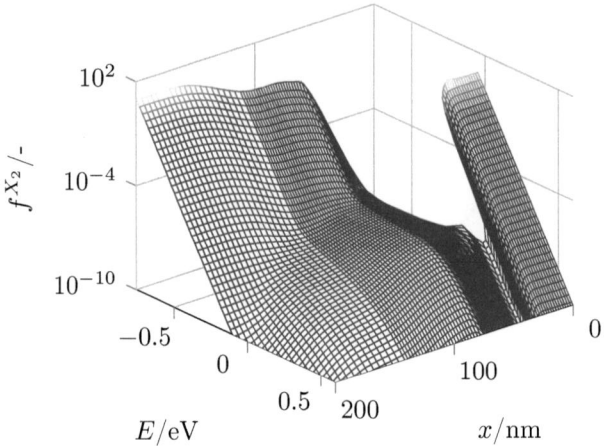

Figure 4.23: Electron distribution function (without Pauli principle) of the X_2-valleys over energy and position at $V_{BE}=0.70\,V$.

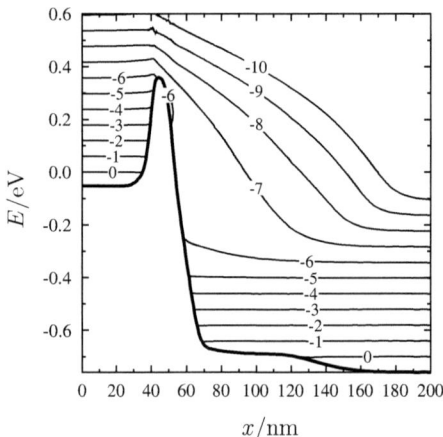

Figure 4.24: Contour lines of the electron distribution function (without Pauli principle) of the X_2-valleys over energy and position at $V_{BE}=0.70\,V$. The decadic logarithm is used for the contour lines.

Sige HBT at its physical limits

Figure 4.25 illustrates the doping and GE profile of the HBT published in [57], which represents a possible candidate of the physical limits of the SiGE-HBT technology. The transfer characteristic and normalized transconductance are

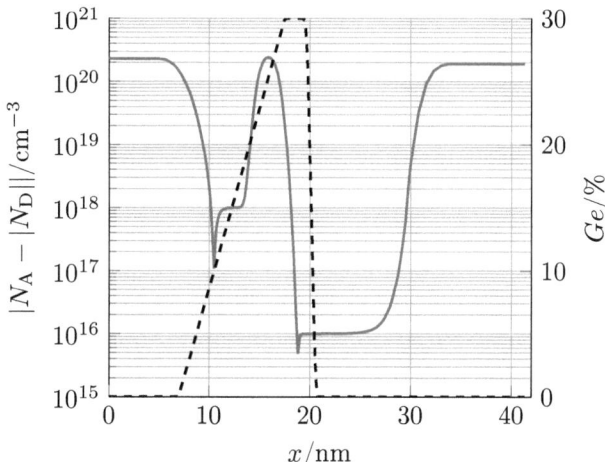

Figure 4.25: Doping and GE-profile of the "limits" SiGE HBT [57].

shown in figure 4.26, and the transit frequency (f_T) is shown in figure 4.27 (for $V_{BC}=0$ V). In terms of transfer characteristic and normalized transconductance, only marginal differences between SHE results and those obtained from the method employed in this work (with/without Pauli principle) are observed (see figure 4.26). In terms of f_T with neglected Pauli principle, small deviations between the peak values obtained by SHE and the method of this work are seen. Considering the Pauli principle, a higher peak value of f_T is obtained. Due to the good agreement obtained for the transfer characteristic and the normalized transconductance, the accumulated hole transit time $\tau_p(x)$, shown in figure 4.28, is used to localize the origin of the small deviations in f_T [25] (see equation (4.2.2-24) and (4.2.2-25), with the elementary charge q, the transconductance g_m, hole density p and the total length L of the HBT).

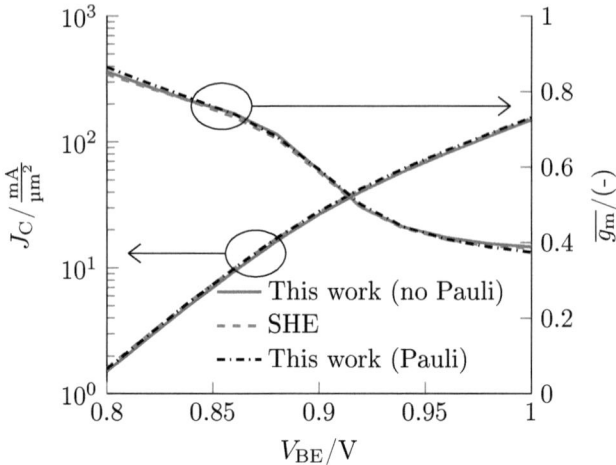

Figure 4.26: Transfer characteristic and normalized transconductance of the "limits" SiGe HBT [57] at $V_{BC}=0\,\mathrm{V}$.

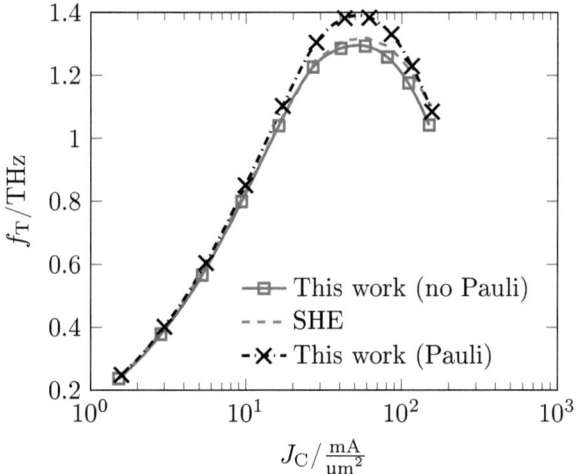

Figure 4.27: Transit frequency of the "limits" SiGe HBT [57] at $V_{BC}=0\,\mathrm{V}$.

$$\tau_p(x) = \frac{q}{g_m} \int_0^{\acute{x}=x} \frac{\partial p(\acute{x})}{\partial V_{BE}}\bigg|_{V_{CE}=\text{const.}} d\acute{x} \qquad (4.2.2\text{-}24)$$

$$f_T = \frac{1}{2\pi\tau_p(L)} \qquad (4.2.2\text{-}25)$$

Besides a too narrow emitter region and thus a visible increase in $\tau_p(x)$

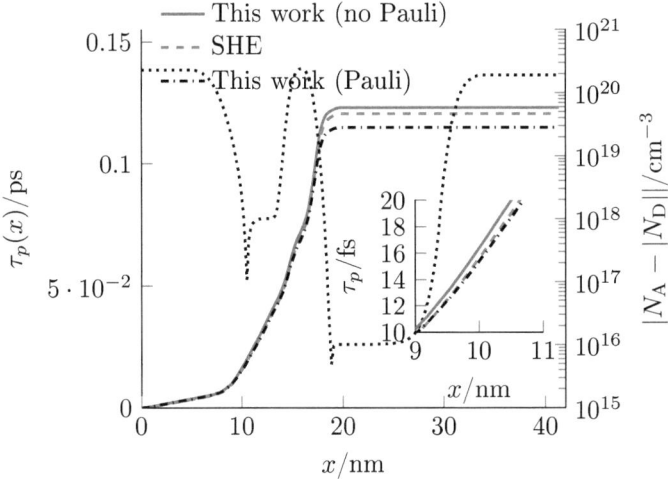

Figure 4.28: Accumulated hole transit time at $f_{\mathrm{T,peak}}$.

already close to the emitter contact at $x=0$ nm, slight deviations in the accumulated hole transit time (figure 4.28) are seen in the simulation results. Especially in the emitter and for the most part in the base region, slightly different slopes of $\tau_p(x)$ are observed. Since the collector region does not contribute to a further increase of $\tau_p(x)$, the different f_{T} values obtained are stemming from the slopes in the regions before. Focusing on results without Pauli principle, marginal deviations in $\tau_p(x)$ obtained by SHE and the method used in this work are obtained. Due to the missing output of material related physical quantities in SHE (e.g. band gap), a distinct origin of the small deviations in $\tau_p(x)$ can not be identified. Since for all simulations an identical intrinsic carrier density is imposed, the deviations are most likely to originate from material related physical models, comparable to discrepancies in $\psi_{\mathrm{equ.}}$, shown for the IHP SiGe HBT in the previous section.

In contrast to the small deviations in $\tau_p(x)$ with neglected Pauli principle, the consideration of the Pauli principle leads to a more pronounced reduction of $\tau_p(x)$ and thus to a higher peak value of f_{T} (see figure 4.27). As the BTE is solved for electrons only, the electron densities are compared in figure 4.29 for different operation points. According to figure 4.29, the Pauli principle leads

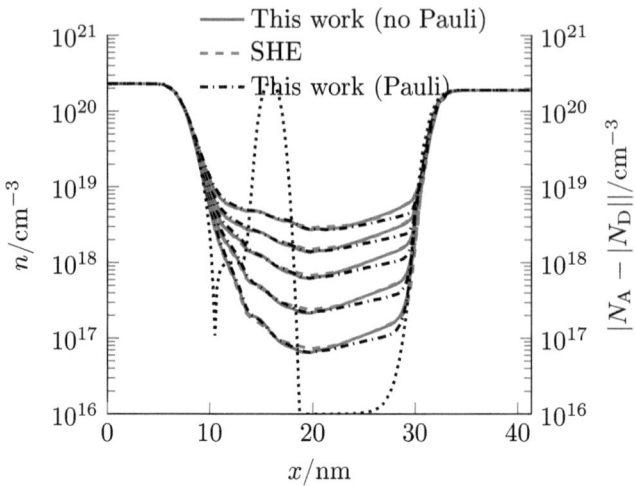

Figure 4.29: Electron densities of the limits SiGe HBT [57] for $V_{BE}=0.82\,\text{V}$, $0.86\,\text{V}$, $0.90\,\text{V}$, $0.94\,\text{V}$ and $0.98\,\text{V}$ at $V_{BC}=0\,\text{V}$.

to smaller electron densities in the collector region. As the Pauli principle prevents the electron distribution function to exceed values larger than one, the distribution function exhibits a plateau-like shape near the conduction band edge for highly doped regions (emitter and collector contact regions), similarly to the BULK case shown in figure 4.14. Thus, with considered Pauli principle, a less pronounced diffusion into the low doped collector region – due to fully occupied states in the low doped collector region – takes place. Consequently, a lower electron density compared to case with neglected Pauli principle is seen. In figure 4.30 and 4.31 the contour lines of the distribution function within the X_4-valleys are shown w/o and w/ Pauli principle, respectively. The main deviations in the distribution functions are located around $x < 30\,\text{nm}$, which coincides with the deviations in the electron densities shown in figure 4.29. The distribution function in figure 4.31 is about one order of magnitude smaller in the collector transition region than the one without Pauli principle in figure 4.30.

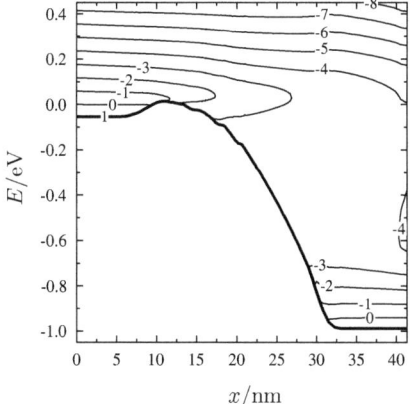

Figure 4.30: Electron distribution function (without Pauli principle) of the X_4-valleys over energy and position at $f_{T,peak}$. The decadic logarithm is used for the contour lines.

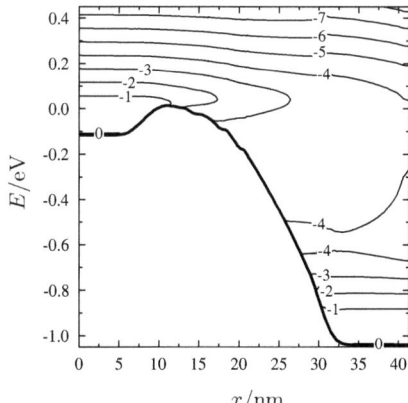

Figure 4.31: Electron distribution function (with Pauli principle) of the X_4-valleys over energy and position at $f_{T,peak}$. The decadic logarithm is used for the contour lines.

With an almost identical collector current density J_C (see figure 4.26) and smaller electron densities in the collector transition region (with Pauli principle in figure 4.29), an increase in the electron velocity is needed for current conservation. The increase in the electron velocity (with Pauli principle) is shown in figure 4.32. Due to the demand of charge neutrality, the decreasing

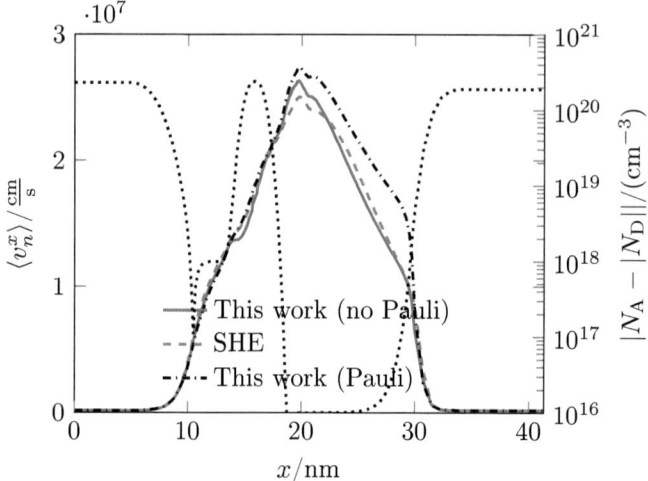

Figure 4.32: Average electron velocities in x-direction at $f_{T,\text{peak}}$ (identical V_{BE}, and almost identical J_C in accordance with figure 4.26).

electron density and increasing electron velocity affect $\tau_p(x = L)$:

$$
\begin{aligned}
\tau_p(L) &= \frac{q}{g_m} \int_0^{\acute{x}=L} \left.\frac{\partial p(\acute{x})}{\partial V_{BE}}\right|_{V_{CE}=\text{const.}} d\acute{x} \\
&= \underbrace{\frac{q}{g_m} \int_0^{\acute{x}=L} \left.\frac{\partial n(\acute{x})}{\partial V_{BE}}\right|_{V_{CE}=\text{const.}} d\acute{x}}_{\tau_n(L)} \\
&\simeq \int_0^{\acute{x}=L} \frac{1}{\langle v_n^x(\acute{x})\rangle} \left.\left(1 - \frac{V_T}{g_m}\frac{1}{\langle v_n^x(\acute{x})\rangle}\frac{\partial \langle v_n^x(\acute{x})\rangle}{\partial V_{BE}}\right)\right|_{V_{CE}=\text{const.}} d\acute{x}\,,
\end{aligned}
$$

$$(4.2.2\text{-}26)$$

with the transconductance g_m, the thermal voltage V_T, the normalized transconductance $\overline{g_{\mathrm{m}}} = \frac{g_{\mathrm{m}} V_{\mathrm{T}}}{J_{\mathrm{C}}}$ and the elementary charge q. A detailed derivation of (4.2.2-26) can be found in appendix C.1. Based on equation (4.2.2-26), one can estimate that an increase in $\langle v_n^x(\hat{x}) \rangle$ yields a decrease of $\tau_{\mathrm{p,n}}(L)$ and therefore a higher f_{T} (cf. equation (4.2.2-25)). Compared to the case of neglected Pauli principle, less but faster electrons are obtained within the low doped collector region and higher f_{T} values are reached. The described correlation is shown in figure 4.33. For the sake of clarity, the results of the method employed in this work are shown only.

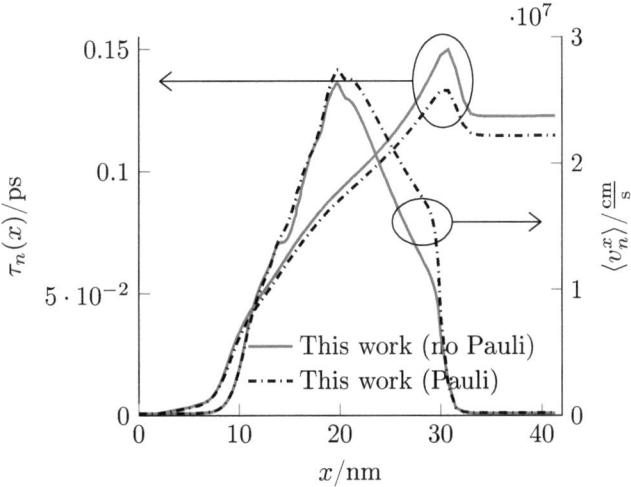

Figure 4.33: Average electron velocities in x-direction at $f_{\mathrm{T,peak}}$.

Despite of some minor deviations between SHE and the method of this work, a good overall-agreement is obtained for both internal and external (terminal) characteristics. As mentioned before, the exact origins of the minor deviations could not be identified due to limited set of output quantities provided by SHE. Nevertheless, simulation results typically vary slightly depending on the simulation software used and the observed deviations are well within the associated spread of experimental data.

4.3 InP/InGaAs

In this section, hetero-structures fabricated in the INP, INGAAS and INP/ INGAAS material systems are investigated. These material systems are categorized as so called III-V semiconductors due to the position of the involved elements in the periodic table. Since no access to another BTE-solver for such materials is available, the simulation results of the developed method are compared with scarcely available experimental and simulation results published in literature. For all simulations, a two-valley model is used. The first valley represents the Γ-valley in center of the first Brillouin zone and the second one the L-valleys (see section 2.2, figure 2.7 and figure 2.9). In the case of the Γ-valley, the valley degeneracy equals one. The second valley represents the eight L-valleys, each contributing half to the first Brillouin zone (see figure 2.10(b) in section 2.2). Thus, for convenience and for comparison purpose, a valley degeneracy factor of four is assigned to the L-valleys. Although the equi-energy surfaces of the L-valleys exhibit an ellipsoidal shape (see figure 2.9 in section 2.2), their anisotropy is usually neglected. Instead, isotropic and thus spherical equi-energy surfaces are assumed [19, 20, 46, 69]. Figure 4.34 summarizes the described simplifications.

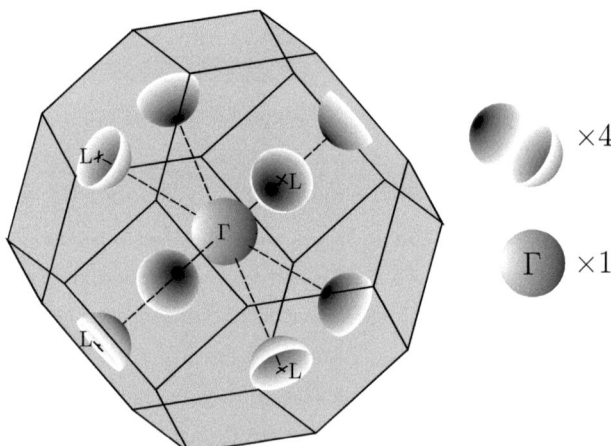

Figure 4.34: Schematic illustration of the 1st *Brillouin*-zone together with the valleys and their degeneracy as considered for the III-V structures INP and INGAAS.

Quantity	Symbol	Value	Unit
General material parameters			
Permittivity (static)	ε_{st}	$12.5+1.4\,x_C$	-
Permittivity (HF)	ε_{hf}	$9.61+2.0\,x_C$	-
Sound velocity	v_s	$5130\text{-}1030\,x_C$	m/s
Mass density	ρ	$4810+690\,x_C$	kg/m^3
Band parameters			
Electron affinity	χ_r (INP)	4.38	eV
Electron affinity change	$\Delta\chi$	$0.4078\,x_C\text{-}0.1657\,x_C^2$	eV
Energetic valley offset	ΔE_0^ν	$\Gamma:0.0$ $L:0.586-0.126\,x_C$	eV
Valley degeneracy factor	Z^ν	$\Gamma:1$ $L:4$	-
Non-parabolicity factor	α_ν	$\Gamma:0.84+0.4670\,x_C$ $L:0.23+0.4610\,x_C$	eV^{-1}
Isotropic masses	m_ν^*	$\Gamma:0.078-0.037\,x_C$ $L:0.260+0.031\,x_C$	-
Scattering parameters			
Acoustic deform. pot.	D_A	INP : 8.0 INGAAS : 9.2	eV
Polar optical phonons	$\hbar\omega_{POP}$	INP : 42.3 INGAAS : 31.5	meV
Intravalley phonon scat.		INP INGAAS	
ODP type	D_O	$6.7\cdot10^{10}$ $5.0\cdot10^{10}$	eV/m
L-valley only	$\hbar\omega_O$	43.0 39.0	meV
Intervalley phonon scat.		INP INGAAS	
ODP type	D_{ij}	$1.0\cdot10^{11}$ $11.3\cdot10^{10}$	eV/m
$\Gamma \leftrightarrow L$	$\hbar\omega_{ij}$	27.8 24.3	meV
ODP type	D_{ij}	$1.0\cdot10^{11}$ $4.44\cdot10^{10}$	eV/m
$L \leftrightarrow L$, $Z^{\nu'} = Z^L - 1$	$\hbar\omega_{ij}$	29.0 23.8	meV

Table 4.2: INP (x_C=0) and INGAAS (x_C=1 $\widehat{=}$ IN$_{0.53}$GA$_{0.47}$AS) parameters [53].

Table 4.2 summarizes the parameters of the physical models used for the simulations. Like for SI/SIGE materials, also here the isotropic approximation of impurity scattering is employed. In addition, piezoelectric scattering is neglected, since this scattering mechanism only plays an important role at low lattice temperatures [51]. In contrast to the simulations of SI/SIGE materials, a position dependent dispersion relation is taken into account due to the composition dependent non-parabolicity factors α_ν and effective masses m_ν^* (cf. table 4.2). For the composition dependent material parameters in table 4.2, a linear interpolation between INP and INGAAS is employed. Note, although the differences in the electron affinity between INP and INGAAS are about 120 meV according to literature [43], a difference of 242 meV is used instead. Latter discrepancy stems from the fact that the difference in the electron affinity reported in [43] does not agree with the values of the conduction band discontinuity between INP and INGAAS commonly used in simulations [15, 22]. For the calculation of the conduction band edges, the equations (4.2.0-17) and (4.2.0-19) given in section 4.2 are also valid here. For equation (4.2.0-17), the doping dependent shift of the conduction band is needed. Following [35], the latter is obtained by the doping dependent band gap narrowing

$$\Delta E_{\mathrm{G}} = A\,G^{\frac{1}{3}} + B\,G^{\frac{1}{4}} + C\,G^{\frac{1}{2}} \; , \tag{4.3.0-27}$$

with $G = N_{\mathrm{A}} + |N_{\mathrm{D}}|$ and, as suggested in [59], by partitioning the band gap narrowing to the conduction and valence band edge by

$$\frac{\Delta E_{\mathrm{C},0,\mathrm{dop.}}}{m_n} = \frac{\Delta E_{\mathrm{V},0,\mathrm{dop.}}}{m_p} \; ,$$
$$\rightarrow \Delta E_{\mathrm{C},0,\mathrm{dop.}} = \frac{\Delta E_{\mathrm{G}}}{1 + \frac{m_p}{m_n}} \; , \tag{4.3.0-28}$$

with the effective electron (hole) mass m_n (m_p), respectively. Note, m_p denotes the average of light and heavy hole masses [35,59]. However, the effective masses m_n and m_p in (4.3.0-28) must not be mistaken with the masses involved in the dispersion relation, e.g. m_ν^* in table 4.2. The effective masses in (4.3.0-28) are responsible of retrieving the effective density of states N_C and N_V, respectively, used for the electron and hole density calculation based on Boltzmann distributions. Behind the concept of N_C and N_V, a parabolic dispersion relation is assumed. In contrast to m_ν^* in table 4.2, describing the curvature of the dispersion relation for valley ν, m_n combines from a macroscopic point of view the isotropic masses of each, as parabolic assumed, valley. The parameters involved in (4.3.0-27) and (4.3.0-28) are listed in table 4.3. In contrast to the simulations for SI/SIGE, the Pauli principle is considered for all simulations, here.

Symbol	Value	Unit
Doping dependent conduction band shift/band gap narrowing		
n-type		
A	$6.826 \cdot 10^{-8} - 6.826 \cdot 10^{-8} x_C$	eVcm
B	10^{-20}	$\mathrm{eVcm}^{\frac{3}{4}}$
C	$1.6383 \cdot 10^{-12} + 4.3407 \cdot 10^{-11} x_C$	$\mathrm{eVcm}^{\frac{3}{2}}$
p-type		
A	$1.03 \cdot 10^{-8} - 1.10 \cdot 10^{-9} x_C$	eVcm
B	$4.43 \cdot 10^{-7} - 8.60 \cdot 10^{-8} x_C$	$\mathrm{eVcm}^{\frac{3}{4}}$
C	$3.38 \cdot 10^{-12} + 2.70 \cdot 10^{-13} x_C$	$\mathrm{eVcm}^{\frac{3}{2}}$
Effective electron/hole masses		
m_n	$0.08\text{-}0.03895 x_C$	-
m_p	$0.6\text{-}0.143 x_C$	-

Table 4.3: INP ($x_C=0$) and INGAAS ($x_C=1 \,\widehat{=}\, \mathrm{I}_{N0.53}\mathrm{G}_{A0.47}\mathrm{A}_S$) parameters for (4.3.0-27)-(4.3.0-28) [53].

4.3.1 BULK simulations

Due to the scarcely available experimental data, the BULK simulation results are mainly compared with the data provided in [43]. Figure 4.35 shows the electron low field mobility of INP and INGAAS as function of the total doping concentration $G = N_A + |N_D|$. In the case of the BTE-results, both the minority and the majority case is shown. The data provided in [43], abbreviated by Lit., are available for INGAAS, only. Therefore and for comparison purposes, the model and its parameters given in [64], which were adjusted to experimental data, are also shown in figure 4.35 and abbreviated by Mod. From the qualitative point of view, acceptable agreement is obtained for both

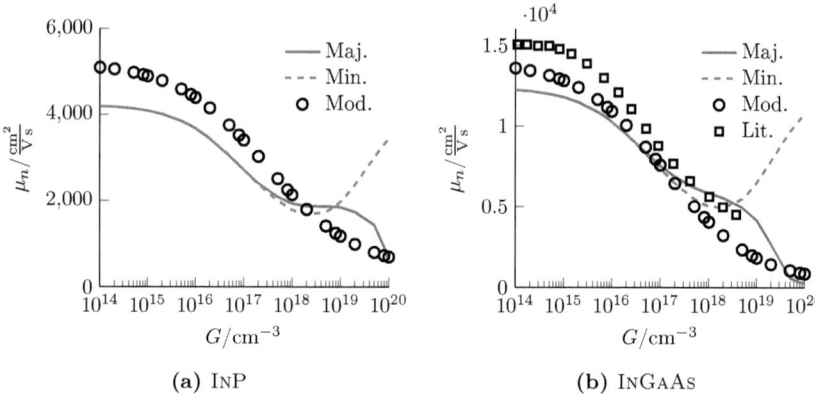

(a) INP

(b) INGAAS

Figure 4.35: INP and INGAAS electron low field mobility at $T_L=300$ K.

INP and INGAAS without spending any adjustment effort. As pointed out for SI in section 4.2.1, for low doping concentrations phonon scattering is dominating scattering process. Thus, the deviations for $G < 1 \times 10^{15}$ cm^{-3} are mainly due to the phonon scattering parameters used for the simulations (see table 4.2) which should be revised therefore. Remaining deviations, after a revision of the phonon scattering parameters, can be reduced by introducing doping dependent fit- functions (for the minority and the majority case), like used for SI and SIGE (cf. appendix B.5).

Figure 4.36 illustrates the electron drift velocities over applied electric field for different lattice temperatures. Here, the simulation results are compared with the simulated data given in [18] for unscreened electron-phonon interaction. A doping concentration of $|N_D|=1 \times 10^{15}$ cm^{-3} is considered for the simulations, in accordance with the range given in [18]. Therefore, negligible impurity scattering can be assumed (see figure 4.35). All curves shown in figure 4.36 exhibit the characteristical *negative differential mobility* (NDM), stemming from the electron transfer from the Γ- to the L-valley [46,69]. However,

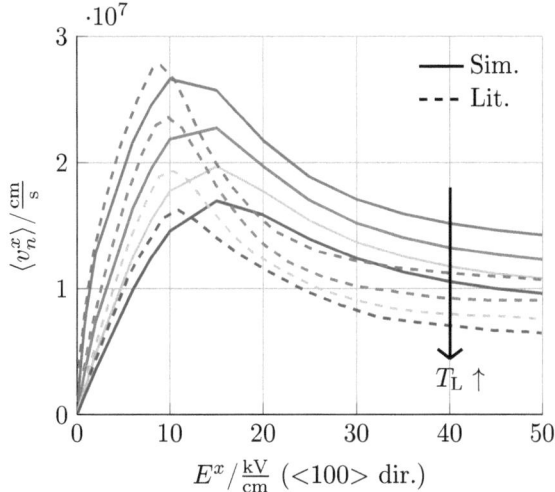

Figure 4.36: INP electron drift velocities over electric field at T_L=200 K, 300 K, 400 K and 500 K. The literature data are taken from [18].

already for low electric fields (E^x <5 kV/cm), deviations between simulation results and literature data are visible. In figure 4.36 and for E^x <5 kV/cm, the slope of the simulated electron drift velocity (low field mobility) is smaller than the one of the literature data, which is in accordance with the underestimation of the low field electron mobility in figure 4.35(a) obtained by simulation. Despite of the good agreement in the peak electron drift velocities over temperature, the corresponding field strengths are, compared to the data in [18], slightly overestimated by the simulations. In figure 4.37, the valley occupancy obtained by the simulations (Orig.) is shown for T_L=300 K. The onset of an increase in the L-valley occupancy (critical field) is directly linked to the velocity peaks, seen in figure 4.36.

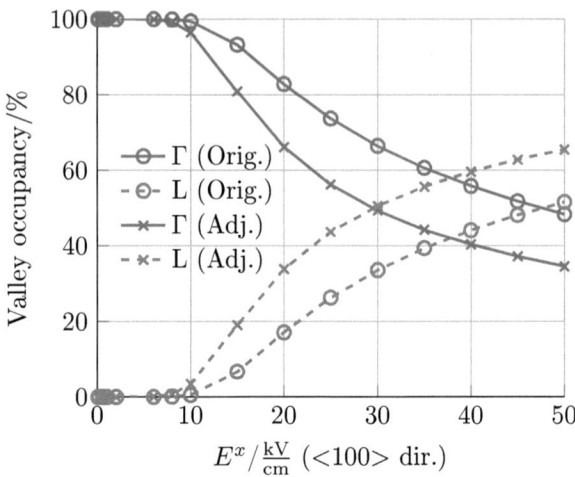

Figure 4.37: INP valley occupancy over electric field at T_L=300 K. **Orig.**: parameters of table 4.2; **Adj.**: band parameters given in [18].

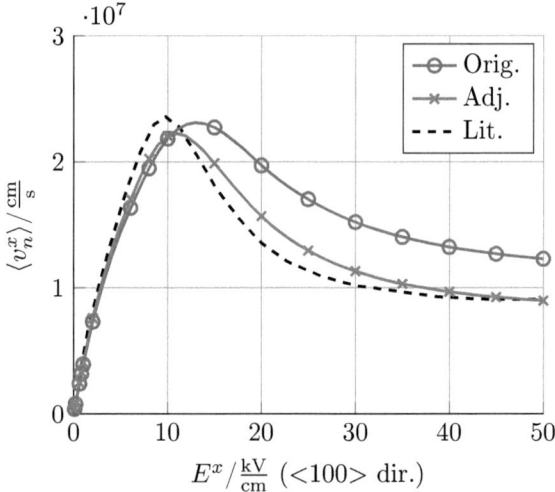

Figure 4.38: INP electron drift velocity over electric field at T_L=300 K. **Orig.**: parameters of table 4.2; **Adj.**: band parameters given in [18]; **Lit.**: curve in [18].

Beside discrepancies in the phonon scattering parameters, the main source of the deviations seen in figure 4.36 are the parameters used for the dispersion relation. Figure 4.38 compares the electron velocities for $T_\text{L}=300\,\text{K}$ after adjusting the dispersion relation parameters (Adj.) according to the values given in [18]. Besides some small discrepancies due to different scattering parameters, a better agreement is obtained according to figure 4.38. The discrepancies in the electron drift velocities shown for INP reflect the variation of the III-V model parameters found in literature exemplary and demonstrate the necessity for adjusting simulation parameters. The most recent parameters found agree with the ones given in table 4.2 and are therefore used for further investigations.

Like for INP, the electron drift velocity over field characteristic for INGAAS is shown in figure 4.39 for $T_\text{L}=300\,\text{K}$. The literature data shown in figure 4.39 are taken from [43, 67]. Although the qualitative shape of the velocity-

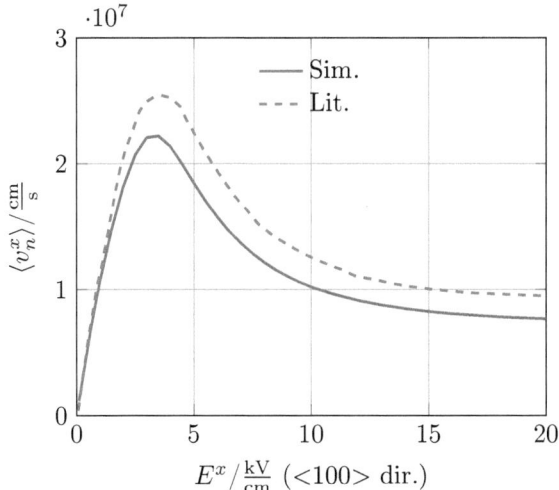

Figure 4.39: INGAAS electron drift velocity over electric field at $T_\text{L}=300\,\text{K}$. The literature results are taken from [43, 67].

field characteristics and the field strength of the velocity maximum is well reproduced by the simulations, quantitative deviations are also seen here. In contrast to the previously considered INP characteristics, a smaller electron velocity is predicted by the simulation, here. Unfortunately, the dispersion relation parameters used for the literature results were not given in [67] and the references therein. Thus, in order to estimate the impact of the dispersion relation parameters on the velocity-field characteristic, the effective masses (see figure 4.40), the non-parabolicity factor of the Γ-valley α_Γ and the valley separation $\Delta E_{\Gamma L}$ (see figure 4.41) are varied by ±10%.

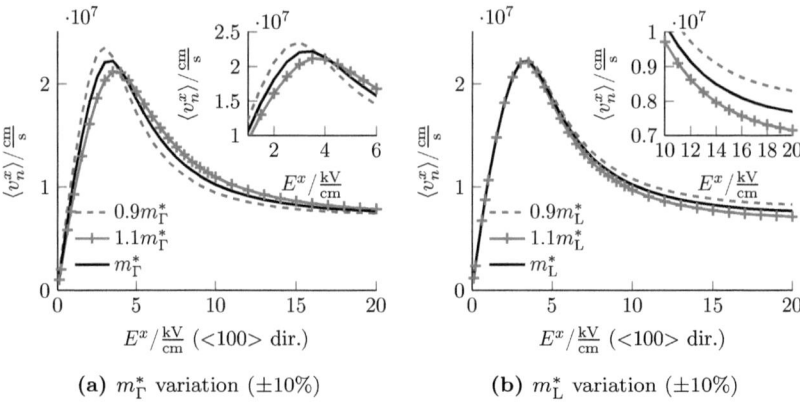

(a) m_Γ^* variation (±10%)   (b) m_L^* variation (±10%)

Figure 4.40: INGAAS: impact of the effective masses on the electron drift velocity at T_L=300 K.

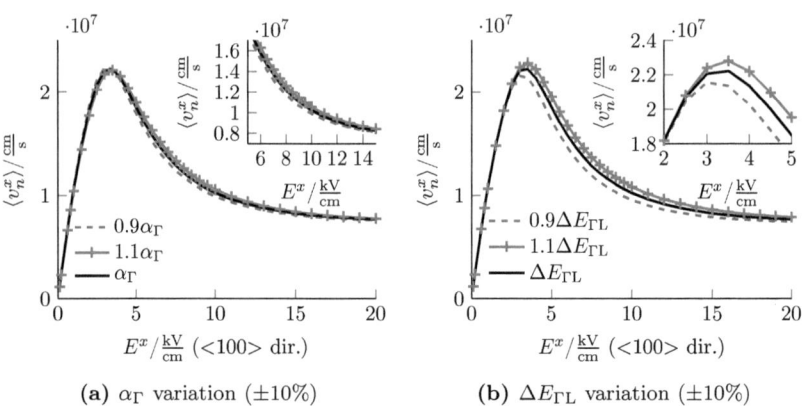

(a) α_Γ variation (±10%)   (b) $\Delta E_{\Gamma L}$ variation (±10%)

Figure 4.41: INGAAS: impact of the non-parabolicity factor α_Γ and the valley separation $\Delta E_{\Gamma L}$ on the electron drift velocity at T_L=300 K.

According to figure 4.40(a), an increase in the effective mass of the Γ-valley (m_Γ^*) leads to a decrease in the peak drift velocity. Thus, the slope prior the velocity peak and therefore the low field mobility is decreased. Additionally, the critical field is shifted to higher field strengths with increasing m_Γ^*. Below the critical field, most particles are occupying states within the Γ-valley, like exemplary shown for INP by figure 4.37 and 4.38. By assigning those particles a higher effective mass results for an identical electric field (identical force acting on them) in a lower average velocity, which can simply be approximated by using Newton's laws of motion. The same argumentation holds for fields higher than the critical field. Due to the valley separation $\Delta E_{\Gamma L}$, this amount of kinetic energy (excluding the phonon energy) transforms into potential energy during the particle transfer. Additionally, as the effective mass of the L-valleys is higher compared to those of the Γ-valley, the increase in the carrier velocity is less pronounced. Thus, the average carrier velocity, taken over both the Γ- and L-valley, is decreasing. This effect is amplified by increasing the mass of the L-valley as seen in figure 4.40(b).

The variation of the non-parabolicity factor of the Γ-valley has only a minor impact on the velocity-field characteristic. In the case of the L-valley, the variation did show an almost negligible impact and is therefore not shown.

By varying the valley separation $\Delta E_{\Gamma L}$, the position of the critical field and the peak velocity can be altered. With increasing $\Delta E_{\Gamma L}$, the particles in the Γ-valley have to gain a higher kinetic energy before the valley transfer is enabled by intervalley scattering. Thus, due to the higher kinetic energy needed for the transfer, an increase in the peak drift velocity is seen. As higher kinetic energies are accompanied with stronger electric fields, the critical field is shifted to higher field strengths.

Figure 4.42 compares the INGAAS velocity-field characteristic found in literature [43, 67] with simulation results. The latter are obtained by adjusting the dispersion relation parameters roughly, according to the trends shown in the figures 4.40 and 4.41. The good agreement in figure 4.42 was obtained by decreasing the effective mass of the Γ-/L-valley by 7.5%/25%, respectively. Additionally, the valley separation $\Delta E_{\Gamma L}$ had to be increased by 50%, which seems to be unrealistically high. One reason for the pronounced change in $\Delta E_{\Gamma L}$ might be the neglect of the L-valley anisotropy in the simulations. However, the discrepancies in curves or parameters already seen for INP among the literature and their rare availability prevent a clear judgment about the quality of the used parameters. Therefore, as a first step, the L-valley anisotropy should be considered in consecutive works, which unfortunately demands the discretization of the variable φ (see section 2.4.2) and thus eliminates the rotational symmetry depicted in section 2.4.4.

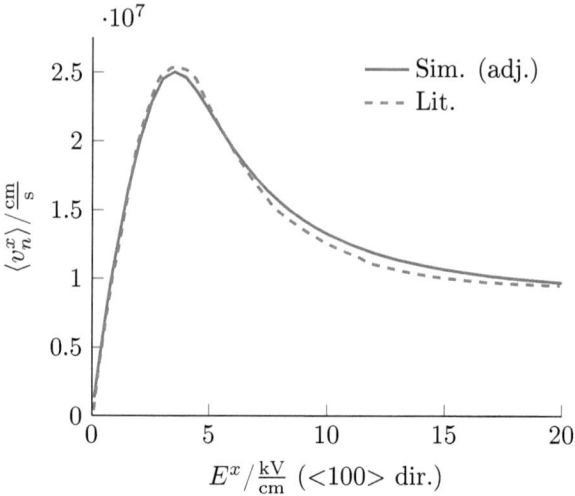

Figure 4.42: INGAAS electron drift velocity-field characteristic at $T_L = 300\,\mathrm{K}$: simulation results with adjusted dispersion relation parameters and literature results [43, 67].

4.3.2 Device simulations

In terms of III-V device simulation, an INP/INGAAS DHBT published in [16] is investigated. The doping and composition profile is shown in figure 4.43. For convenience, the doping profile used for the simulations is modified, compared to the layer structure described in [16]. In order to reflect the fabricated profile more realistically, smoothed instead of abrupt doping transitions are considered. In addition, a 20 nm INGAASP layer within the collector region – in front of the collector doping hill shown in figure 4.43 – is replaced by graded composition profile (INGAAS→ INP). In [16], the INGAASP layer is used to avoid current blocking at the BC junction stemming from the conduction band discontinuity of INP/INGAAS material transitions. However, in order to ease the amount of models and parameters needed for the simulation, the INGAASP layer is replaced by the graded composition profile, which has the same principle effect on the conduction band edge. The simulated transfer characteristic of the HBT is shown in figure 4.44 and compared with measurement results from which the 1D-characteristics were extracted. The latter were obtained based on a extracted HICUM compact model parameter set and by setting all external and peripheral elements to zero. The simulated and measured f_{T} characteristics are shown in figure 4.45.

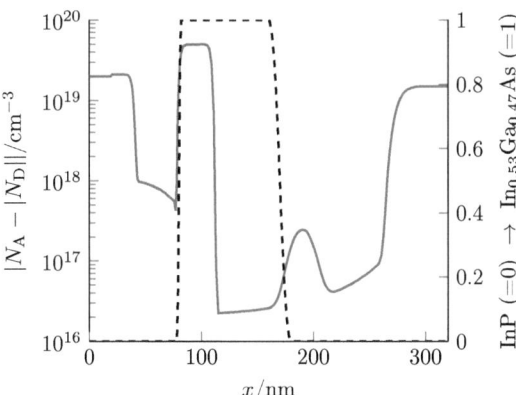

Figure 4.43: Doping and composition profile of the INP/INGAAS HBT used for the simulations based on [16].

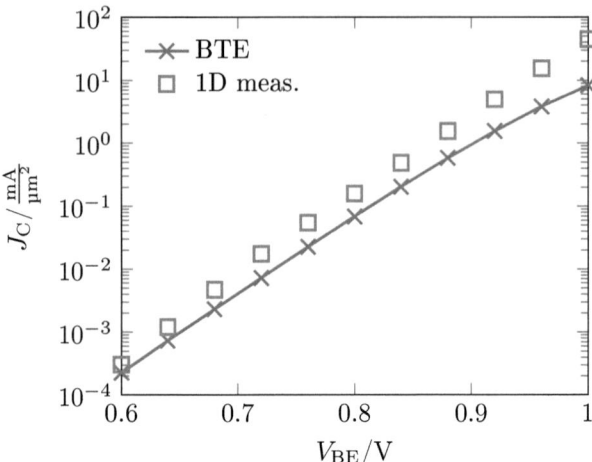

Figure 4.44: Simulated and measured (1D data extracted) transfer characteristic of the INP/INGAAS HBT at T_L=300 K and V_{BC}=0 V.

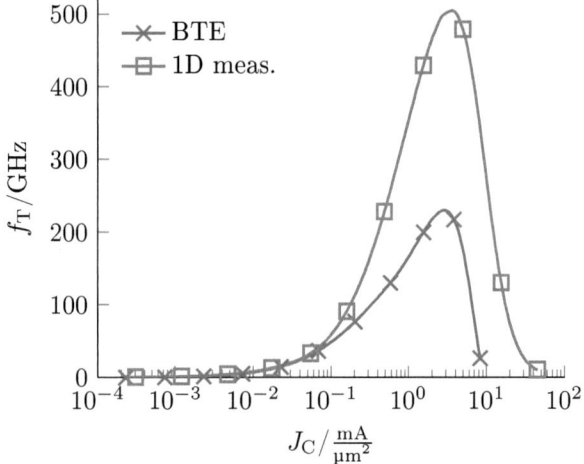

Figure 4.45: Simulated and measured (1D data extracted) transit frequency (f_T) characteristic of the INP/INGAAS HBT at T_L=300 K and V_{BC}=0 V.

Besides the moderate agreement between the measured and simulated transfer characteristic, major deviations, especially at low J_C and in the peak values, are obtained for f_T. The BTE results predict a peak value of f_T, which is about 45% of the measured 1D f_T peak value. To identify the origin of the discrepancies in f_T, the accumulated electron transit time at peak f_T is drawn in figure 4.46. Here, a major increase of the total electron transit time arising from the L-valley portion near the metallurgical BE junction is observed. Due to the composition profile, conduction band discontinuities for both the

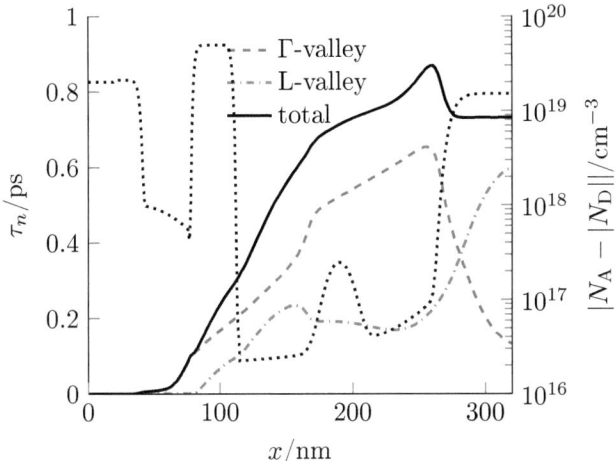

Figure 4.46: Accumulated electron transit time of the INP/INGAAS HBT at the simulated peak value of f_T (T_L=300 K and V_{BC}=0 V).

Γ- and the L-valley edge are seen near the metallurgical BE junction (cf. figure 4.47). With the formulation for the effective field $\vec{E}^{\nu}_{\text{eff.}}$ (2.4.0-108) (cf. section 2.4), the conduction band discontinuities result in an almost abrupt decrease of $\vec{E}^{\nu}_{\text{eff.}}$, as shown in figure 4.48. Thus, the particle distribution function in the Γ-valley is shifted to higher kinetic energies, enabling a pronounced Γ-to L-valley transfer, as seen in figure 4.49 by the valley occupancy. Due to the higher L-valley's effective mass, most particles are populating states at the bottom of the L-valley having a low kinetic energy. Consequently, the average electron velocity, shown in figure 4.50, decreases abruptly at the rising edge of the composition profile near the metallurgical BE junction.

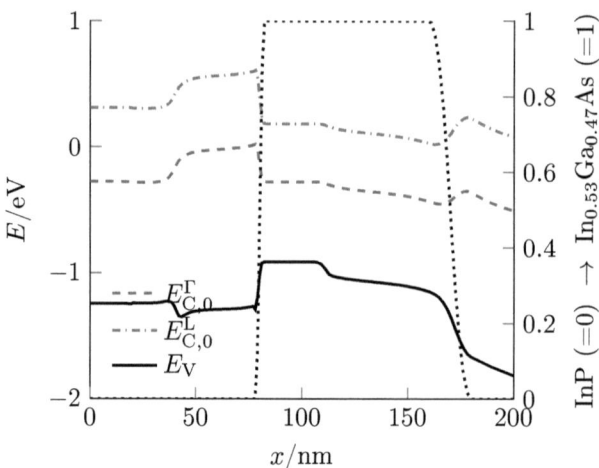

Figure 4.47: Conduction and valence band edges of the InP/InGaAs HBT at the simulated peak value of f_T ($T_\mathrm{L}=300\,\mathrm{K}$ and $V_\mathrm{BC}=0\,\mathrm{V}$).

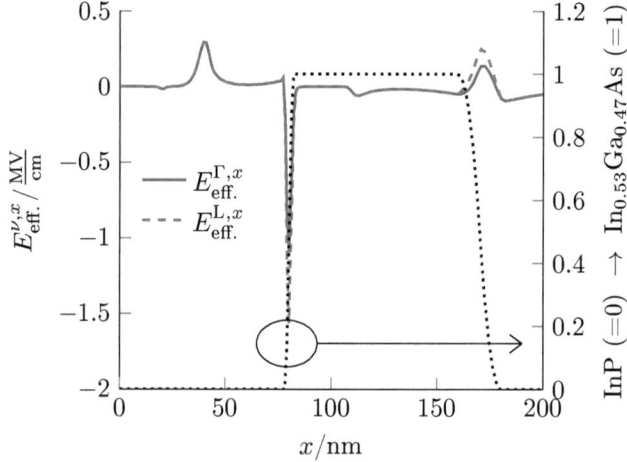

Figure 4.48: Effective fields $\vec{E}_\mathrm{eff.}^{\nu}$ (x-components) of the InP/InGaAs HBT at the simulated peak value of f_T ($T_\mathrm{L}=300\,\mathrm{K}$ and $V_\mathrm{BC}=0\,\mathrm{V}$).

With the formula of the accumulated electron or hole transit time (4.2.2-26) as function of the electron velocity, the abrupt decrease of $\langle v_n^x \rangle$ leads to an

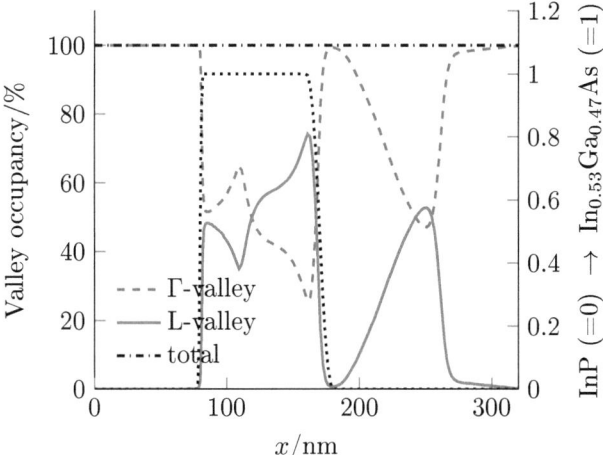

Figure 4.49: Γ- and L-valley occupancy of the InP/InGaAs HBT at simulated peak value of f_T (T_L=300 K and V_{BC}=0 V).

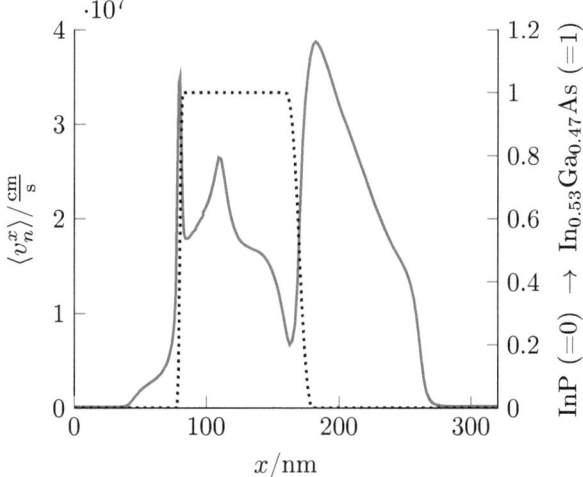

Figure 4.50: Average electron velocity in x-direction of the InP/InGaAs HBT at the simulated peak value of f_T (T_L=300 K and V_{BC}=0 V).

increasing τ_n and consequently degrades f_T. This effect stems from the treatment of InP/InGaAs hetero-junctions within the framework of the BTE.

187

Despite of the fact that the developed BTE solver can handle almost abrupt changes in its driving force $E^{\vec{\nu}}_{\text{eff.}}$ after (2.4.0-108), the shown results appear to be nonphysical. The nonphysical appearance of the results are not linked to numerical issues, since convergence was obtained for all considered operating points. However, the physical appearance of the results can be improved by:

1) increasing the valley separation $\Delta E_{\Gamma\text{L}}$ as function of the composition and/or

2) grading the composition profile at the BE junction (flattening of the composition slope m_{c}).

An increase in $\Delta E_{\Gamma\text{L}}$ suppresses the Γ- to L-valley transfer artificially. Thus more particles remain in the *fast* Γ-valley, leading to a higher average electron velocity and thus to a higher f_{T}. The second option affects the driving force $E^{\vec{\nu}}_{\text{eff.}}$ at the BE junction directly. With a smaller m_{c}, $E^{\vec{\nu}}_{\text{eff.}}$ is weakened and thus prevents an excessive shift of the Γ-particles to kinetic energies, where the transfer to the L-valley is enabled. The impact of the two listed options on the transfer characteristic and f_{T} is shown in figure 4.51 and 4.52, respectively.

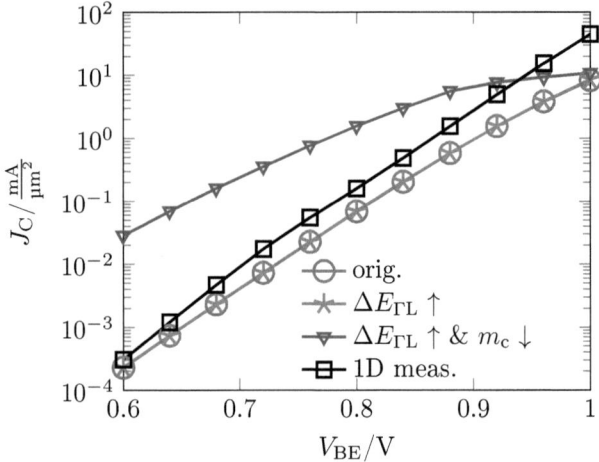

Figure 4.51: Impact of $\Delta E_{\Gamma\text{L}}$ and m_{c} on the transfer characteristics $J_{\text{C}}(V_{\text{BE}})$ ($T_{\text{L}}=300\,\text{K}$ and $V_{\text{BC}}=0\,\text{V}$).

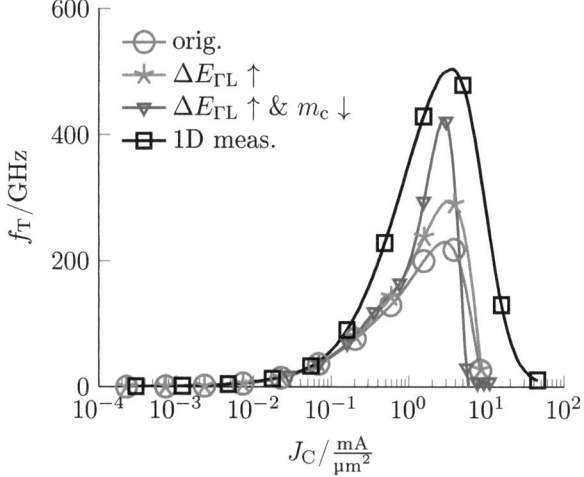

Figure 4.52: Impact of $\Delta E_{\Gamma L}$ & m_c on the transit frequency f_T (T_L=300 K and V_{BC}=0 V).

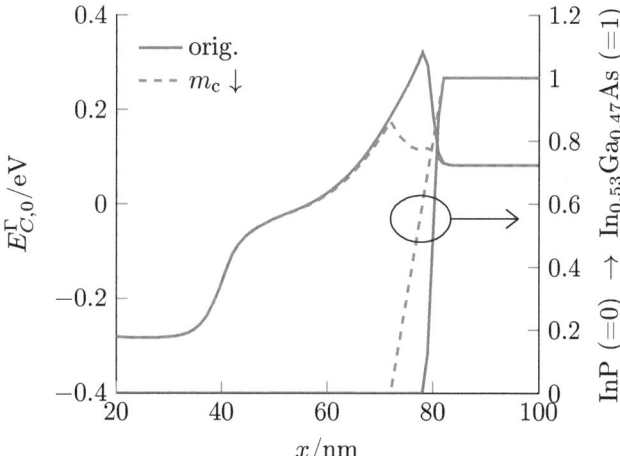

Figure 4.53: Conduction band edge of the Γ-valley (T_L=300 K, V_{BE}=0.6 V and V_{BC}=0 V).

For the results shown in figure 4.51 and 4.52, the valley separation $\Delta E_{\Gamma L}$ is increased by 14% within the InGaAs layer, which is caused by halving the parameter of ΔE_0^L responsible for its composition dependence (cf. table 4.2). The composition slope m_c is decreased from 0.25 $\frac{1}{nm}$ to 0.1 $\frac{1}{nm}$, where the first discretized INGAAS point (in the base region) is left unchanged (see figure 4.53). Despite of improvements in the simulated peak values of f_T w.r.t. to 1D meas. data, especially for decreasing m_c, variations of $\Delta E_{\Gamma L}$ and m_c are not capable of achieving a better agreement between the simulated and experimentally determined transfer characteristics (cf. figure 4.51). Contrary to f_T, a decreasing m_c leads to increasing discrepancies in J_C w.r.t. the 1D meas. data, because of the fundamentally different shape of the BE conduction band barrier, shown in figure 4.53. Therefore, an adjustment of $\Delta E_{\Gamma L}$ and/or m_c is not a constructive workaround for handling almost abrupt band discontinuities within the BTE.

In [54, 55] an interface condition is formulated for the BTE, which is specially designed for conduction band discontinuities at hetero-junctions. The basic idea behind the interface condition is to describe the electron transport at the discontinuity by means of thermionic transmission/reflection. High energetic electrons, which can overcome the discontinuity by thermionic transmission, are transferred, while the remaining electrons are reflected. In addition, the components for the k-vector in-plane of the interface are conserved, where the normal component varies during the interface passage in order to conserve the total energy. The interface condition is schematically illustrated by figure 4.54. Thus, the interface condition published in [54, 55] and described before ensures that electrons entering the INGAAS base from the INP emitter will not undergo an abrupt transition to the L-valley. Thus, an abrupt decrease of the average electron velocity, like shown in figure 4.50, will be suppressed. This will in turn decrease the accumulated electron transit time and therefore result in a higher f_T. Unfortunately, the interface condition could not be implemented and tested within the frame of this work due to time constraints.

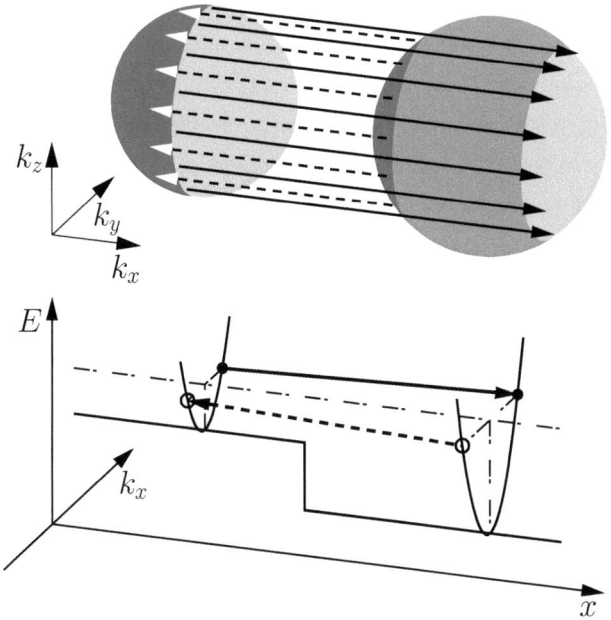

Figure 4.54: Schematic illustration of the interface condition for III-V hetero-junctions (e.g. InP→InGaAs interface). The spheres are sketching equi-energy surfaces of kinetic energy. Below, the total energy is sketched. Here, a conduction band edge (potential energy) with a discontinuity and parabolas, indicating the kinetic energy, are shown. As the interface condition conserves the in-plane k-vectors, only the variation of k_x is considered below. The solid/dashed arrows mark the passage of particles in positive/negative x-direction, respectively. Due to conservation of the in-plane k-vectors, only the light/dark shaded parts of the spheres participate to the particle transfer. Thus, reflection is assigned to the medium grey shaded part of the right sphere.

4.4 Fields of application

Deterministic solvers of the BTE, such as the one developed in this work, are currently out of scope of being suitable for the everyday industrial work. Due to their long simulations times and huge memory consumption, compared to DD/HD simulators, those simulation techniques loose attractiveness, although they are faster than MC and provide a more profound physical insight into the device under investigation compared to DD/HD. The disadvantage of the currently higher computational costs might get put into a new perspective with the ongoing development of faster CPUs and cheaper memory.

To combine both the physical accuracy of a deterministic BTE-solver and the smaller simulation times of DD/HD simulators, the BTE is used for the calibration and development of physical material models and for the adjustment of the HD transport model parameters.

In terms of physical material models, e.g. electron mobility or band-gap, less computationally expensive BULK simulations can be utilized. Based on those simulation results (see section 4.2.1 and 4.3.1), physical model equations for the considered material can either be calibrated or newly be developed. As an example, figure 4.55 compares the low field mobility of electrons for InP (electrons as majorities) obtained by BTE-BULK simulations and the calibrated Caughey-Thomas mobility model.

Regarding device simulations, the BTE results for advanced HBTs serve as reference for calibrating the HD transport model parameters f^{td}, f^{tc} and f^{ec} [73]. With earlier adjusted physical material models, the HD transport model parameters can be calibrated in order to match at least the terminal quantities obtained from the BTE. Since BTE simulations are needed to obtain the best suitable set of HD transport model parameters, the BTE is solved for exemplary profiles for each technology node only. Once a suitable set of parameters is found, HD instead of BTE simulations are employed within a technology node to predict the device performance. However, this strategy is only valid as long as no major profile changes are made. Thus, it is recommended to regularly revise and cross-check the HD transport model parameters, e.g. after profile optimizations, to preserve the predictive capability of the HD simulations. Figure 4.56 compares exemplary the transfer and f_T characteristic obtained by BTE (without Pauli principle) and HD (calibrated f^{td}, f^{tc} and f^{ec}) of the IHP SiGe HBT in section 4.2.2. Although good agreement regarding the transfer characteristic is obtained, f_T is underestimated by HD of about $80\,\mathrm{GHz}$ ($\approx 15\%$). This is a side-effect of the adjustment strategy depicted in [73] and stems from the demand for a positive (physically reasonable) Early-voltage. A better agreement between the f_Ts can be obtained at the cost of meaningful output characteristics.

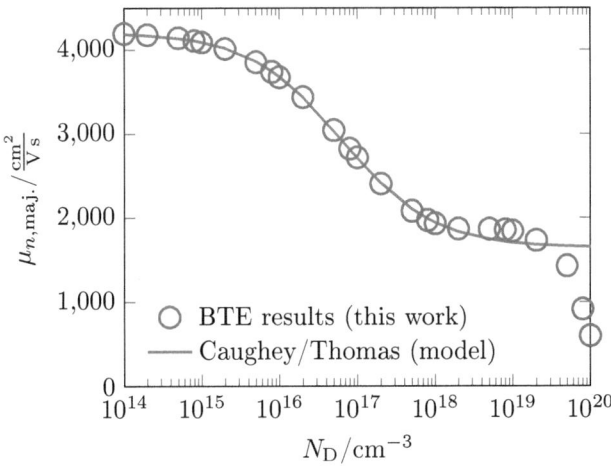

Figure 4.55: Electron majority low field mobility for InPat T_L=300 K: BTE-BULK results and calibrated Chaughey-Thomas model [68].

In addition to the aforementioned material and transport model calibration, the developed BTE solver, including the interface condition described at the end of section 4.3.2, is crucial for the simulation of III-V devices. Although not explicitly published in literature, the standard one-valley DD and HD simulators fail to simulate III-V devices. The issues related to III-V devices stem from the high-field mobility models (*transferred electron* or *negative differential mobility* (NDM) model), which imitate the Γ- to L-valley electron transition seen for the electron drift velocity characteristics (cf. section 4.3.1). The usage of these models results in convergence problems for DD and HD simulations but are, nevertheless, implemented in both commer-

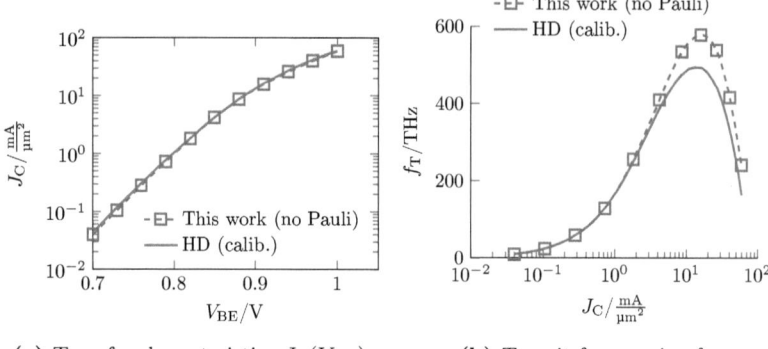

(a) Transfer characteristics $J_C(V_{BE})$.

(b) Transit frequencies f_T.

Figure 4.56: BTE and HD terminal characteristics for the IHP SiGeHBT from section 4.2.2.

cial [61, 65] and academic [32] simulators. The most often available model is the one of Barnes [3]. In order to indicate the problems arising from the NDM-models, an InP $n^+ - n - n^+$ structure, shown in figure 4.57(a), is investigated. The corresponding J–V characteristics (BTE, DD and HD) are illustrated in figure 4.57(b), where the left contact at x=0 µm is grounded and the bias is applied at the right contact (at x=1.5 µm). The DD/HD results

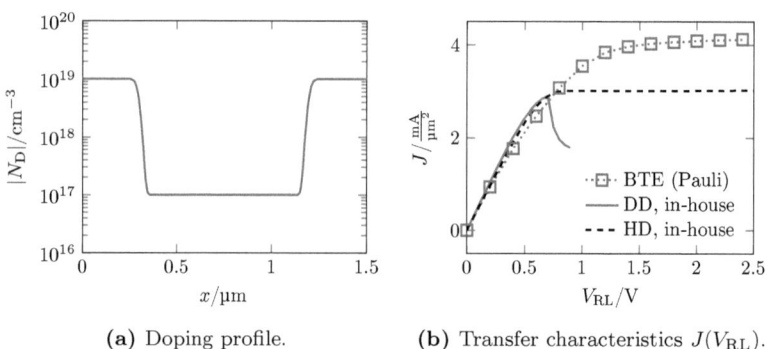

(a) Doping profile.

(b) Transfer characteristics $J(V_{RL})$.

Figure 4.57: Doping and transfer characteristic (BTE, DD and HD) of the InP $n^+ - n - n^+$ structure.

shown in figure 4.57(b) are obtained by the in-house simulator, called CHIEF, based on the work [72]. Focusing first on the DD results, in can be taken from figure 4.57(b) that no results beyond V_{RL}=0.9 V are shown, since the simulator failed to converge. In addition, a nonphysical drop of the current density J is seen starting at V_{RL}=0.7 V. This behavior is originating from the NDM-model. Figure 4.58 shows the effective DD field (gradient of the electron quasi-Fermi potentials φ_n), the NDM-model, electron density and the resulting electron velocities for the *converged* operating point V_{RL}=0.8 V. Although the DD system converged from the mathematical point of view,

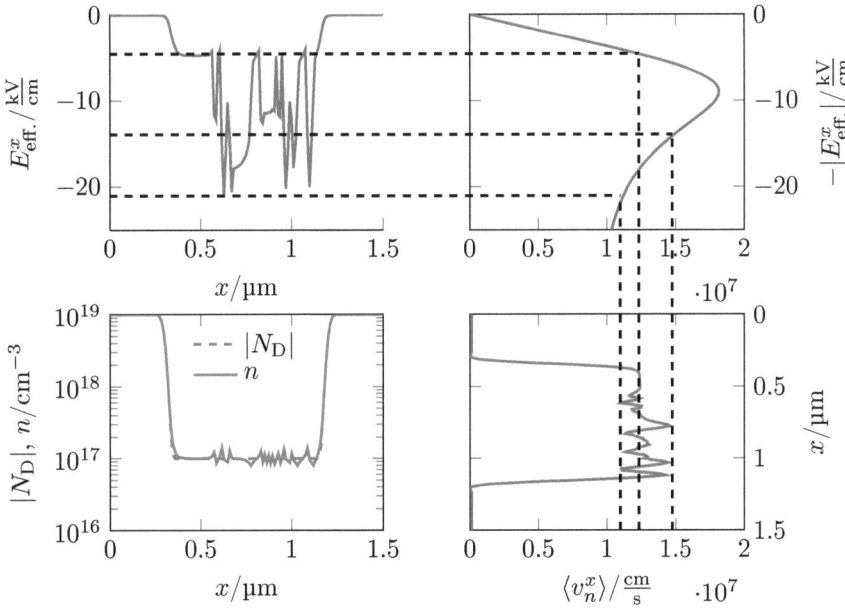

Figure 4.58: Effective DD field, NDM-model, electron density and electron velocities of the $n^+ - n - n^+$ structure for V_{RL}=0.8 V.

nonphysical oscillations in the effective field, electron density and thus in the electron velocities are seen within the low doped n-region.

Due to the continuity equation for electrons, the electron current density within $n^+ - n - n^+$ structure has to be spatially independent. Thus, the 1D DD system converges (besides the fulfillment of the *Poisson* equation and continuity equation of holes), if

$$J_n^x = q\,n\,\mu_n(|E_{\text{eff.}}^x|)E_{\text{eff.}}^x = q\,n\,\langle v_n^x(E_{\text{eff.}}^x)\rangle = \text{const.,} \qquad (4.4.0\text{-}29)$$

with $E_{\text{eff.}}^x = -\frac{\partial \varphi_n}{\partial x}$, is fulfilled. The highly doped n$^+$-regions are almost charge neutral, except for the transition regions to the n-region. Thus, $n^+ = |N_{\text{D}}|$ is valid over wide ranges in the n$^+$-regions. Therefore, the problem (4.4.0-29) can be reformulated for the n$^+$-regions to: Find the correct electron velocity for fulfilling the electron continuity equation. The charge neutrality of the n$^+$-regions leads to electron mobilities, which equal the doping dependent low field electron mobility $\mu_{n,G}$ (cf. figure 4.59) and the electron velocity can be approximated by

$$\langle v_n^{x,+}(E_{\text{eff.}}^{x,+})\rangle \approx \left.\frac{\partial v_n^{x,+}}{\partial E_{\text{eff.}}^{x,+}}\right|_{E_{\text{eff.}}^{x,+}=0\,\frac{\text{V}}{\text{cm}}} \qquad E_{\text{eff.}}^{x,+} \approx \mu_{n,G}^+ E_{\text{eff.}}^{x,+}, \qquad (4.4.0\text{-}30)$$

with the +-signs specifying the n$^+$-regions.

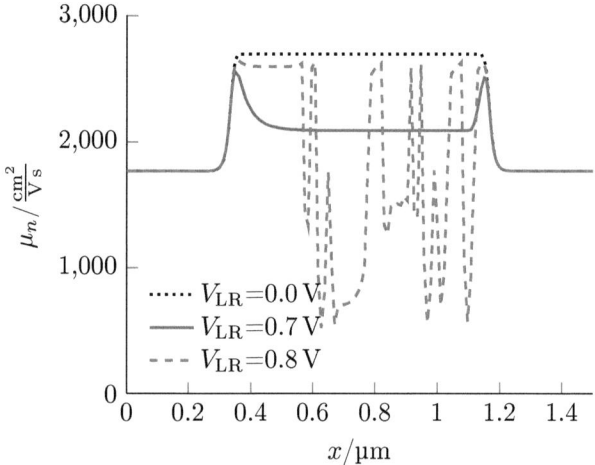

Figure 4.59: INP $n^+ - n - n^+$ structure: Electron mobilities μ_n at V_{RL}=0.0 V, 0.7 V and 0.8 V.

The low doped n-region is sufficiently wide for exhibiting a neutral region

$(n^- = |N_\mathrm{D}|, V_\mathrm{RL} \leq 0.7\,\mathrm{V})$. Likewise, the problem (4.4.0-29) for the n-region can also be reformulated to: Find the correct electron velocity for fulfilling the electron continuity equation. However, due to the lower doping concentration compared to the n^+-regions, the voltage mainly occurs across the n-region. Thus, the field dependent decrease of the electron mobility is seen within the n-region (cf. figure 4.59) and the electron velocity becomes

$$\langle v_n^{x,-}(E_\mathrm{eff.}^{x,-})\rangle \approx \mu_n^-(|E_\mathrm{eff.}^{x,-}|)E_\mathrm{eff.}^{x,-}\ , \tag{4.4.0-31}$$

with the $-$-signs specifying the n-region. In (4.4.0-31), $\mu_n(|E_\mathrm{eff.}^{x,-}|)$ is defined by the NDM-model. With (4.4.0-29) and (4.4.0-30), equation (4.4.0-31) is

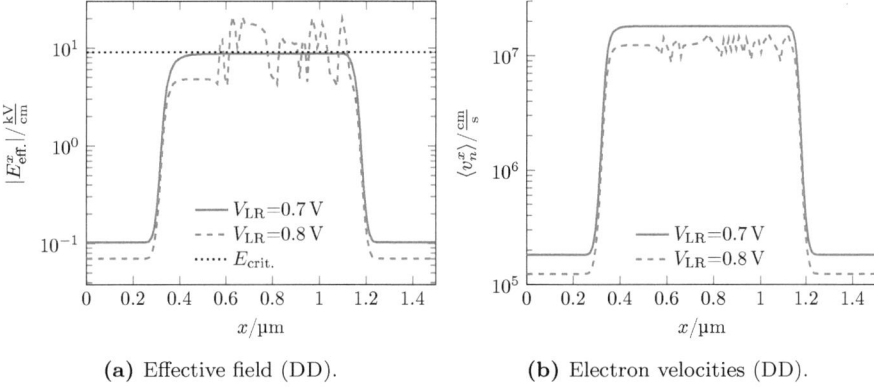

(a) Effective field (DD).   (b) Electron velocities (DD).

Figure 4.60: InP $n^+ - n - n^+$ structure: Effective field (DD) & electron velocities for V_RL=0.7 V and V_RL=0.8 V.

approximated by

$$\langle v_n^{x,-}(E_\mathrm{eff.}^{x,-})\rangle \approx \frac{n^+}{n^-}\mu_{n,G}^+ E_\mathrm{eff.}^{x,+}\ , \tag{4.4.0-32}$$

with $\frac{n^+}{n^-} \approx 100$ due to the doping profile (cf. figure 4.60(b)). Equation (4.4.0-32) is in accordance with the simulation results, as long as $\langle v_n^{x,-}(E_\mathrm{eff.}^{x,-})\rangle$ is increasing for increasing $E_\mathrm{eff.}^{x,-}$. This is the case up to V_RL=0.7 V, which is the last operating point for which the effective field within the n-region does not exceed the critical field (see figure 4.60(a)).

For higher applied voltages ($V_{RL} > 0.7\,$V), $\langle v_n^{x,-}(E_{\text{eff.}}^{x,-})\rangle$ decreases for increasing $E_{\text{eff.}}^{x,-}$ due to the NDM-model. Approximating the electron density by $n^- \approx |N_D|$ (see figure 4.58), the electron current density J has to decrease, too. With the continuity equation and (4.4.0-30), the effective fields $E_{\text{eff.}}^{x,+}$ within the n$^+$-regions are re-adjusted to lower values (cf. figure 4.60(a)) to provide the smaller current density. In combination with the Poisson-equation, the lower electron current density demands in turn a slightly higher conduction band barrier close to the n$^+$ $-$ n transition (x=0.3 µm, see figure 4.61). Consequently, for V_{RL}=0.8 V less electrons diffuse into the n-region

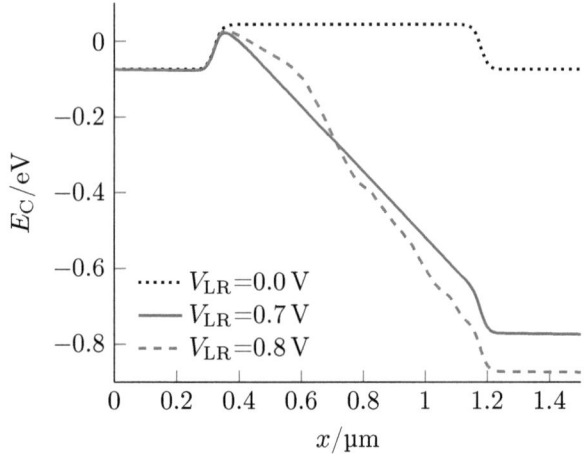

Figure 4.61: InP n$^+$ $-$ n $-$ n$^+$ structure: Conduction band edge (DD) at V_{RL}=0.0 V, 0.7 V and 0.8 V..

compared to V_{RL}=0.7 V enforcing a longer neutral n-region and a smaller effective field $E_{\text{eff.}}^{x,-}$ (here up to $x\approx$0.5 µm at V_{RL}=0.8 V, cf. figure 4.60(a) and 4.61). The current density within that particular range of the n-region dictates the current density for the rest of the n-region via (4.4.0-29). Another consequence is that the portion of the voltage drop over the remaining part of the n-region is increased compared to V_{RL}=0.7 V. Due to the NDM-model sketched in figure 4.58, two effective fields for one electron velocity are possible. Thus, during the solution of the DD-system, both the effective field $E_{\text{eff.}}^{x,-}$ and the electron density change abruptly to handle the increase in the proportional voltage drop and to conserve the current (see equation (4.4.0-29)), respectively. Therefore, oscillations are seen in both the effective field $E_{\text{eff.}}^{x,-}$ and the electron density, which lead in the end to convergence problems.

The issues described above do not occur for homogeneously doped homostructures. The effective field $E_{\text{eff.}}^{x}$ is spatially invariant in this case. Thus, all discretization points pass through the curve given by the NDM-model likewise.

However, for non-homogeneous devices the NDM-model exhibits different shapes due to the doping dependent low field mobility. Thus, the region of the device with an effective field beyond the critical field forces the adjacent regions to down-regulate the current density artificially (by the continuity equation) and leads to nonphysical oscillations/results.

In the case of HD simulations a similar problem might occur depending on the employed formulation of the effective field $E_{\text{eff.}}^{x}$. For example, the effective field obtained by the simplified energy balance equation under homogeneous BULK conditions [24, 73] reads

$$E_{\text{eff.}}^{x} = \sqrt{\frac{3}{2} \frac{k_b}{q} \frac{T_n - T_{\text{L}}}{\tau_{\omega,n} \mu_n(E_{\text{eff.}}^{x})}} \ , \tag{4.4.0-33}$$

with the energy relaxation time $\tau_{\omega,n}$ and the electron/lattice temperature T_n/T_{L}, respectively. Equation (4.4.0-33) can be reformulated to

$$E_{\text{eff.}}^{x} = \frac{3}{2} \frac{k_b}{q} \frac{T_n - T_{\text{L}}}{\tau_{\omega,n} \langle v_n^x(E_{\text{eff.}}^{x}) \rangle} \ . \tag{4.4.0-34}$$

Equation (4.4.0-34) gives implicitly the dependence of the $E_{\text{eff.}}^{x}$ as function of the electron temperature. Since the electron velocity modeled by the NDM-model is involved in (4.4.0-34), a unique solution might not be found, if $E_{\text{eff.}}^{x}$ is close to or higher than the critical field, similar to the DD-case. Thus oscillations are already occurring at the calculation of $E_{\text{eff.}}^{x}$, which degrades the convergence behavior of the HD system. In CHIEF and for HD-NDM-model only, this problem is circumvented by using the electron saturation drift velocity $v_{n,\text{sat.}}$ in (4.4.0-34), which tremendously improves the convergence behavior at the expense of physical basis.

Thus, the standard DD and HD simulators in combination with the NDM-model are not suitable for simulating III-V devices, since the Γ- to L-valley electron transition is modeled by a numerically unsuitable electron high-field mobility (velocity) equation. The intention of these models is to imitate the result of the valley transfer but keeping all electrons within one valley. A possible way to circumvent this issue is the extension of DD and HD by multiple valleys. In this context, the developed BTE solver is able to support both the development of the material and transport models involved. The commercial simulator SDEVICE [65] offers the possibility to artificially enlarge the velocity peak. Thus, a velocity plateau is forming which enhances the convergential behavior. Nevertheless, the shape of the NDM-model turns qualitatively into the well-known SI velocity-field characteristics and thus does not resemble the curves seen for III-V material anymore.

CHAPTER 5

Summary and Outlook

In this work, the deterministic solution method for the Boltzmann transport equation based on the work [19,20] has been discussed and extended. The necessary fundamentals are briefly reviewed and a detailed view on the numerical treatment is given before simulation results of semiconductor devices suitable for high-frequency applications are presented. The mathematical basis of the pioneering work [19,20] is reviewed and explained in detail. In addition, most limitations of the original method are removed.

For example, with demand of charge neutrality for BULK simulations and by applying an inflow/outflow boundary condition to device simulations, stationary solutions of the BTE can be directly obtained.

Furthermore, by application of the Herring-Vogt transformation, the solver developed in this work is capable to consider aniostropic valleys.

Compared to [19,20], with a slightly reformulated solution variable and a few additional terms, the method is extended for spatially dependent dispersion relations.

In addition, the non-linear collision term, due to the Pauli principle, is taken into account.

Also, the original constraint of an equi-distant real-space discretization has been removed.

In [19,20] Dirac δ-functions were used in order to cope with the anisotropic scattering mechanisms for III-V materials, like POP or PZ scattering. However, in this work the anisotropic scattering mechanisms are as far as possible handled by analytic equations, which do not violate the particle conservation and do not exhibit the necessity of a different set of basis functions depending on the material under investigation.

Simulation results for two most prominent material combinations, SI-SIGE and INP-INGAAS are given. For both material combinations, BULK simulation results are compared with already existing data from literature. In terms of SI-SIGE good agreement with published results is obtained. Motivated by the good agreement obtained, a brief investigation regarding the influence of the Pauli principle on the doping and GE-dependent band gap is given. In terms of device simulations, the simulation results for a current cutting-edge technology from IHP and for a SIGE HBT at the presently known physical limit are presented. Those results are compared with the ones obtained by the deterministic BTE solver (SHE) based on the work in [31]. Despite of some small deviations due to the material models involved, here also good agreement between the two simulators is obtained verifying the correctness of the developed solver. Regarding INP-INGAAS BULK simulations, discrepancies between simulation and published results are shown. In most cases, the discrepancies could be explained by the usage of different model parameters, e.g. for the band structure or dispersion relation. As far as published, results with parameters in accordance to literature are shown or at least corrective actions are described. However, in terms of HBT simulations, major deviations between simulation and measured 1D results are observed. Further investigations are pointing to the deficiency of the BTE of handling almost abrupt hetero-junctions accurately. A possible corrective implementation of an interface boundary condition is described, but could not be pursued in this work due to time constraints.

The work so far is completed by listing the possible areas of application for the developed BTE solver. Here, the focus is put on assisting the every-day engineering, work as well as research and development applications. However, at least for III-V materials, which exhibit a pronounced intervalley electron transfer seen in the velocity-field characteristics, the mis-modeling of such effects by commercial and academic DD and HD simulators and the resulting bad convergence behavior is described.

Compared to the SHE solver [31], the solver developed in this work currently exhibits simulations times, which are about a factor of 10 larger than the ones of SHE. Beside the non-optimized prototype status of the developed solver, which therefore offers room for optimization, a non-equidistant discretization of the reciprocal space appears to reduce the simulation time drastically. A rough estimation indicates the speed-up potential. Assuming that the time needed for the sparse matrix solver grows quadratically with the dimension of the matrix, a reduction of the points needed for the reciprocal space by a factor of three up to four is needed. From the author's experience, such reductions, especially in terms of the kinetic energy, seem to be feasible. Figure 5.1 exemplary illustrated the percentage of the total electron density over kinetic energy for the "limits" HBT (see section 4.2.2) at $f_{T,peak}$ for $V_{BC}=0$ V. It can be taken from figure 5.1 that for most regions 95% of the

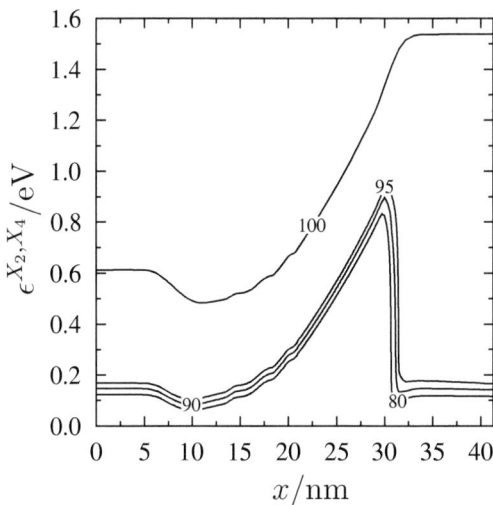

Figure 5.1: Percentage of the electron density over kinetic energy for the "limits" HBT at $f_{T,peak}$ for $V_{BC}=0$ V, Pauli principle considered.

total electron density is located at kinetic energies below 200 meV. Only in the collector region, where the high energetic tail of the distribution function is located (cf. figure 4.31), the contour line for 95% exhibits values around 900 meV. However, in that region a plateau is forming (negligible gradient of the distribution function w.r.t. the kinetic energy – figure 4.31), which can also be captured by a coarser energy discretization. Thus, by employing a fine energy discretization for energies below 200 meV followed by a continuous coarsening of the energy discretization, a reduction of the grid points by a factor of at least 3 seems to be plausible.

Once the non-equidistant discretization of the reciporcal space is implemented, the interface boundary condition for INP-INGAAS should be considered, since its actual implementation and evaluation strongly depends on the discretization of the kinetic energy. In order to reduce the work load (analytic preprocessing and revision of existing implementations) and to provide more time efficient simulations, it is recommended to consider this only after extensions regarding the energy discretization have been conducted.

Another point, which should be mentioned, is the rare event of oscillations by the WENO method. These oscillations disturb the convergence behavior and thus lead to increasing simulation times. Some counteractions are already given in literature, e.g. WENO-M [29] or WENO-Z [7] method, and were also tested during the course of this work. Although the WENO-Z method seems to be the more robust method, a complete avoidance of oscillations is not observed. However, the mentioned oscillations are sometimes seen for fine step sizes of the energy discretization and at high kinetic energies, where values of the solution variable are in the range of the relative floating-point accuracy ($\approx 2 \times 10^{-16}$). Since these oscillations disappeared by coarsening the grid, it is supposed that their occurrence will be suppressed by employing a non-uniform energy discretization. Since the WENO method is, compared to SHE, a relatively new method, research activities w.r.t. its improvement are still ongoing. Therefore, the literature should be regularly checked about the latest improvements of the WENO and related methods, like the WLS-ENO method [44].

Although the method depicted in this work is only applied to materials, which exhibit a 3D reciprocal space, it is also possible to consider 1D materials, like CNTs. For example, the work [1] shows simulation results of a deterministic BTE-solver for metallic CNTs, where in [28] results for semi-conducting CNTs are shown. Figure 5.2 shows exemplary the distribution functions for an infinitely long semiconducting CNT (BULK) with two considered sub-valleys. Here, the results obtained by both the MC-method and a prototype solver, derived from the scheme depicted in this work, are shown. Thus, the method presented in this work is not restricted to materials with

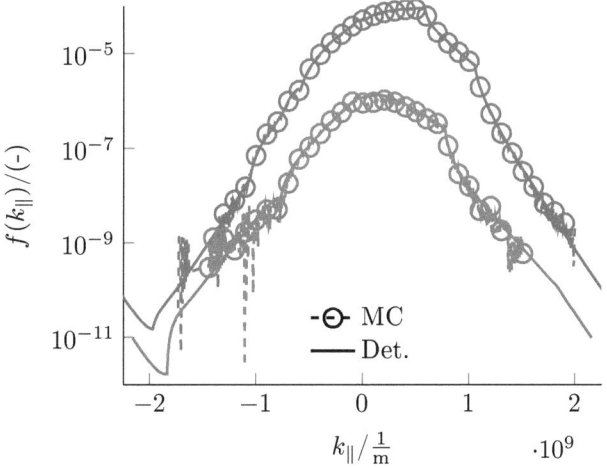

Figure 5.2: Electron distribution function of a semiconducting CNT for two considered sub-valleys. Since a pure 1D transport is considered, $k_\parallel$ states the reciprocal vectors along the CNT.

3D reciprocal space, but can be also applied, after some revisions, to 1D and 2D materials like CNTs and Graphene, respectively.

APPENDIX A

Supplementary material for chapter 2

A.1 A detailed view on the Herring-Vogt transformation

As mentioned in section 2.2, the Herring-Vogt transformation is responsible of mapping ellipsoidal equi-energy surfaces onto spheres. Following the example of (2.2.0-45), an ellipsoid with the semi-axes $\sqrt{m_x^\nu}$, $\sqrt{m_y^\nu}$ and $\sqrt{m_z^\nu}$ in the $\vec{k}^\nu$-space is considered. Thus, in order to obtain a sphere, each axis in $\vec{k}^\nu$-space is stretched or compressed, so that the ellipsoid becomes a sphere. The stretched/compressed $\vec{k}^\nu$-space forms the $\vec{\tilde{k}}^\nu$-space. The scaling of the axes is schematically shown in figure A.1 and is mathematically described by the Herring-Vogt transformation matrix (2.2.0-46) (see section 2.2)

$$
\vec{\tilde{k}}^\nu \;=\; \underbrace{\begin{pmatrix} \sqrt{\dfrac{m_\nu^*}{m_x^\nu}} & 0 & 0 \\[2mm] 0 & \sqrt{\dfrac{m_\nu^*}{m_y^\nu}} & 0 \\[2mm] 0 & 0 & \sqrt{\dfrac{m_\nu^*}{m_z^\nu}} \end{pmatrix}}_{T^{\mathrm{HV},\nu}} \vec{k}^\nu \;, \qquad \text{(A.1.0-1)}
$$

with the anisotropic effective masses $m^\nu_{x,y,z}$ of a valley ν and resulting isotropic effective mass m^*_ν after the transformation. In vector notation the component-

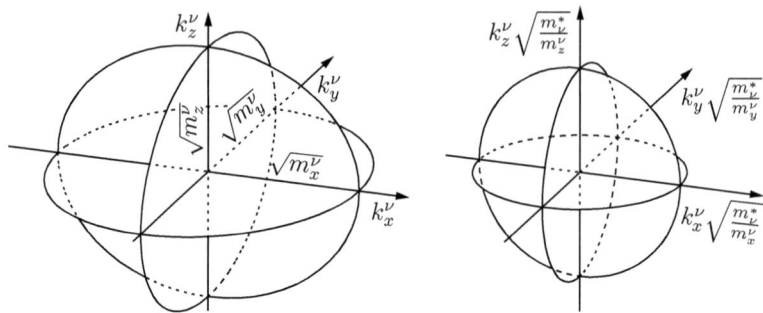

Figure A.1: Schematic illustration of axis scaling due to the Herring-Vogt transformation.

wise scaling of $\vec{k}^\nu$ (after (2.2.0-46) or (A.1.0-1)) becomes more clear:

$$\begin{pmatrix} \tilde{k}^\nu_x \\ \tilde{k}^\nu_y \\ \tilde{k}^\nu_z \end{pmatrix} = \begin{pmatrix} \sqrt{\frac{m^*_\nu}{m^\nu_x}}k^\nu_x \\ \sqrt{\frac{m^*_\nu}{m^\nu_y}}k^\nu_y \\ \sqrt{\frac{m^*_\nu}{m^\nu_z}}k^\nu_z \end{pmatrix} = \begin{pmatrix} T^{\mathrm{HV},\nu}_{1,1}k^\nu_x \\ T^{\mathrm{HV},\nu}_{2,2}k^\nu_y \\ T^{\mathrm{HV},\nu}_{3,3}k^\nu_z \end{pmatrix} , \begin{pmatrix} k^\nu_x \\ k^\nu_y \\ k^\nu_z \end{pmatrix} = \begin{pmatrix} 1/T^{\mathrm{HV},\nu}_{1,1}\tilde{k}^\nu_x \\ 1/T^{\mathrm{HV},\nu}_{2,2}\tilde{k}^\nu_y \\ 1/T^{\mathrm{HV},\nu}_{3,3}\tilde{k}^\nu_z \end{pmatrix} ,\text{(A.1.0-2)}$$

Since the transformation does not affect the unit vectors

$$\vec{e}_{\tilde{k}^\nu_d} = \frac{\frac{\partial \vec{k}^\nu}{\partial k^\nu_d}}{h_{\tilde{k}^\nu_d}} \text{ , with } d = \{x,y,z\} \text{ and } h_{\tilde{k}^\nu_d} = \left|\frac{\partial \vec{k}^\nu}{\partial \tilde{k}^\nu_d}\right| \text{ , } h_{\tilde{k}^\nu_x} = \frac{1}{T^{\mathrm{HV},\nu}_{1,1}} \text{ , } h_{\tilde{k}^\nu_y} = \frac{1}{T^{\mathrm{HV},\nu}_{2,2}} \text{ , }$$

$$h_{\tilde{k}^\nu_z} = \frac{1}{T^{\mathrm{HV},\nu}_{3,3}} \text{ , } \vec{e}_{\tilde{k}^\nu_x} = \begin{pmatrix} 1 \\ 0 \\ 0 \end{pmatrix} = \vec{e}_{k^\nu_x} + 0\vec{e}_{k^\nu_y} + 0\vec{e}_{k^\nu_z} = \vec{e}_{k^\nu_x} \text{ , }$$

$$\vec{e}_{\tilde{k}^\nu_y} = \begin{pmatrix} 0 \\ 1 \\ 0 \end{pmatrix} = 0\vec{e}_{k^\nu_x} + \vec{e}_{k^\nu_y} + 0\vec{e}_{k^\nu_z} = \vec{e}_{k^\nu_y} \text{ , } \vec{e}_{\tilde{k}^\nu_z} = \begin{pmatrix} 0 \\ 0 \\ 1 \end{pmatrix} = 0\vec{e}_{k^\nu_x} + 0\vec{e}_{k^\nu_y} + \vec{e}_{k^\nu_z} = \vec{e}_{k^\nu_z} \text{ , }$$

$$\text{(A.1.0-3)}$$

the transformation matrices for vector fields $\vec{F}$ become identity matrices

$$
\begin{pmatrix} F_{\tilde{k}^\nu_x}(\vec{\tilde{k}}^\nu) \\ F_{\tilde{k}^\nu_z}(\vec{\tilde{k}}^\nu) \\ F_{\tilde{k}^\nu_z}(\vec{\tilde{k}}^\nu) \end{pmatrix} = \begin{pmatrix} \vec{e}^{\mathsf{T}}_{\tilde{k}^\nu_x} \\ \vec{e}^{\mathsf{T}}_{\tilde{k}^\nu_y} \\ \vec{e}^{\mathsf{T}}_{\tilde{k}^\nu_z} \end{pmatrix} \cdot \begin{pmatrix} F_{k^\nu_x}(\vec{\tilde{k}}^\nu) \\ F_{k^\nu_z}(\vec{\tilde{k}}^\nu) \\ F_{k^\nu_z}(\vec{\tilde{k}}^\nu) \end{pmatrix}
$$

$$
= \begin{pmatrix} 1 & 0 & 0 \\ 0 & 1 & 0 \\ 0 & 0 & 1 \end{pmatrix} \cdot \begin{pmatrix} F_{k^\nu_x}(\vec{\tilde{k}}^\nu) \\ F_{k^\nu_z}(\vec{\tilde{k}}^\nu) \\ F_{k^\nu_z}(\vec{\tilde{k}}^\nu) \end{pmatrix} = \begin{pmatrix} F_{k^\nu_x}(\vec{\tilde{k}}^\nu) \\ F_{k^\nu_z}(\vec{\tilde{k}}^\nu) \\ F_{k^\nu_z}(\vec{\tilde{k}}^\nu) \end{pmatrix} \quad \text{(A.1.0-4)}
$$

for the transformation of a vector field in $\vec{k}^\nu$-space to the $\vec{\tilde{k}}^\nu$-space and

$$
\begin{pmatrix} F_{k_x}(\vec{k}^\nu) \\ F_{k_z}(\vec{k}^\nu) \\ F_{k_z}(\vec{k}^\nu) \end{pmatrix} = \begin{pmatrix} \vec{e}_{\tilde{k}^\nu_x} & \vec{e}_{\tilde{k}^\nu_y} & \vec{e}_{\tilde{k}^\nu_z} \end{pmatrix} \cdot \begin{pmatrix} F_{\tilde{k}_x}(\vec{k}^\nu) \\ F_{\tilde{k}_z}(\vec{k}^\nu) \\ F_{\tilde{k}_z}(\vec{k}^\nu) \end{pmatrix}
$$

$$
= \begin{pmatrix} 1 & 0 & 0 \\ 0 & 1 & 0 \\ 0 & 0 & 1 \end{pmatrix} \cdot \begin{pmatrix} F_{\tilde{k}_x}(\vec{k}^\nu) \\ F_{\tilde{k}_z}(\vec{k}^\nu) \\ F_{\tilde{k}_z}(\vec{k}^\nu) \end{pmatrix} = \begin{pmatrix} F_{\tilde{k}_x}(\vec{k}^\nu) \\ F_{\tilde{k}_z}(\vec{k}^\nu) \\ F_{\tilde{k}_z}(\vec{k}^\nu) \end{pmatrix} \quad , \quad \text{(A.1.0-5)}
$$

vice versa. Thus, the only thing to be done is to express the components of a vector field $\vec{F}$ in terms of the k-vectors of that space, where the field has to be transformed to. This also holds of scalar functions. Since the $\tilde{k}^\nu$ coordinate system is orthogonal ($\vec{e}_{\tilde{k}^\nu_x} \cdot \vec{e}_{\tilde{k}^\nu_y} = \vec{e}_{\tilde{k}^\nu_x} \cdot \vec{e}_{\tilde{k}^\nu_z} = \vec{e}_{\tilde{k}^\nu_y} \cdot \vec{e}_{\tilde{k}^\nu_z} = 0$) the gradient and divergence of a vector field $\vec{F}$ in the $\vec{k}^\nu$-space can be transformed by [8]

$$
\text{grad}_{\tilde{k}^\nu}\left(F(\vec{k}^\nu)\right) = \frac{1}{h_{\tilde{k}^\nu_x}}\frac{\partial F(\vec{\tilde{k}}^\nu)}{\partial \tilde{k}^\nu_x}\vec{e}_{\tilde{k}^\nu_x} + \frac{1}{h_{\tilde{k}^\nu_y}}\frac{\partial F(\vec{\tilde{k}}^\nu)}{\partial \tilde{k}^\nu_y}\vec{e}_{\tilde{k}^\nu_y} + \frac{1}{h_{\tilde{k}^\nu_z}}\frac{\partial F(\vec{\tilde{k}}^\nu)}{\partial \tilde{k}^\nu_z}\vec{e}_{\tilde{k}^\nu_z}
$$

$$
= T^{\text{HV},\nu}_{1,1}\frac{\partial F(\vec{\tilde{k}}^\nu)}{\partial \tilde{k}^\nu_x}\vec{e}_{\tilde{k}^\nu_x} + T^{\text{HV},\nu}_{2,2}\frac{\partial F(\vec{\tilde{k}}^\nu)}{\partial \tilde{k}^\nu_y}\vec{e}_{\tilde{k}^\nu_y} + T^{\text{HV},\nu}_{3,3}\frac{\partial F(\vec{\tilde{k}}^\nu)}{\partial \tilde{k}^\nu_z}\vec{e}_{\tilde{k}^\nu_z} \quad \text{(A.1.0-6)}
$$

and

$$
D = h_{\tilde{k}^\nu_x}h_{\tilde{k}^\nu_y}h_{\tilde{k}^\nu_z} = \frac{1}{T^{\text{HV},\nu}_{1,1}}\frac{1}{T^{\text{HV},\nu}_{2,2}}\frac{1}{T^{\text{HV},\nu}_{3,3}} \quad \text{(A.1.0-7)}
$$

$$
\operatorname{div}_{\tilde{k}^{\nu}}\left(\vec{F}(\vec{k}^{\nu})\right) = \frac{1}{D}\left[\frac{\partial}{\partial \tilde{k}_x^{\nu}}\left(h_{\tilde{k}_y^{\nu}} h_{\tilde{k}_z^{\nu}} F_{k_x^{\nu}}(\vec{k}^{\nu})\right) + \frac{\partial}{\partial \tilde{k}_y^{\nu}}\left(h_{\tilde{k}_x^{\nu}} h_{\tilde{k}_z^{\nu}} F_{k_y^{\nu}}(\vec{k}^{\nu})\right)\right.
$$

$$
\left. + \frac{\partial}{\partial \tilde{k}_z^{\nu}}\left(h_{\tilde{k}_x^{\nu}} h_{\tilde{k}_y^{\nu}} F_{k_z^{\nu}}(\vec{k}^{\nu})\right)\right]
$$

$$
= \left[\frac{\partial}{\partial \tilde{k}_x^{\nu}}\left(T_{1,1}^{\mathrm{HV},\nu} F_{k_x^{\nu}}(\vec{k}^{\nu})\right) + \frac{\partial}{\partial \tilde{k}_y^{\nu}}\left(T_{2,2}^{\mathrm{HV},\nu} F_{k_y^{\nu}}(\vec{k}^{\nu})\right)\right.
$$

$$
\left. + \frac{\partial}{\partial \tilde{k}_z^{\nu}}\left(T_{3,3}^{\mathrm{HV},\nu} F_{k_z^{\nu}}(\vec{k}^{\nu})\right)\right], \tag{A.1.0-8}
$$

where for (A.1.0-8) the result of (A.1.0-4) has already been employed. Note, the elements of the Herring-Vogt transformation matrix ((A.1.0-1) or (2.2.0-46)) are independent of $\vec{k}^{\nu}$ and thus $\vec{\tilde{k}}^{\nu}$, which allows to treat them as constants for the transformation of the gradient and divergence. Because of that and due to identical unit vectors in (see (A.1.0-3)) $\vec{k}^{\nu}$- and $\vec{\tilde{k}}^{\nu}$-space, the transformation of gradient and divergence to the $\vec{\tilde{k}}^{\nu}$-space can also be obtained in a simple manner. Due to the identical unit vectors, each component of a vector field in one space can be treated as a scalar function F in the other space and a conversion between them can be achieved by using the relations given by (A.1.0-2). Thus, by obeying the chain-rule of differentiation, one obtains for example

$$
\frac{\partial F(\tilde{k}_x^{\nu}, \tilde{k}_y^{\nu}, \tilde{k}_z^{\nu})}{\partial k_x^{\nu}} = \frac{\partial F(\tilde{k}_x^{\nu}, \tilde{k}_y^{\nu}, \tilde{k}_z^{\nu})}{\partial \tilde{k}_x^{\nu}}\frac{\partial \tilde{k}_x^{\nu}}{\partial k_x^{\nu}} = T_{1,1}^{HV,\nu}\frac{\partial F(\tilde{k}_x^{\nu}, \tilde{k}_y^{\nu}, \tilde{k}_z^{\nu})}{\partial \tilde{k}_x^{\nu}}. \tag{A.1.0-9}
$$

The inverse Jacobian of the Herring-Vogt transformation equals the inverse Herring-Vogt transformation, due to the constant matrix elements:

$$
J_{\mathrm{HV}}^{-1} = \begin{pmatrix} \frac{\partial k_x^{\nu}}{\partial \tilde{k}_x^{\nu}} & \frac{\partial k_x^{\nu}}{\partial \tilde{k}_y^{\nu}} & \frac{\partial k_x^{\nu}}{\partial \tilde{k}_z^{\nu}} \\ \frac{\partial k_y^{\nu}}{\partial \tilde{k}_x^{\nu}} & \frac{\partial k_y^{\nu}}{\partial \tilde{k}_y^{\nu}} & \frac{\partial k_y^{\nu}}{\partial \tilde{k}_z^{\nu}} \\ \frac{\partial k_z^{\nu}}{\partial \tilde{k}_x^{\nu}} & \frac{\partial k_z^{\nu}}{\partial \tilde{k}_y^{\nu}} & \frac{\partial k_z^{\nu}}{\partial \tilde{k}_z^{\nu}} \end{pmatrix} = \begin{pmatrix} \frac{1}{T_{1,1}^{\mathrm{HV},\nu}} & 0 & 0 \\ 0 & \frac{1}{T_{2,2}^{\mathrm{HV},\nu}} & 0 \\ 0 & 0 & \frac{1}{T_{3,3}^{\mathrm{HV},\nu}} \end{pmatrix} = \left\{T^{\mathrm{HV},\nu}\right\}^{-1}.
$$

$$
\tag{A.1.0-10}
$$

Thus the infinitesimal volume element in $\vec{k}^\nu$-space transforms to

$$
\begin{aligned}
dk_x^\nu \, dk_y^\nu \, dk_z^\nu \;\; &= \;\; \left| J_{\mathrm{HV}}^{-1} \right| = \frac{1}{T_{1,1}^{\mathrm{HV},\nu}} \frac{1}{T_{2,2}^{\mathrm{HV},\nu}} \frac{1}{T_{3,3}^{\mathrm{HV},\nu}} d\tilde{k}_x^\nu \, d\tilde{k}_y^\nu \, d\tilde{k}_z^\nu \\
&= \;\; D \, d\tilde{k}_x^\nu \, d\tilde{k}_y^\nu \, d\tilde{k}_z^\nu \; ,
\end{aligned}
\tag{A.1.0-11}
$$

with D after (A.1.0-7). In (A.1.0-11) the scaling of the $\vec{k}^\nu$-space is observed again, if an integration over the $\vec{k}^\nu$-space is performed. In section 2.2, equation (2.2.0-48), D became one due to the demand of an infinitesimal volume element, being invariant w.r.t. the transformation/considered space. Therefore, the isotropic effective mass became

$$
m_\nu^* = \sqrt[3]{m_x^\nu m_y^\nu m_z^\nu} \; .
\tag{A.1.0-12}
$$

Another advantage, besides the identical infinitesimal volume element, is that due to (A.1.0-12) the determinant of the Herring-Vogt matrix and its inverse equals one:

$$
\left| T^{\mathrm{HV},\nu} \right| = \left| \left\{ T^{\mathrm{HV},\nu} \right\}^{-1} \right| = 1 \; ,
\tag{A.1.0-13}
$$

and that the Herring-Vogt matrix is invariant under transposition

$$
\left\{ T^{\mathrm{HV},\nu} \right\}^{\mathsf{T}} = T^{\mathrm{HV},\nu} \; .
\tag{A.1.0-14}
$$

A.2 Conservative form of the BTE

This section describes the steps of deriving the conservative form of the BTE after (2.4.0-110) without any loss of generality. Starting point is the BTE given by (2.4.0-107)

$$
\begin{aligned}
&\frac{\partial f^\nu(\vec{r}, \vec{k}^\nu, t)}{\partial t} + \vec{v}_{\mathrm{g}}^\nu(\vec{r}, \vec{k}^\nu) \mathrm{grad}_{\vec{r}} \left(f^\nu(\vec{r}, \vec{k}^\nu, t) \right) \\
&- \frac{q}{\hbar} \vec{E}_{\mathrm{eff.}}^\nu(\vec{r}, \vec{k}^\nu, t) \mathrm{grad}_{\vec{k}^\nu} \left(f^\nu(\vec{r}, \vec{k}^\nu, t) \right) = C(\vec{r}, f^\nu(\vec{r}, \vec{k}^\nu, t), f^{\nu'}(\vec{r}, \vec{k}^{\nu'}, t)) \; .
\end{aligned}
\tag{A.2.0-15}
$$

Using the product rule, the diffusion term can be rearranged to

$$\vec{v}_{\mathrm{g}}^{\nu}(\vec{r},\vec{k}^{\nu})\mathrm{grad}_{\vec{r}}\left(f^{\nu}(\vec{r},\vec{k}^{\nu},t)\right) = \mathrm{div}_{\vec{r}}\left(\vec{v}_{\mathrm{g}}^{\nu}(\vec{r},\vec{k}^{\nu})f^{\nu}(\vec{r},\vec{k}^{\nu},t)\right)$$
$$- f^{\nu}(\vec{r},\vec{k}^{\nu},t)\mathrm{div}_{\vec{r}}\left(\vec{v}_{\mathrm{g}}^{\nu}(\vec{r},\vec{k}^{\nu})\right) ,$$
(A.2.0-16)

and likewise the drift term

$$-\frac{q}{\hbar}\vec{E}_{\mathrm{eff.}}^{\nu}(\vec{r},\vec{k}^{\nu},t)\mathrm{grad}_{\vec{k}^{\nu}}\left(f^{\nu}(\vec{r},\vec{k}^{\nu},t)\right) =$$
$$-\frac{q}{\hbar}\left(\mathrm{div}_{\vec{k}^{\nu}}\left(\vec{E}_{\mathrm{eff.}}^{\nu}(\vec{r},\vec{k}^{\nu},t)f^{\nu}(\vec{r},\vec{k}^{\nu},t)\right) - f^{\nu}(\vec{r},\vec{k}^{\nu},t)\mathrm{div}_{\vec{k}^{\nu}}\left(\vec{E}_{\mathrm{eff.}}^{\nu}(\vec{r},\vec{k}^{\nu},t)\right)\right) .$$
(A.2.0-17)

With the driving force for electrons after (2.4.0-108), the second term on the right-hand side of (A.2.0-17) is rewritten to

$$\frac{q}{\hbar}f^{\nu}(\vec{r},\vec{k}^{\nu},t)\mathrm{div}_{\vec{k}^{\nu}}\left(\vec{E}_{\mathrm{eff.}}^{\nu}(\vec{r},\vec{k}^{\nu},t)\right)$$
$$= \frac{q}{\hbar}\left(f^{\nu}(\vec{r},\vec{k}^{\nu},t)\mathrm{div}_{\vec{k}^{\nu}}\left(\mathrm{grad}_{\vec{r}}\left(-\psi(\vec{r},t)+\frac{E_{\mathrm{C,0}}^{\nu}(\vec{r})}{q}+\frac{\varepsilon^{\nu}(\vec{r},\vec{k}^{\nu})}{q}\right)\right)\right)$$
$$= \frac{q}{\hbar}\left(f^{\nu}(\vec{r},\vec{k}^{\nu},t)\mathrm{div}_{\vec{r}}\left(\mathrm{grad}_{\vec{k}^{\nu}}\left(-\psi(\vec{r},t)+\frac{E_{\mathrm{C,0}}^{\nu}(\vec{r})}{q}+\frac{\varepsilon^{\nu}(\vec{r},\vec{k}^{\nu})}{q}\right)\right)\right)$$
$$= \frac{q}{\hbar}\left(f^{\nu}(\vec{r},\vec{k}^{\nu},t)\mathrm{div}_{\vec{r}}\left(\mathrm{grad}_{\vec{k}^{\nu}}\left(\frac{\varepsilon^{\nu}(\vec{r},\vec{k}^{\nu})}{q}\right)\right)\right) ,$$
(A.2.0-18)

which evaluates with the definition of the group velocity after (2.4.0-109) to

$$\frac{q}{\hbar}f^{\nu}(\vec{r},\vec{k}^{\nu},t)\mathrm{div}_{\vec{k}^{\nu}}\left(\vec{E}_{\mathrm{eff.}}^{\nu}(\vec{r},\vec{k}^{\nu},t)\right) = f^{\nu}(\vec{r},\vec{k}^{\nu},t)\mathrm{div}_{\vec{r}}\left(\vec{v}_{\mathrm{g}}^{\nu}(\vec{r},\vec{k}^{\nu})\right) . \quad \text{(A.2.0-19)}$$

Thus, adding the diffusion term (A.2.0-16) and the drift term (A.2.0-17) by using the result of (A.2.0-19) gives

$$\vec{v}_{\mathrm{g}}^{\nu}(\vec{r},\vec{k}^{\nu})\mathrm{grad}_{\vec{r}}\left(f^{\nu}(\vec{r},\vec{k}^{\nu},t)\right) - \frac{q}{\hbar}\vec{E}_{\mathrm{eff.}}^{\nu}(\vec{r},\vec{k}^{\nu},t)\mathrm{grad}_{\vec{k}^{\nu}}\left(f^{\nu}(\vec{r},\vec{k}^{\nu},t)\right)$$

$$= \mathrm{div}_{\vec{r}}\left(\vec{v}_{\mathrm{g}}^{\nu}(\vec{r},\vec{k}^{\nu})f^{\nu}(\vec{r},\vec{k}^{\nu},t)\right) - \frac{q}{\hbar}\mathrm{div}_{\vec{k}^{\nu}}\left(\vec{E}_{\mathrm{eff.}}^{\nu}(\vec{r},\vec{k}^{\nu},t)f^{\nu}(\vec{r},\vec{k}^{\nu},t)\right) \ .$$

$$(A.2.0\text{-}20)$$

A.3   Transformed k-space divergence

Starting point for the transformation of the divergence in k-space into the new coordinate system $(\epsilon^{\nu}, \mu, \varphi)$ is equation (2.4.2-152):

$$\mathrm{div}_{\vec{k}^{\nu}}\left(\vec{F}\right) \;=\; \mathrm{grad}_{\vec{k}^{\nu}}\left(\epsilon^{\nu}\right)\cdot\frac{\partial\vec{F}}{\partial\epsilon^{\nu}} + \mathrm{grad}_{\vec{k}^{\nu}}\left(\mu\right)\cdot\frac{\partial\vec{F}}{\partial\mu}$$

$$+\mathrm{grad}_{\vec{k}^{\nu}}\left(\varphi\right)\cdot\frac{\partial\vec{F}}{\partial\varphi} \ . \qquad (A.3.0\text{-}21)$$

Like mentioned in section 2.4.2.3, the gradients $(\mathrm{grad}_{\vec{k}^{\nu}}(\{\epsilon^{\nu},\mu,\varphi\}))$ have to be known for further calculations. Those terms can be directly read off the Jacobian – here inverse of sub-matrix D after (2.4.2-136) and (2.4.2-137) – given in section 2.4.2.2:

$$\mathrm{grad}_{\vec{k}^{\nu}}\left(\epsilon^{\nu}\right) = \frac{1}{\frac{\partial|\vec{k}^{\nu}|}{\partial\epsilon^{\nu}}}\begin{pmatrix}\mu\\\sqrt{1-\mu^{2}}\cos(\varphi)\\\sqrt{1-\mu^{2}}\sin(\varphi)\end{pmatrix} \ , \qquad (A.3.0\text{-}22)$$

$$\mathrm{grad}_{\vec{k}^{\nu}}\left(\mu\right) = \frac{1}{|\vec{k}^{\nu}|}\begin{pmatrix}1-\mu^{2}\\-\mu\sqrt{1-\mu^{2}}\cos(\varphi)\\-\mu\sqrt{1-\mu^{2}}\sin(\varphi)\end{pmatrix} \ , \qquad (A.3.0\text{-}23)$$

$$\mathrm{grad}_{\vec{k}^{\nu}}\left(\varphi\right) = \frac{1}{|\vec{k}^{\nu}|\sqrt{1-\mu^{2}}}\begin{pmatrix}0\\-\sin(\varphi)\\\cos(\varphi)\end{pmatrix} \ . \qquad (A.3.0\text{-}24)$$

Thus, the divergence in (A.3.0-21) becomes

$$
\operatorname{div}_{\vec{k}^\nu}\left(\vec{F}\right) = \frac{1}{\frac{\partial |\vec{k}^\nu|}{\partial \epsilon^\nu}} \begin{pmatrix} \frac{\mu}{\sqrt{1-\mu^2}}\cos(\varphi) \\ \sqrt{1-\mu^2}\sin(\varphi) \end{pmatrix} \cdot \frac{\partial \vec{F}}{\partial \epsilon^\nu}
$$

$$
+\frac{1}{|\vec{k}^\nu|}\begin{pmatrix} \frac{1-\mu^2}{-\mu\sqrt{1-\mu^2}\cos(\varphi)} \\ -\mu\sqrt{1-\mu^2}\sin(\varphi) \end{pmatrix} \cdot \frac{\partial \vec{F}}{\partial \mu} + \frac{1}{|\vec{k}^\nu|\sqrt{1-\mu^2}}\begin{pmatrix} 0 \\ -\sin(\varphi) \\ \cos(\varphi) \end{pmatrix} \cdot \frac{\partial \vec{F}}{\partial \varphi} .
$$

$$(A.3.0\text{-}25)$$

Note, unit vectors for (A.3.0-22)-(A.3.0-24) and $\vec{F}$ in (A.3.0-21) are still the ones of the original k-space ($\vec{e}_{\vec{k}^\nu_x}$, $\vec{e}_{\vec{k}^\nu_y}$ and $\vec{e}_{\vec{k}^\nu_z}$). Thus, each component of $\vec{F}$ is expressed by means of the unit vectors in the new space ($\epsilon^\nu, \mu, \varphi$) with the help of (2.4.2-151)

$$
\begin{pmatrix} F_{\vec{k}^\nu_x} \\ F_{\vec{k}^\nu_y} \\ F_{\vec{k}^\nu_z} \end{pmatrix} = \begin{pmatrix} \mu F_{\epsilon^\nu} + \sqrt{1-\mu^2}F_\mu \\ \sqrt{1-\mu^2}\cos(\varphi)F_{\epsilon^\nu} - \mu\cos(\varphi)F_\mu - \sin(\varphi)F_\varphi \\ \sqrt{1-\mu^2}\sin(\varphi)F_{\epsilon^\nu} - \mu\sin(\varphi)F_\mu + \cos(\varphi)F_\varphi \end{pmatrix} . (A.3.0\text{-}26)
$$

With (A.3.0-26), the derivatives $\frac{\partial \vec{F}}{\partial \epsilon^\nu}$, $\frac{\partial \vec{F}}{\partial \mu}$ and $\frac{\partial \vec{F}}{\partial \varphi}$ become

$$
\frac{\partial \vec{F}}{\partial \epsilon^\nu} = \begin{pmatrix} \mu\frac{\partial F_{\epsilon^\nu}}{\partial \epsilon^\nu} + \sqrt{1-\mu^2}\frac{\partial F_\mu}{\partial \epsilon^\nu} \\ \sqrt{1-\mu^2}\cos(\varphi)\frac{\partial F_{\epsilon^\nu}}{\partial \epsilon^\nu} - \mu\cos(\varphi)\frac{\partial F_\mu}{\partial \epsilon^\nu} - \sin(\varphi)\frac{\partial F_\varphi}{\partial \epsilon^\nu} \\ \sqrt{1-\mu^2}\sin(\varphi)\frac{\partial F_{\epsilon^\nu}}{\partial \epsilon^\nu} - \mu\sin(\varphi)\frac{\partial F_\mu}{\partial \epsilon^\nu} + \cos(\varphi)\frac{\partial F_\varphi}{\partial \epsilon^\nu} \end{pmatrix} , (A.3.0\text{-}27)
$$

$$
\frac{\partial \vec{F}}{\partial \mu} = \left(\begin{array}{l} \mu\frac{\partial F_{\epsilon^\nu}}{\partial \mu} + \sqrt{1-\mu^2}\frac{\partial F_\mu}{\partial \mu} \\ \sqrt{1-\mu^2}\cos(\varphi)\frac{\partial F_{\epsilon^\nu}}{\partial \mu} - \mu\cos(\varphi)\frac{\partial F_\mu}{\partial \mu} - \sin(\varphi)\frac{\partial F_\varphi}{\partial \mu} \\ \sqrt{1-\mu^2}\sin(\varphi)\frac{\partial F_{\epsilon^\nu}}{\partial \mu} - \mu\sin(\varphi)\frac{\partial F_\mu}{\partial \mu} + \cos(\varphi)\frac{\partial F_\varphi}{\partial \mu} \\ +F_{\epsilon^\nu} - \frac{\mu}{\sqrt{1-\mu^2}}F_\mu \\ -\frac{\mu}{\sqrt{1-\mu^2}}\cos(\varphi)F_{\epsilon^\nu} - \cos(\varphi)F_\mu \\ -\frac{\mu}{\sqrt{1-\mu^2}}\sin(\varphi)F_{\epsilon^\nu} - \sin(\varphi)F_\mu \end{array}\right) ,
$$

$$(A.3.0\text{-}28)$$

$$\frac{\partial \vec{F}}{\partial \varphi} = \begin{pmatrix} \mu \frac{\partial F_{\epsilon\nu}}{\partial \varphi} + \sqrt{1-\mu^2} \frac{\partial F_\mu}{\partial \varphi} \\ \sqrt{1-\mu^2} \cos(\varphi) \frac{\partial F_{\epsilon\nu}}{\partial \varphi} - \mu \cos(\varphi) \frac{\partial F_\mu}{\partial \varphi} - \sin(\varphi) \frac{\partial F_\varphi}{\partial \varphi} \\ \sqrt{1-\mu^2} \sin(\varphi) \frac{\partial F_{\epsilon\nu}}{\partial \varphi} - \mu \sin(\varphi) \frac{\partial F_\mu}{\partial \varphi} + \cos(\varphi) \frac{\partial F_\varphi}{\partial \varphi} \\ +0 \\ -\sqrt{1-\mu^2} \sin(\varphi) F_{\epsilon\nu} + \mu \sin(\varphi) F_\mu - \cos(\varphi) F_\varphi \\ +\sqrt{1-\mu^2} \cos(\varphi) F_{\epsilon\nu} - \mu \cos(\varphi) F_\mu - \sin(\varphi) F_\varphi \end{pmatrix} . \qquad \text{(A.3.0-29)}$$

Inserting the derivatives (A.3.0-27)-(A.3.0-29) in (A.3.0-25) yields after some calculations to

$$\text{div}_{\vec{k}^\nu}\left(\vec{F}\right) = \frac{1}{|\vec{k}^\nu| \frac{\partial |\vec{k}^\nu|}{\partial \epsilon^\nu}} \left(\left(|\vec{k}^\nu| \frac{\partial F_{\epsilon\nu}}{\partial \epsilon^\nu} + 2 F_{\epsilon\nu} \frac{\partial |\vec{k}^\nu|}{\partial \epsilon^\nu} \right) \right.$$

$$\left. + \frac{\partial |\vec{k}^\nu|}{\partial \epsilon^\nu} \left(\sqrt{1-\mu^2} \frac{\partial F_\mu}{\partial \mu} - \frac{\mu}{\sqrt{1-\mu^2}} F_\mu \right) + \frac{1}{\sqrt{1-\mu^2}} \frac{\partial |\vec{k}^\nu|}{\partial \epsilon^\nu} \frac{\partial F_\varphi}{\partial \varphi} \right) . \text{(A.3.0-30)}$$

By noticing that

$$\frac{1}{|\vec{k}^\nu|} \frac{\partial}{\partial \epsilon^\nu} \left(|\vec{k}^\nu|^2 F_{\epsilon\nu} \right) = \left(|\vec{k}^\nu| \frac{\partial F_{\epsilon\nu}}{\partial \epsilon^\nu} + 2 F_{\epsilon\nu} \frac{\partial |\vec{k}^\nu|}{\partial \epsilon^\nu} \right) , \qquad \text{(A.3.0-31)}$$

$$\frac{\partial}{\partial \mu} \left(\sqrt{1-\mu^2} F_\mu \right) = \left(\sqrt{1-\mu^2} \frac{\partial F_\mu}{\partial \mu} - \frac{\mu}{\sqrt{1-\mu^2}} F_\mu \right) , \text{(A.3.0-32)}$$

allows to rewrite the equation (A.3.0-30) to

$$\text{div}_{\vec{k}^\nu}\left(\vec{F}\right) = \frac{1}{|\vec{k}^\nu|^2 \frac{\partial |\vec{k}^\nu|}{\partial \epsilon^\nu}} \left(\frac{\partial}{\partial \epsilon^\nu} \left(|\vec{k}^\nu|^2 F_{\epsilon\nu} \right) + \frac{\partial}{\partial \mu} \left(|\vec{k}^\nu| \frac{\partial |\vec{k}^\nu|}{\partial \epsilon^\nu} \sqrt{1-\mu^2} F_\mu \right) \right.$$

$$\left. + \frac{\partial}{\partial \varphi} \left(|\vec{k}^\nu| \frac{\partial |\vec{k}^\nu|}{\partial \epsilon^\nu} \frac{F_\varphi}{\sqrt{1-\mu^2}} \right) \right) . \qquad \text{(A.3.0-33)}$$

Using the transformed DOS after (2.4.2-156) and h_{ϵ^ν}, h_μ, h_φ from (2.4.2-147)-(2.4.2-149), the divergence becomes finally

$$
\mathrm{div}_{\vec{k}^\nu}\left(\vec{F}\right) \;=\; \frac{1}{\mathrm{dos}^\nu(\vec{r},\epsilon^\nu)}\left[\frac{\partial}{\partial\epsilon^\nu}\left(\frac{\mathrm{dos}^\nu(\vec{r},\epsilon^\nu)}{h_{\epsilon^\nu}}F_{\epsilon^\nu}\right) + \frac{\partial}{\partial\mu}\left(\frac{\mathrm{dos}^\nu(\vec{r},\epsilon^\nu)}{h_\mu}F_\mu\right)\right.
$$
$$
\left.+\frac{\partial}{\partial\varphi}\left(\frac{\mathrm{dos}^\nu(\vec{r},\epsilon^\nu)}{h_\varphi}F_\varphi\right)\right]\,, \tag{A.3.0-34}
$$

which is identical to the formulation (2.4.2-157).

APPENDIX B

Supplementary material for chapter 3

B.1 Normalization factors

As sketched in section 3.1, the normalization factor K_N of the collision term is involved in both the one for the time t (t_N after (3.1.0-17)) and the one for the real-space dimensions (L_N after (3.1.0-20)). The unit of the collision term is associated with the following expression

$$[C] = \left[\sum_X \frac{\Omega}{8\pi^3} \int_{\mathrm{BZ}} S^X \mathrm{dos}^{\nu'} d\acute{\epsilon}^{\nu'} d\mu' d\varphi' \right] , \qquad (\text{B.1.0-1})$$

where X labels the considered scattering mechanisms and the square brackets represent a function, which returns the unit of the expression enclosed. As sketched in section 2.3.5, the transition rate can be expressed by a prefactor A and a delta-function δ for the energy conservation (see equation (2.3.5-87) for the elastic and (2.3.5-88) for the inelastic case, respectively). Thus, S^X in

(B.1.0-1) is replaced by

$$S^X = A^X \delta\left(\acute{\epsilon}^{\nu'} - \epsilon^{\nu} \pm \Delta\epsilon^X\right), \text{ and} \qquad \text{(B.1.0-2)}$$

$$\tilde{S}^X = \frac{\Omega A^X}{8\pi^3}\delta\left(\acute{\epsilon}^{\nu'} - \epsilon^{\nu} \pm \Delta\epsilon^X\right), \qquad \text{(B.1.0-3)}$$

where $\Delta\epsilon^X$ accounts for an inelastic scattering process and equals zero for an elastic one. Using the normalization factors E_N after (3.1.0-10) and D_N after (3.1.0-12), equation (B.1.0-1) becomes

$$[C] =$$

$$\left[\sum_X \int_{BZ} \underbrace{\frac{\Omega A^X}{8\pi^3}}_{\tilde{A}^X} \frac{1}{E_N} \delta\left(\acute{w}^{\nu'} - w^{\nu} \pm \Delta w^X\right) \frac{D_N}{E_N} \frac{(\sqrt{m^*_{\nu'}})^3}{2} s^{\nu}(\acute{w}^{\nu'}) E_N d\acute{w}^{\nu'} d\mu' d\varphi'\right]$$

$$= \left[\frac{D_N}{E_N}\sum_X \int_{BZ} \tilde{A}^X \underbrace{f_{\text{aniso.}}(\vec{k}^{\nu}, \vec{k}^{\nu'})}_{[...]=1} \underbrace{\delta\left(\acute{w}^{\nu'} - w^{\nu} \pm \Delta w^X\right)}_{[...]=1} \right.$$

$$\left. \underbrace{\frac{(\sqrt{m^*_{\nu'}})^3}{2}}_{[...]=1} \underbrace{s^{\nu}(\acute{w}^{\nu'})}_{[...]=1} d\acute{w}^{\nu'} d\mu' d\varphi'\right] = \left[\frac{D_N}{E_N}\tilde{A}^X\right] = \left[\frac{D_N}{E_N}K^X_N\right], \qquad \text{(B.1.0-4)}$$

with the identity for the the δ-function

$$\delta(\epsilon) = \delta(E_N w) = \frac{1}{E_N}\delta(w), \qquad \text{(B.1.0-5)}$$

and $f_{\text{aniso.}}(\vec{k}^{\nu}, \vec{k}^{\nu'})$ is a function taking the possible anisotropy of the currently considered scattering mechanism into account. According to (B.1.0-4), the unit of the collision term consists of the normalization factor of the DOS (see (3.1.0-13)) and the normalization factor of the considered transition rate

$$K^X_N = \tilde{A}^X. \qquad \text{(B.1.0-6)}$$

In order to find a common normalization factor for all transition rates, one can, for example, use the maximum normalization factor of all transition rates

$$K_{\mathrm{N}} = \max_X \left(K_{\mathrm{N}}^X \right) = \max_X \left(\tilde{A}^X \right) = \max_X \left(\frac{\Omega}{8\pi^3} A \right) . \qquad \text{(B.1.0-7)}$$

With K_{N} after (B.1.0-7) it is guaranteed that the normalized transition rates involved in the collision term C are within the range of $[0, 1]$. This choice is advantageous, since the distances between real numbers represented on a computer are the closest ones in that range. Although cancellation is not completely avoided, it is at least suppressed in a optimal manner. Based on (3.1.0-18)/(B.1.0-7), the normalization factor for the time t becomes

$$t_{\mathrm{N}} = \frac{E_{\mathrm{N}}}{D_{\mathrm{N}} K_{\mathrm{N}}} . \qquad \text{(B.1.0-8)}$$

For the normalization factor of the real-space dimensions, the formulation of the BTE after (3.0.0-1) is employed and the transient and diffusion terms are considered, only. First, the transient term becomes with (3.1.0-17)/(B.1.0-8)

$$\frac{\partial f^\nu}{\partial t} = \frac{1}{t_{\mathrm{N}}} \frac{\partial f^\nu}{\partial \bar{t}} = \frac{D_{\mathrm{N}} K_{\mathrm{N}}}{E_{\mathrm{N}}} \frac{\partial f^\nu}{\partial \bar{t}} , \qquad \text{(B.1.0-9)}$$

where $\bar{t}$ represents the normalized (and thus dimensionless) time. Focusing on the diffusion term, without the correction term regarding a position dependent band structure, and assuming a (up to now unknown) normalization length L_{N}, one obtains

$$
\begin{aligned}
\frac{1}{\hbar} \frac{\partial}{\partial x_{\mathrm{T}}} \left(\frac{f^\nu}{\frac{\partial |\vec{k}^\nu|}{\partial \epsilon^\nu}} \sqrt{\frac{m_\nu^*}{m_x^\nu}} \mu \right) &= \frac{1}{\hbar L_{\mathrm{N}}} \frac{\partial}{\partial \overline{x_{\mathrm{T}}}} \left(\frac{f^\nu}{\frac{k_{\mathrm{N}}}{E_{\mathrm{N}}} \frac{\partial |\vec{k}^\nu|}{\partial w^\nu}} \sqrt{\frac{m_\nu^*}{m_x^\nu}} \mu \right) \\
&= \frac{E_{\mathrm{N}}}{\hbar L_{\mathrm{N}} k_{\mathrm{N}}} \frac{\partial}{\partial \overline{x_{\mathrm{T}}}} \left(\frac{f^\nu}{\frac{\partial |\vec{k}^\nu|}{\partial w^\nu}} \sqrt{\frac{m_\nu^*}{m_x^\nu}} \mu \right) . \quad \text{(B.1.0-10)}
\end{aligned}
$$

Since the normalization factor $\frac{1}{t_N}$ has to be valid for each term in the BTE (3.0.0-1), L_N can be constructed as

$$\frac{1}{t_N} = \frac{D_N K_N}{E_N} \overset{!}{=} \frac{E_N}{\hbar L_N k_N} ,$$

$$\rightarrow L_N = \frac{E_N^2}{\hbar D_N k_N K_N} . \qquad \text{(B.1.0-11)}$$

Note, the result (3.1.0-20)/(B.1.0-11) for L_N is strongly connected to t_N after (3.1.0-17)/(B.1.0-8), and thus to K_N after (3.1.0-18)/(B.1.0-7).

An alternative approach for L_N is to use the reciprocal value of k_N after (3.1.0-18). In this case, L_N and t_N read

$$L_N = \frac{1}{k_N} , \qquad \text{(B.1.0-12)}$$

$$t_N = \frac{\hbar}{E_N} \text{ and} \qquad \text{(B.1.0-13)}$$

$$K_N = \frac{E_N^2}{\hbar D_N} . \qquad \text{(B.1.0-14)}$$

However, this approach has not been pursued, in order to suppress numerical issues arising from K_N after (B.1.0-14) in combination with the transition rates in the collision term.

B.2 Rearranged and normalized diffusion term

For the BTE in chapter 3 (equation (3.0.0-4)), the terms associated with the diffusion term are rearranged, so that the derivatives w.r.t. x, ϵ^ν and μ are appearing in a more conservative form (divergence of a flux and a corrective dot-product). Employing the product rule one obtains

$$
\frac{\mathrm{dos}^\nu}{\hbar} \mathrm{div}_{\vec{r}_T} \left(\frac{f^\nu}{\frac{\partial |\vec{k}^\nu|}{\partial \epsilon^\nu}} \underbrace{\begin{pmatrix} T_{1,1}^{\mathrm{HV},\nu} \mu \\ T_{2,2}^{\mathrm{HV},\nu} \sqrt{1-\mu^2}\cos(\varphi) \\ T_{3,3}^{\mathrm{HV},\nu} \sqrt{1-\mu^2}\sin(\varphi) \end{pmatrix}}_{\vec{v}} \right)
$$

$$
= \frac{1}{\hbar} \left(\mathrm{div}_{\vec{r}_T} \left(\frac{f^\nu \mathrm{dos}^\nu}{\frac{\partial |\vec{k}^\nu|}{\partial \epsilon^\nu}} \vec{v} \right) - \frac{f^\nu}{\frac{\partial |\vec{k}^\nu|}{\partial \epsilon^\nu}} \vec{v} \cdot \mathrm{grad}_{\vec{r}_T}(\mathrm{dos}^\nu) \right)
$$

$$
= \frac{1}{\hbar} \left(\mathrm{div}_{\vec{r}_T} \left(\frac{F^\nu}{\frac{\partial |\vec{k}^\nu|}{\partial \epsilon^\nu}} \vec{v} \right) - \frac{F^\nu}{\frac{\partial |\vec{k}^\nu|}{\partial \epsilon^\nu}} \vec{v} \cdot \frac{1}{\mathrm{dos}^\nu} \mathrm{grad}_{\vec{r}_T}(\mathrm{dos}^\nu) \right)
$$

$$
\overset{\mathrm{1D}}{=} \frac{1}{\hbar} \left(\frac{\partial}{\partial x_T} \left(\frac{F^\nu}{\frac{\partial |\vec{k}^\nu|}{\partial \epsilon^\nu}} T_{1,1}^{\mathrm{HV},\nu} \mu \right) - \frac{F^\nu}{\frac{\partial |\vec{k}^\nu|}{\partial \epsilon^\nu}} T_{1,1}^{\mathrm{HV},\nu} \mu \frac{1}{\mathrm{dos}^\nu} \frac{\partial \mathrm{dos}^\nu}{\partial x_T} \right) \quad , \text{(B.2.0-15)}
$$

with F^ν after (3.0.0-2).

With the normalized version of F^ν (Φ^ν after (3.1.0-16)) and the normalization factors introduced in section 3.1, (B.2.0-15) becomes

$$
\frac{1}{\hbar} \left(\frac{\partial}{\partial x_T} \left(\frac{F^\nu}{\frac{\partial |\vec{k}^\nu|}{\partial \epsilon^\nu}} T_{1,1}^{\mathrm{HV},\nu} \mu \right) - \frac{F^\nu}{\frac{\partial |\vec{k}^\nu|}{\partial \epsilon^\nu}} T_{1,1}^{\mathrm{HV},\nu} \mu \frac{1}{\mathrm{dos}^\nu} \frac{\partial \mathrm{dos}^\nu}{\partial x_T} \right)
$$

$$
= \frac{1}{\hbar} \frac{D_N}{2 E_N L_N} \left(\frac{\partial}{\partial \overline{x_T}} \left(\frac{\Phi^\nu T_{1,1}^{\mathrm{HV},\nu} \mu}{\frac{k_N}{E_N} \frac{\partial |\vec{k}^\nu|}{\partial w^\nu}} \right) - \frac{\frac{E_N}{k_N} \Phi^\nu T_{1,1}^{\mathrm{HV},\nu} \mu}{\frac{\partial |\vec{k}^\nu|}{\partial w^\nu} \left(\sqrt{m_\nu^*}\right)^3 s^\nu(w^\nu)} \frac{\partial \left(\sqrt{m_\nu^*}\right)^3 s^\nu(w^\nu)}{\partial \overline{x_T}} \right)
$$

$$
= \frac{1}{\hbar} \frac{D_N}{2 L_N k_N} \left(\frac{\partial}{\partial \overline{x_T}} \left(\frac{\Phi^\nu T_{1,1}^{\mathrm{HV},\nu} \mu}{\frac{\partial |\vec{k}^\nu|}{\partial w^\nu}} \right) - \frac{\Phi^\nu T_{1,1}^{\mathrm{HV},\nu} \mu}{\frac{\partial |\vec{k}^\nu|}{\partial w^\nu} \left(\sqrt{m_\nu^*}\right)^3 s^\nu(w^\nu)} \frac{\partial \left(\sqrt{m_\nu^*}\right)^3 s^\nu(w^\nu)}{\partial \overline{x_T}} \right) ,
$$

$$
\text{(B.2.0-16)}
$$

with the dos$^\nu$ (3.1.0-13) already inserted in (B.2.0-16). Next, the missing derivatives are given for an assumed/considered non-parabolic dispersion relation in normalized form:

$$\frac{\partial |\vec{k^\nu}|}{\partial w^\nu} = \frac{\sqrt{m_\nu^*}}{2} \frac{1 + 2a_\nu w^\nu}{\sqrt{w^\nu (1 + a_\nu w^\nu)}}, \tag{B.2.0-17}$$

$$\frac{\partial \left(\sqrt{m_\nu^*}\right)^3 s^\nu (w^\nu)}{\partial x_\mathrm{T}} = \frac{\sqrt{m_\nu^*}}{2} \left\{ 3 \frac{\partial m_\nu^*}{\partial x_\mathrm{T}} \sqrt{w^\nu (1 + a_\nu w^\nu)} \, (1 + 2a_\nu w^\nu) \right.$$

$$\left. + m_\nu^* \frac{w^{\nu 2} (5 + 6a_\nu w^\nu)}{\sqrt{w^\nu (1 + a_\nu w^\nu)}} \frac{\partial a_\nu}{\partial x_\mathrm{T}} \right\} . \tag{B.2.0-18}$$

With the expressions (B.2.0-17) and (B.2.0-18) one obtains for (B.2.0-16):

$$\frac{1}{\hbar} \left(\frac{\partial}{\partial x_\mathrm{T}} \left(\frac{F^\nu}{\frac{\partial |\vec{k^\nu}|}{\partial \epsilon^\nu}} T_{1,1}^{\mathrm{HV},\nu} \mu \right) - \frac{F^\nu}{\frac{\partial |\vec{k^\nu}|}{\partial \epsilon^\nu}} T_{1,1}^{\mathrm{HV},\nu} \mu \frac{1}{\mathrm{dos}^\nu} \frac{\partial \mathrm{dos}^\nu}{\partial x_\mathrm{T}} \right)$$

$$= BTE_\mathrm{N} \left\{ \frac{\partial}{\partial x_\mathrm{T}} \left(\frac{2\sqrt{w^\nu (1 + a_\nu w^\nu)}}{\sqrt{m_\nu^*} (1 + 2a_\nu w^\nu)} \Phi^\nu T_{1,1}^{\mathrm{HV},\nu} \mu \right) \right.$$

$$\left. - \frac{\Phi^\nu T_{1,1}^{\mathrm{HV},\nu} \mu}{\left(\sqrt{m_\nu^*}\right)^3 (1 + 2a_\nu w^\nu)^2} \left(3 \frac{\partial m_\nu^*}{\partial x_\mathrm{T}} s^\nu (w^\nu) + m_\nu^* \frac{w^{\nu 2} (5 + 6a_\nu w^\nu)}{\sqrt{w^\nu (1 + a_\nu w^\nu)}} \frac{\partial a_\nu}{\partial x_\mathrm{T}} \right) \right\} .$$

$$\tag{B.2.0-19}$$

The extension of (B.2.0-19) towards the 2D/3D case can be performed analogously to the 1D case considered here. One only has to take the *full* vector $\vec{v}$ in (B.2.0-15) into account and needs to replace the derivatives $\frac{\partial \dots}{\partial x_\mathrm{T}}$ in (B.2.0-18) by the considered components of the real space gradient.

In next stage, the terms associated with the position dependent dispersion relation are considered (see section 2.4.2.1, equation (2.4.2-125)). First, the terms associated with $\mathrm{grad}_{\vec{r}}(\epsilon^{\nu})$ are investigated and normalized. For the diffusion term one obtains

$$
\frac{\mathrm{dos}^{\nu}}{\hbar} \left[\underbrace{\begin{pmatrix} T_{1,1}^{\mathrm{HV},\nu} \mu \\ T_{2,2}^{\mathrm{HV},\nu} \sqrt{1-\mu^2}\cos(\varphi) \\ T_{3,3}^{\mathrm{HV},\nu} \sqrt{1-\mu^2}\sin(\varphi) \end{pmatrix} \cdot \begin{pmatrix} \frac{\partial \epsilon^{\nu}}{\partial x} \\ \frac{\partial \epsilon^{\nu}}{\partial y} \\ \frac{\partial \epsilon^{\nu}}{\partial z} \end{pmatrix} }_{\vec{v}} \right] \frac{\partial}{\partial \epsilon^{\nu}} \left(\frac{F^{\nu}}{\mathrm{dos}^{\nu} \frac{\partial |\vec{k}^{\nu}|}{\partial \epsilon^{\nu}}} \right)
$$

$$
= \frac{\mathrm{dos}^{\nu}}{\hbar} \left[\vec{v} \cdot \mathrm{grad}_{\vec{r}}(\epsilon^{\nu}) \right] \frac{\partial}{\partial \epsilon^{\nu}} \left(\frac{F^{\nu}}{\mathrm{dos}^{\nu} \frac{\partial |\vec{k}^{\nu}|}{\partial \epsilon^{\nu}}} \right)
$$

$$
= \frac{1}{\hbar} \left\{ \frac{\partial}{\partial \epsilon^{\nu}} \left(\frac{F^{\nu}}{\frac{\partial |\vec{k}^{\nu}|}{\partial \epsilon^{\nu}}} \left[\vec{v} \cdot \mathrm{grad}_{\vec{r}}(\epsilon^{\nu}) \right] \right) - \frac{F^{\nu}}{\mathrm{dos}^{\nu} \frac{\partial |\vec{k}^{\nu}|}{\partial \epsilon^{\nu}}} \frac{\partial}{\partial \epsilon^{\nu}} \left(\mathrm{dos}^{\nu} \left[\vec{v} \cdot \mathrm{grad}_{\vec{r}}(\epsilon^{\nu}) \right] \right) \right\}
$$

$$
\overset{\mathrm{1D}}{=} \frac{1}{\hbar} \left\{ \frac{\partial}{\partial \epsilon^{\nu}} \left(\frac{F^{\nu} T_{1,1}^{\mathrm{HV},\nu} \mu}{\frac{\partial |\vec{k}^{\nu}|}{\partial \epsilon^{\nu}}} \frac{\partial \epsilon^{\nu}}{\partial x} \right) - \frac{F^{\nu}}{\mathrm{dos}^{\nu} \frac{\partial |\vec{k}^{\nu}|}{\partial \epsilon^{\nu}}} \frac{\partial}{\partial \epsilon^{\nu}} \left(\mathrm{dos}^{\nu} T_{1,1}^{\mathrm{HV},\nu} \mu \frac{\partial \epsilon^{\nu}}{\partial x} \right) \right\}.
$$

$$(\text{B.2.0-20})$$

Using the normalization factors listed in section 3.1 (table 3.1 and 3.2), equation (B.2.0-20) can be rearranged to

$$
\frac{1}{\hbar} \left\{ \frac{\partial}{\partial \epsilon^{\nu}} \left(\frac{F^{\nu} T_{1,1}^{\mathrm{HV},\nu} \mu}{\frac{\partial |\vec{k}^{\nu}|}{\partial \epsilon^{\nu}}} \frac{\partial \epsilon^{\nu}}{\partial x} \right) - \frac{F^{\nu}}{\mathrm{dos}^{\nu} \frac{\partial |\vec{k}^{\nu}|}{\partial \epsilon^{\nu}}} \frac{\partial}{\partial \epsilon^{\nu}} \left(\mathrm{dos}^{\nu} T_{1,1}^{\mathrm{HV},\nu} \mu \frac{\partial \epsilon^{\nu}}{\partial x} \right) \right\}
$$

$$
= \frac{1}{\hbar} \frac{D_{\mathrm{N}}}{2 k_{\mathrm{N}} L_{\mathrm{N}}} \left\{ \frac{\partial}{\partial w^{\nu}} \left(\frac{\Phi^{\nu} T_{1,1}^{\mathrm{HV},\nu} \mu}{\frac{\partial |\vec{k}^{\nu}|}{\partial w^{\nu}}} \frac{\partial w^{\nu}}{\partial \overline{x}} \right) \right.
$$

$$
\left. - \frac{\Phi^{\nu}}{\left(\sqrt{m_{\nu}^{*}}\right)^3 s^{\nu}(w^{\nu}) \frac{\partial |\vec{k}^{\nu}|}{\partial w^{\nu}}} \frac{\partial}{\partial w^{\nu}} \left(\left(\sqrt{m_{\nu}^{*}}\right)^3 s^{\nu}(w^{\nu}) T_{1,1}^{\mathrm{HV},\nu} \mu \frac{\partial w^{\nu}}{\partial \overline{x}} \right) \right\}.
$$

$$(\text{B.2.0-21})$$

Inserting (B.2.0-17) for $\frac{\partial |\vec{k}^\nu|}{\partial w^\nu}$, equation (B.2.0-21) finally becomes

$$\frac{1}{\hbar}\left\{\frac{\partial}{\partial\epsilon^\nu}\left(\frac{F^\nu T_{1,1}^{\mathrm{HV},\nu}\mu}{\frac{\partial|\vec{k}^\nu|}{\partial\epsilon^\nu}}\frac{\partial\epsilon^\nu}{\partial x}\right) - \frac{F^\nu}{\mathrm{dos}^\nu\frac{\partial|\vec{k}^\nu|}{\partial\epsilon^\nu}}\frac{\partial}{\partial\epsilon^\nu}\left(\mathrm{dos}^\nu T_{1,1}^{\mathrm{HV},\nu}\mu\frac{\partial\epsilon^\nu}{\partial x}\right)\right\}$$

$$= BTE_{\mathrm{N}}\left[\frac{\partial}{\partial w^\nu}\left(\frac{2\sqrt{w^\nu\left(1+a_\nu w^\nu\right)}}{\sqrt{m_\nu^*}\left(1+2a_\nu w^\nu\right)}\Phi^\nu T_{1,1}^{\mathrm{HV},\nu}\mu\frac{\partial w^\nu}{\partial\overline{x}}\right)\right.$$

$$\left.-\frac{2\Phi^\nu T_{1,1}^{\mathrm{HV},\nu}\mu}{\sqrt{m_\nu^*}\left(1+2a_\nu w^\nu\right)^2}\frac{\partial}{\partial w^\nu}\left(s^\nu(w^\nu)\frac{\partial w^\nu}{\partial\overline{x}}\right)\right]. \qquad (\text{B.2.0-22})$$

Note, although formulas for $\frac{\partial w^\nu}{\partial d}$ ($d = \{x, y, z\}$) are listed in section 2.4.2.2, equation (2.4.2-140) (in non-normalized form), the derivatives are intentionally not inserted. As will be seen later, most of the terms, where $\frac{\partial w^\nu}{\partial d}$ is involved, will cancel out with the terms associated with the drift term. Therefore, and for the sake of readability, these derivatives are not further resolved. Again, like for (B.2.0-19), the extension regarding a 2D/3D real space simulation domain can be performed analogously. Here, only the derivative $\frac{\partial w^\nu}{\partial\overline{x}}$ has to be replaced by the normalized real space gradient $\mathrm{grad}_{\overline{r}}(w^\nu)$.

The last remaining part to be considered for the diffusion term is the one associated with the gradient $\mathrm{grad}_{\overline{r}}(\mu)$ (see section 2.4.2.1, equation (2.4.2-125)). In the case of the diffusion term (see equation (2.4.3-159) in section 2.4.3) one obtains

$$\frac{1}{\hbar}\frac{1}{\frac{\partial|\vec{k}^\nu|}{\partial\epsilon^\nu}}\left[\begin{pmatrix}T_{1,1}^{\mathrm{HV},\nu}(\frac{\partial F^\nu}{\partial\mu}\mu+F^\nu)\\T_{2,2}^{\mathrm{HV},\nu}\sqrt{1-\mu^2}\cos(\varphi)(\frac{\partial F^\nu}{\partial\mu}-F^\nu\frac{\mu}{1-\mu^2})\\T_{3,3}^{\mathrm{HV},\nu}\sqrt{1-\mu^2}\sin(\varphi)(\frac{\partial F^\nu}{\partial\mu}-F^\nu\frac{\mu}{1-\mu^2})\end{pmatrix}\cdot\begin{pmatrix}\frac{\partial\mu}{\partial x}\\\frac{\partial\mu}{\partial y}\\\frac{\partial\mu}{\partial z}\end{pmatrix}\right]$$

$$=\frac{1}{\hbar}\frac{D_{\mathrm{N}}}{2k_{\mathrm{N}}L_{\mathrm{N}}}\frac{1}{\frac{\partial|\vec{k}^\nu|}{\partial w^\nu}}\left[\begin{pmatrix}T_{1,1}^{\mathrm{HV},\nu}(\frac{\partial\Phi^\nu}{\partial\mu}\mu+\Phi^\nu)\\T_{2,2}^{\mathrm{HV},\nu}\sqrt{1-\mu^2}\cos(\varphi)(\frac{\partial\Phi^\nu}{\partial\mu}-\Phi^\nu\frac{\mu}{1-\mu^2})\\T_{3,3}^{\mathrm{HV},\nu}\sqrt{1-\mu^2}\sin(\varphi)(\frac{\partial\Phi^\nu}{\partial\mu}-\Phi^\nu\frac{\mu}{1-\mu^2})\end{pmatrix}\cdot\begin{pmatrix}\frac{\partial\mu}{\partial\overline{x}}\\\frac{\partial\mu}{\partial\overline{y}}\\\frac{\partial\mu}{\partial\overline{z}}\end{pmatrix}\right]$$

$$\overset{\mathrm{1D}}{=} BTE_{\mathrm{N}}\frac{T_{1,1}^{\mathrm{HV},\nu}}{\frac{\partial|\vec{k}^\nu|}{\partial w^\nu}}\frac{\partial\mu}{\partial\overline{x}}\frac{\partial\Phi^\nu\mu}{\partial\mu}. \qquad (\text{B.2.0-23})$$

Using the product rule, (B.2.0-23) is rearranged to

$$
BTE_N \frac{T_{1,1}^{HV,\nu}}{\frac{\partial |\vec{k}^\nu|}{\partial w^\nu}} \frac{\partial \mu}{\partial \overline{x}} \frac{\partial \Phi^\nu \mu}{\partial \mu}
$$

$$
= BTE_N \left\{ \frac{\partial}{\partial \mu} \left(\frac{\Phi^\nu T_{1,1}^{HV,\nu} \mu}{\frac{\partial |\vec{k}^\nu|}{\partial w^\nu}} \frac{\partial \mu}{\partial \overline{x}} \right) - \Phi^\nu \mu \frac{\partial}{\partial \mu} \left(\frac{T_{1,1}^{HV,\nu}}{\frac{\partial |\vec{k}^\nu|}{\partial w^\nu}} \frac{\partial \mu}{\partial \overline{x}} \right) \right\}
$$

$$
= BTE_N \left\{ \frac{\partial}{\partial \mu} \left(\frac{2\sqrt{w^\nu (1 + a_\nu w^\nu)}}{\sqrt{m_\nu^*} (1 + 2a_\nu w^\nu)} \Phi^\nu T_{1,1}^{HV,\nu} \mu \frac{\partial \mu}{\partial \overline{x}} \right) \right.
$$

$$
\left. - \frac{2\sqrt{w^\nu (1 + a_\nu w^\nu)}}{\sqrt{m_\nu^*} (1 + 2a_\nu w^\nu)} \Phi^\nu T_{1,1}^{HV,\nu} \mu \frac{\partial}{\partial \mu} \left(\frac{\partial \mu}{\partial \overline{x}} \right) \right\} , \qquad \text{(B.2.0-24)}
$$

where in (B.2.0-24) $\overline{\frac{\partial |\vec{k}^\nu|}{\partial w^\nu}}$ after (B.2.0-17) has been employed.

B.3 Diffusion and Drift terms

Based on the relations given in B.2, the terms required in a discretized representation for the device BTE are given here. Adding the diffusion terms after (3.1.0-23), (3.1.0-25)-(3.1.0-26) and the drift term after (3.1.0-27) results in

$$
\frac{\partial}{\partial \overline{x}_{\mathrm{T}}} \left(\frac{\Phi^{\nu}}{\sqrt{m_{\nu}^{*}}} \frac{2\sqrt{w^{\nu}\left(1 + a_{\nu}w^{\nu}\right)}}{1 + 2a_{\nu}w^{\nu}} T_{1,1}^{\mathrm{HV},\nu} \mu \right)
$$

$$
- \frac{\Phi^{\nu} T_{1,1}^{\mathrm{HV},\nu} \mu}{\sqrt{m_{\nu}^{*}}(1 + 2a_{\nu}w^{\nu})^{2}} \left(\frac{3}{m_{\nu}^{*}} \frac{\partial m_{\nu}^{*}}{\partial \overline{x}_{\mathrm{T}}} s^{\nu}(w^{\nu}) + \frac{w^{\nu 2}(6a_{\nu}w^{\nu} + 5)}{\sqrt{w^{\nu}\left(1 + a_{\nu}w^{\nu}\right)}} \frac{\partial a_{\nu}}{\partial \overline{x}_{\mathrm{T}}} \right)
$$

$$
+ \frac{\partial}{\partial w^{\nu}} \left(\frac{2\sqrt{w^{\nu}(1 + a_{\nu}w^{\nu})}}{\sqrt{m_{\nu}^{*}}(1 + 2a_{\nu}w^{\nu})} \Phi^{\nu} T_{1,1}^{\mathrm{HV},\nu} \mu \frac{\partial w^{\nu}}{\partial \overline{x}} \right)
$$

$$
- \frac{2\Phi^{\nu} T_{1,1}^{\mathrm{HV},\nu} \mu}{\sqrt{m_{\nu}^{*}}(1 + 2a_{\nu}w^{\nu})^{2}} \frac{\partial}{\partial w^{\nu}} \left(s^{\nu}(w^{\nu}) \frac{\partial w^{\nu}}{\partial \overline{x}} \right)
$$

$$
+ \frac{\partial}{\partial \mu} \left(\frac{2\sqrt{w^{\nu}(1 + a_{\nu}w^{\nu})}}{\sqrt{m_{\nu}^{*}}(1 + 2a_{\nu}w^{\nu})} \Phi^{\nu} T_{1,1}^{\mathrm{HV},\nu} \mu \frac{\partial \mu}{\partial \overline{x}} \right)
$$

$$
- \frac{2\sqrt{w^{\nu}(1 + a_{\nu}w^{\nu})}}{\sqrt{m_{\nu}^{*}}(1 + 2a_{\nu}w^{\nu})} \Phi^{\nu} T_{1,1}^{\mathrm{HV},\nu} \mu \frac{\partial}{\partial \mu} \left(\frac{\partial \mu}{\partial \overline{x}} \right)
$$

$$
+ \frac{\partial}{\partial w^{\nu}} \left(-\frac{2\Phi^{\nu} \sqrt{w^{\nu}(1 + a_{\nu}w^{\nu})}}{\sqrt{m_{\nu}^{*}} \left(1 + 2a_{\nu}w^{\nu}\right)} \mu T_{1,1}^{\mathrm{HV},\nu} \overline{E_{\mathrm{eff.}}^{\nu,x}} \right)
$$

$$
+ \frac{\partial}{\partial \mu} \left(-\frac{\Phi^{\nu}(1 - \mu^{2})}{\sqrt{m_{\nu}^{*}} \sqrt{w^{\nu}(1 + a_{\nu}w^{\nu})}} T_{1,1}^{\mathrm{HV},\nu} \overline{E_{\mathrm{eff.}}^{\nu,x}} \right) = T_{\mathrm{diff.}} + T_{\mathrm{drift.}} \, . \qquad \text{(B.3.0-25)}
$$

In equation (B.3.0-25), the first six lines are associated to the rearranged diffusion term (section B.2), where the last two lines are the components of the drift term in w^{ν} and μ direction, respectively. Using the normalized version of the driving force for the BTE (see equation (2.4.0-108))

$$
\overline{E_{\mathrm{eff.}}^{\nu,x}} = -\frac{\partial \overline{\psi}}{\partial \overline{x}} + \frac{\partial \overline{V_{\mathrm{C},0}^{\nu}}}{\partial \overline{x}} + \frac{\partial w^{\nu}}{\partial \overline{x}} = \overline{E_{\mathrm{eff., pot.}}^{\nu,x}} + \frac{\partial w^{\nu}}{\partial \overline{x}} \, , \qquad \text{(B.3.0-26)}
$$

and due to the rearrangement of the diffusion terms, (B.3.0-25) is further simplified. Due to a position dependent dispersion relation, the BULK drift term is modified by the derivatives $\frac{\partial}{\partial w^\nu}(\dots)$ and $\frac{\partial}{\partial \mu}(\dots)$ stemming from the rearranged diffusion term (see section B.2). Lumping those terms together gives

$$
T_{\text{diff.}} + T_{\text{drift}} = \frac{\partial}{\partial \overline{x_T}} \left(\frac{\Phi^\nu}{\sqrt{m_\nu^*}} \frac{2\sqrt{w^\nu (1 + a_\nu w^\nu)}}{1 + 2a_\nu w^\nu} T_{1,1}^{\text{HV},\nu} \mu \right)
$$
$$
+ \frac{\partial}{\partial w^\nu} \left(-\frac{2\sqrt{w^\nu(1 + a_\nu w^\nu)}}{\sqrt{m_\nu^*}(1 + 2a_\nu w^\nu)} \Phi^\nu T_{1,1}^{\text{HV},\nu} \mu \overline{E_{\text{eff., pot.}}^{\nu,x}} \right)
$$
$$
+ \frac{\partial}{\partial \mu} \left(-\frac{\Phi^\nu (1 - \mu^2)}{\sqrt{m_\nu^*}\sqrt{w^\nu(1 + a_\nu w^\nu)}} T_{1,1}^{\text{HV},\nu} \overline{E_{\text{eff.}}^{\nu,x}} + \frac{2\sqrt{w^\nu(1 + a_\nu w^\nu)}}{\sqrt{m_\nu^*}(1 + 2a_\nu w^\nu)} \Phi^\nu T_{1,1}^{\text{HV},\nu} \mu \frac{\partial \mu}{\partial \overline{x}} \right)
$$
$$
- \frac{\Phi^\nu T_{1,1}^{\text{HV},\nu} \mu}{\sqrt{m_\nu^*}} \left[\frac{1}{(1 + 2a_\nu w^\nu)^2} \left(\frac{3}{m_\nu^*} \frac{\partial m_\nu^*}{\partial \overline{x_T}} s^\nu(w^\nu) + \frac{w^{\nu 2}(6a_\nu w^\nu + 5)}{\sqrt{w^\nu (1 + a_\nu w^\nu)}} \frac{\partial a_\nu}{\partial \overline{x_T}} \right) \right.
$$
$$
\left. + \frac{2}{(1 + 2a_\nu w^\nu)^2} \frac{\partial}{\partial w^\nu} \left(s^\nu(w^\nu) \frac{\partial w^\nu}{\partial \overline{x}} \right) + \frac{2\sqrt{w^\nu(1 + a_\nu w^\nu)}}{(1 + 2a_\nu w^\nu)} \frac{\partial}{\partial \mu} \left(\frac{\partial \mu}{\partial \overline{x}} \right) \right] .
$$

$$(B.3.0\text{-}27)$$

Equation (B.3.0-27) emphasizes on the influence of a position dependent dispersion relation on the resulting drift term. Note, due to the driving force after (B.3.0-26), the gradient of the potential energy $\overline{E_{\text{eff., pot.}}^{\nu,x}}$ is involved in the flux in w^ν direction only, where for the one in μ direction an additional term $\frac{\partial \mu}{\partial \overline{x}}$ associated with a position dependent dispersion relation occurs. The last two lines of (B.3.0-27) contain the remaining portions of the rewritten diffusion term (see section B.2). In section 2.4.2.2, the derivatives $\frac{\partial w^\nu}{\partial \overline{x}}$ (2.4.2-140) and $\frac{\partial \mu}{\partial \overline{x}}$ (2.4.2-142) are given in non-normalized form. Both derivatives ($\frac{\partial w^\nu}{\partial \overline{x}}$ and $\frac{\partial \mu}{\partial \overline{x}}$) are according to h_x ($d = x$) (2.4.2-141) φ-dependent. However, for the considered 1D transport of carriers in x-direction and since no other φ-dependence is introduced to the BTE, h_x effectively reduces to

$$
\overline{h_x} \overset{\text{1D}}{=} \frac{1}{2\pi} \int_0^{2\pi} h_x d\varphi = \frac{1}{2} \left(\frac{\partial \ln \left(T_{2,2}^{\text{HV},\nu} \right)}{\partial \overline{x}} + \frac{\partial \ln \left(T_{3,3}^{\text{HV},\nu} \right)}{\partial \overline{x}} \right) , \quad (B.3.0\text{-}28)
$$

which is also obtained by integrating all terms of the BTE (and using the original h_x (2.4.2-141)) over $\varphi \in [0, 2\pi]$ and dividing afterwards all terms by 2π to avoid an additional scaling of the BTE. Thus and for the 1D transport of carriers in x-direction, the derivative $\frac{\partial w^\nu}{\partial \overline{x}}$ becomes with (B.3.0-28)

$$
\frac{\partial w^\nu}{\partial \overline{x}} = \frac{w^\nu (1 + a_\nu w^\nu)}{1 + 2a_\nu w^\nu} \left[(1 - \mu^2) \left(\frac{\partial \ln \left(T_{2,2}^{\text{HV},\nu} \right)}{\partial \overline{x}} + \frac{\partial \ln \left(T_{3,3}^{\text{HV},\nu} \right)}{\partial \overline{x}} \right) \right.
$$
$$
\left. + 2\mu^2 \frac{\partial \ln \left(T_{1,1}^{\text{HV},\nu} \right)}{\partial \overline{x}} - \left(\frac{1}{m_\nu^*} \frac{\partial m_\nu^*}{\partial \overline{x}} + \frac{w^{\nu 2}}{w^\nu (1 + a_\nu w^\nu)} \frac{\partial a_\nu}{\partial \overline{x}} \right) \right]
$$

$$(B.3.0\text{-}29)$$

and analogously for $\frac{\partial \mu}{\partial \overline{x}}$

$$
\frac{\partial \mu}{\partial \overline{x}} = \mu \left(1 - \mu^2 \right) \left[\frac{\partial \ln \left(T_{1,1}^{\text{HV},\nu} \right)}{\partial \overline{x}} - \frac{1}{2} \left(\frac{\partial \ln \left(T_{2,2}^{\text{HV},\nu} \right)}{\partial \overline{x}} + \frac{\partial \ln \left(T_{3,3}^{\text{HV},\nu} \right)}{\partial \overline{x}} \right) \right] .
$$

$$(B.3.0\text{-}30)$$

With $\frac{\partial w^\nu}{\partial \overline{x}}$ (B.3.0-29) and $\frac{\partial \mu}{\partial \overline{x}}$ (B.3.0-30), the following relations needed for further simplifications of (B.3.0-27) are considered

$$
\frac{\partial}{\partial \mu} \left(\frac{\partial \mu}{\partial \overline{x}} \right) =
$$
$$
(1 - 3\mu^2) \left[\frac{\partial \ln \left(T_{1,1}^{\text{HV},\nu} \right)}{\partial \overline{x}} - \frac{1}{2} \left(\frac{\partial \ln \left(T_{2,2}^{\text{HV},\nu} \right)}{\partial \overline{x}} + \frac{\partial \ln \left(T_{3,3}^{\text{HV},\nu} \right)}{\partial \overline{x}} \right) \right]
$$
$$(B.3.0\text{-}31)$$

and

$$
\frac{\partial}{\partial w^\nu} \left(s^\nu (w^\nu) \frac{\partial w^\nu}{\partial \overline{x}} \right) = -\frac{1}{2} \frac{w^{\nu 2} (6a_\nu w^\nu + 5)}{\sqrt{w^\nu (1 + a_\nu w^\nu)}} \frac{\partial a_\nu}{\partial \overline{x}}
$$
$$
+ \frac{3}{2} s^\nu (w^\nu) \left[(1 - \mu^2) \left(\frac{\partial \ln \left(T_{2,2}^{\text{HV},\nu} \right)}{\partial \overline{x}} + \frac{\partial \ln \left(T_{3,3}^{\text{HV},\nu} \right)}{\partial \overline{x}} \right) \right.
$$
$$
\left. + 2\mu^2 \frac{\partial \ln \left(T_{1,1}^{\text{HV},\nu} \right)}{\partial \overline{x}} - \frac{1}{m_\nu^*} \frac{\partial m_\nu^*}{\partial \overline{x}} \right] . \quad (B.3.0\text{-}32)
$$

Inserting the equation (B.3.0-31) and (B.3.0-32) in (B.3.0-27) leads to

$$
T_{\text{diff.}} + T_{\text{drift}} = \frac{\partial}{\partial \overline{x_T}} \left(\frac{\Phi^\nu}{\sqrt{m_\nu^*}} \frac{2\sqrt{w^\nu \left(1 + a_\nu w^\nu\right)}}{1 + 2a_\nu w^\nu} T_{1,1}^{\text{HV},\nu} \mu \right)
$$

$$
+ \frac{\partial}{\partial w^\nu} \left(-\frac{2\sqrt{w^\nu \left(1 + a_\nu w^\nu\right)}}{\sqrt{m_\nu^*}\left(1 + 2a_\nu w^\nu\right)} \Phi^\nu T_{1,1}^{\text{HV},\nu} \mu \overline{E_{\text{eff., pot.}}^{\nu,x}} \right)
$$

$$
+ \frac{\partial}{\partial \mu} \left(-\frac{\Phi^\nu \left(1 - \mu^2\right)}{\sqrt{m_\nu^*}\sqrt{w^\nu \left(1 + a_\nu w^\nu\right)}} T_{1,1}^{\text{HV},\nu} \overline{E_{\text{eff.}}^{\nu,x}} + \frac{2\sqrt{w^\nu \left(1 + a_\nu w^\nu\right)}}{\sqrt{m_\nu^*}\left(1 + 2a_\nu w^\nu\right)} \Phi^\nu T_{1,1}^{\text{HV},\nu} \mu \frac{\partial \mu}{\partial \overline{x}} \right)
$$

$$
- \frac{2\sqrt{w^\nu \left(1 + a_\nu w^\nu\right)}}{\sqrt{m_\nu^*}\left(1 + 2a_\nu w^\nu\right)} \Phi^\nu T_{1,1}^{\text{HV},\nu} \mu \left\{ \frac{\partial \ln \left(T_{1,1}^{\text{HV},\nu}\right)}{\partial \overline{x}} + \frac{\partial \ln \left(T_{2,2}^{\text{HV},\nu}\right)}{\partial \overline{x}} + \frac{\partial \ln \left(T_{3,3}^{\text{HV},\nu}\right)}{\partial \overline{x}} \right\} .
$$

$$(\text{B.3.0-33})$$

Finally, with the definition of the Herring-Vogt matrix elements (see equation (2.2.0-46) in section 2.2 or appendix A.1, equation (A.1.0-1)) the last line in (B.3.0-33) becomes zero, since

$$
\frac{\partial \ln \left(T_{1,1}^{\text{HV},\nu}\right)}{\partial \overline{x}} + \frac{\partial \ln \left(T_{2,2}^{\text{HV},\nu}\right)}{\partial \overline{x}} + \frac{\partial \ln \left(T_{3,3}^{\text{HV},\nu}\right)}{\partial \overline{x}} = 0 , \qquad (\text{B.3.0-34})
$$

with $m_\nu^* = \sqrt[3]{m_x^\nu m_y^\nu m_z^\nu}$ after (2.2.0-48)/(A.1.0-12). Thus, a position dependent dispersion relation in combination with the solution variable Φ^ν after (3.1.0-16) (section 3.1) does not result in any additional corrective term, but alters the fluxes associated with the drift term. For the sake of completeness, the final form of (B.3.0-33) evaluates to

$$
T_{\text{diff.}} + T_{\text{drift}} = \frac{\partial}{\partial \overline{x_T}} \left(\frac{\Phi^\nu}{\sqrt{m_\nu^*}} \frac{2\sqrt{w^\nu \left(1 + a_\nu w^\nu\right)}}{1 + 2a_\nu w^\nu} T_{1,1}^{\text{HV},\nu} \mu \right)
$$

$$
+ \frac{\partial}{\partial w^\nu} \left(-\frac{2\sqrt{w^\nu \left(1 + a_\nu w^\nu\right)}}{\sqrt{m_\nu^*}\left(1 + 2a_\nu w^\nu\right)} \Phi^\nu T_{1,1}^{\text{HV},\nu} \mu \overline{E_{\text{eff., pot.}}^{\nu,x}} \right)
$$

$$
+ \frac{\partial}{\partial \mu} \left(-\frac{\Phi^\nu \left(1 - \mu^2\right)}{\sqrt{m_\nu^*}\sqrt{w^\nu \left(1 + a_\nu w^\nu\right)}} T_{1,1}^{\text{HV},\nu} \overline{E_{\text{eff.}}^{\nu,x}} + \frac{2\sqrt{w^\nu \left(1 + a_\nu w^\nu\right)}}{\sqrt{m_\nu^*}\left(1 + 2a_\nu w^\nu\right)} \Phi^\nu T_{1,1}^{\text{HV},\nu} \mu \frac{\partial \mu}{\partial \overline{x}} \right) .
$$

$$(\text{B.3.0-35})$$

Note, a definition of Φ^ν being different than the one after (3.1.0-16) will result in an additional corrective term in (B.3.0-35), which might be disadvantageous for the numerical treatment. Basically, a spatial variation of the dispersion relation is inherently included in the solution variable Φ^ν (and thus in its numerical equivalent $n_{i,j,k}^{\nu,\tau}$) due to m_ν^* and a_ν in $s^\nu(w^\nu)$, which makes a corrective term unnecessary. Therefore, only $\overline{E_{\text{eff., pot.}}^{\nu,x}}$ is taken into account for the flux in w^ν direction, since a position dependent m_ν^* or a_ν (which results in a spatial dependent w^ν) is already included by the definition of Φ^ν after (3.1.0-16). In contrast to the flux in w^ν direction, the complete driving force of the BTE after (2.4.0-108)/(B.3.0-26) plus a corrective term associated to $\frac{\partial \mu}{\partial \bar{x}}$ is needed for the flux in μ direction. By the term associated to $\frac{\partial \mu}{\partial \bar{x}}$, a position dependent anisotropy of the considered valley for a fixed w^ν is accounted for (see section 2.4.2.2, equation (2.4.2-142)). The complete driving force of the BTE is also included in the flux in μ direction, since a spatial dependent isotropic dispersion relation yields a spatially dependent μ. Latter becomes obvious by considering (2.4.2-122) (section 2.4.2), for a fixed k_x^ν and $T^{HV,\nu} = I_3$ (I_3 is the identity matrix of size 3, isotropic band structure):

$$\mu = \frac{k_x^\nu}{|\vec{k}^\nu(\epsilon^\nu, \vec{r})|} \ . \tag{B.3.0-36}$$

Thus, for an isotropic and spatially varying $|\vec{k}^\nu|$ (or spatial dependent w^ν via the dispersion relation), μ varies w.r.t. the real space position, which is considered by $\overline{E_{\text{eff.}}^{\nu,x}}$ associated with the μ-flux in (B.3.0-35).

B.4 Collision term without Pauli principle

In this section, the collision term without Pauli principle is considered (see section 2.4.1, equation (2.4.1-116)). Therefore, the normalized collision term within the employed numerical framework (**BIM**) after (3.2.3-49) (section 3.2.3) simplifies to

$$
(\sqrt{m_\nu^*})^3 s^\nu(w^\nu)\overline{C} = \left[\sum_X \int_{\mathrm{BZ}} \frac{(\sqrt{m_\nu^*})^3 s^\nu(w^\nu)}{2} \left(\underbrace{\overline{S^X}(\{\acute{w}^{\nu'},\mu',\varphi'\},\{w^\nu,\mu,\varphi\})}_{\overline{S^{X,\mathrm{in}}}} f^{\nu'} \right.\right.
$$
$$
\left.\left. - \underbrace{\overline{S^X}(\{w^\nu,\mu,\varphi\},\{\acute{w}^{\nu'},\mu',\varphi'\})}_{\overline{S^{X,\mathrm{out}}}} f^\nu \right)(\sqrt{m_{\nu'}^*})^3 s^{\nu'}(\acute{w}^{\nu'})d\acute{w}^{\nu'}d\mu'd\varphi' \right] . \quad \text{(B.4.0-37)}
$$

Following the procedure depicted in section 3.2.3, the equation to be considered for the numerical treatment becomes

$$
\int_{w_{i-\frac{1}{2}}^\nu}^{w_{i+\frac{1}{2}}^\nu} \int_{\mu_{j-\frac{1}{2}}}^{\mu_{j+\frac{1}{2}}} (\sqrt{m_\nu^*})^3 s^\nu(w^\nu)\overline{C}d\mu\, dw^\nu = \int_{w_{i-\frac{1}{2}}^\nu}^{w_{i+\frac{1}{2}}^\nu} dw^\nu \int_{\mu_{j-\frac{1}{2}}}^{\mu_{j+\frac{1}{2}}} d\mu
$$
$$
\left[\sum_X \sum_{\nu'=1}^{N_{\mathrm{val.}}} \frac{(\sqrt{m_\nu^*})^3 s^\nu(w^\nu)}{2} \right.
$$
$$
\left. \int_0^{2\pi} \int_{-1}^1 \int_0^{w_{\max}} \left(\overline{S^{X,\mathrm{in}}}\, \Phi^{\nu'} - \overline{S^{X,\mathrm{out}}} f^\nu (\sqrt{m_{\nu'}^*})^3 s^{\nu'}(\acute{w}^{\nu'}) \right)d\acute{w}^{\nu'}d\mu'd\varphi' \right] ,
$$
$$
\text{(B.4.0-38)}
$$

with the total number of considered valleys $N_{\mathrm{val.}}$ and X represents all scattering mechanisms to be accounted for.

As depicted in section 2.4.1, different results for the discretized collision term with/without Pauli principle are obtained only for inelastic scattering mechanisms. Therefore, only inelastic scattering mechanisms are discussed in the subsequent subsections.

B.4.1 Intervalley optical phonon scattering

With the transition rate of intervalley phonon scattering given in section 3.2.3.2, equation (3.2.3-60)-(3.2.3-63), the collision term with neglected Pauli principle after (B.4.0-37) reads

$$
(\sqrt{m_\nu^*})^3 s^\nu(w^\nu)\overline{C^{\text{NOP}}} = (\sqrt{m_\nu^*})^3 s^\nu(w^\nu) N_{\text{ph}}^\pm \sum_{\nu'} \left\{ K^{\text{NOP},\nu'} \pi \right.
$$

$$
\left[\int_{-1}^{1} d\mu' \int_{0}^{w_{\max}} d\acute{w}^{\nu'} \left(\delta\left(\acute{w}^{\nu'} - w^\nu \mp \Delta w_{\text{ph}} - \Delta w_0^{\nu-\nu'}\right) \Phi^{\nu'}(\acute{w}^{\nu'},\mu') \right. \right.
$$

$$
\left. \left. \left. - \delta\left(\acute{w}^{\nu'} - w^\nu \pm \Delta w_{\text{ph}} - \Delta w_0^{\nu-\nu'}\right) f^\nu(\sqrt{m_{\nu'}^*})^3 s^{\nu'}(\acute{w}^{\nu'}) \right) \right] \right\} , \quad \text{(B.4.1-39)}
$$

where the upper/lower sign of the superscripts $\pm$ in (B.4.1-39) corresponds to an emission/absorption process like for (3.2.3-60) and (3.2.3-62) in section 3.2.3.2. Following the procedure sketched in section 3.2.3.2 for taking an arbitrary Δw_{ph} (normalized phonon energy) and $\Delta w_0^{\nu-\nu'}$ (energetic difference of the potential valley energy) into account, equation (B.4.1-39) becomes

$$
(\sqrt{m_\nu^*})^3 s^\nu(w^\nu)\overline{C^{\text{NOP}}} = (\sqrt{m_\nu^*})^3 s^\nu(w^\nu) \pi N_{\text{ph}}^\pm \sum_{\nu'} \left\{ K^{\text{NOP},\nu'} \right.
$$

$$
\left. \times \left[\sum_{j'=1}^{N_\mu} \frac{n_{i'_{\text{in}}+\acute{r}-2,j'}^{\nu',\tau}}{\Delta w} - 2f^\nu(\sqrt{m_{\nu'}^*})^3 s^{\nu'}(w^\nu \mp \Delta w_{\text{ph}} + \Delta w_0^{\nu-\nu'}) \right] \right\} , \quad \text{(B.4.1-40)}
$$

with i'_{in} after (3.2.3-69) and $\acute{r} \in [1,2]$ counts the two adjacent control volumes in w^ν direction. Performing the transition to the discretized solution variable, (B.4.0-38) reads for intervalley phonon scattering

$$
\sum_{\acute{r}=1}^{2} \int_{w_l^\nu(\acute{r})}^{w_u^\nu(\acute{r})} \int_{\mu_{j-\frac{1}{2}}}^{\mu_{j+\frac{1}{2}}} (\sqrt{m_\nu^*})^3 s^\nu(w^\nu)\overline{C^{\text{NOP}}} d\mu\, dw^\nu = \frac{\pi}{\Delta w} N_{\text{ph}}^\pm \sum_{\nu'} \left\{ K^{\text{NOP},\nu'} \sum_{\acute{r}=1}^{2} \left[\right. \right.
$$

$$
\left. \left. \left(\Delta\mu(\sqrt{m_\nu^*})^3 S^\nu(w^\nu)\big|_{w_l^\nu(\acute{r})}^{w_u^\nu(\acute{r})} \right) \sum_{j'=1}^{N_\mu} n_{i'_{\text{in}}+\acute{r}-2,j'}^{\nu',\tau} - 2n_{i,j}^{\nu,\tau}(\sqrt{m_{\nu'}^*})^3 S^{\nu'}(\acute{w}^{\nu'})\big|_{\acute{w}_l^{\nu'}(\acute{r})}^{\acute{w}_u^{\nu'}(\acute{r})} \right] \right\} ,
$$

$$
\text{(B.4.1-41)}
$$

where the integration limits after (3.2.3-73) and (3.2.3-74) have been used. In contrast to (3.2.3-71), it can clearly be seen from equation (B.4.1-41) that the collision term becomes a linear function of the discretized solution variables $n^{\nu',\tau}_{i'_{\text{in}}+\acute{r}-2,j'}$ and $n^{\nu,\tau}_{i,j}$ if the Pauli principle is neglected.

B.4.2 Intravalley optical phonon scattering

Compared to **inter**valley optical phonon scattering, for **intra**valley optical phonon scattering the summation over all considered valleys is skipped. Therefore, an almost identical discretized collision term like for intervalley optical phonon scattering is obtained (cf. section 3.2.3.2, equation (3.2.3-71) and section 3.2.3.3, equation (3.2.3-78)). Proceeding as for intervalley optical phonon scattering in B.4.1, one obtains

$$\sum_{\acute{r}=1}^{2} \int_{w^{\nu}_{\text{l}}(\acute{r})}^{w^{\nu}_{\text{u}}(\acute{r})} \int_{\mu_{j-\frac{1}{2}}}^{\mu_{j+\frac{1}{2}}} (\sqrt{m^*_\nu})^3 s^\nu(w^\nu)\overline{C^{\text{OP}}} d\mu\, dw^\nu = \frac{\pi(\sqrt{m^*_\nu})^3}{\Delta w} N^{\pm}_{\text{ph}} K^{\text{OP}} \sum_{\acute{r}=1}^{2} \Bigg[$$

$$\left(\Delta\mu\ S^\nu(w^\nu)\big|_{w^\nu_{\text{l}}(\acute{r})}^{w^\nu_{\text{u}}(\acute{r})}\right) \sum_{j'=1}^{N_\mu} n^{\nu,\tau}_{i'_{\text{in}}+\acute{r}-2,j'} - 2n^{\nu,\tau}_{i,j}\ S^\nu(\acute{w}^\nu)\big|_{\acute{w}^\nu_{\text{l}}(\acute{r})}^{\acute{w}^\nu_{\text{u}}(\acute{r})}\Bigg] , \qquad (\text{B.4.2-42})$$

with K^{OP} after (3.2.3-76), $N^{\pm}_{\text{ph}}$ after (3.2.3-77), the integration limits (3.2.3-73) and (3.2.3-74) and $\acute{r}$ for counting the two adjacent w^ν intervals.

B.4.3 Polar optical phonon scattering

Like in section 3.2.3.4, here also the anisotropic version and its isotropic approximation of POP scattering is considered. In addition, an isotropic dispersion relation is assumed, like for the numerical treatment in section 3.2.3.4.

Anisotropic version

Based on the considerations concerning the anisotropy and the corresponding equations (3.2.3-81)-(3.2.3-82), (B.4.0-37) becomes here

$$(\sqrt{m_\nu^*})^3 s^\nu(w^\nu)\overline{C^{POP}} =$$

$$\frac{\sqrt{m_\nu^*}K^{POP}\pi}{\Delta w \Delta \mu} N_{ph}^\pm \left(s^\nu(w^\nu) \sum_{j'=1}^{N_\mu} \left\{ n_{i'_{in}+\acute{r}-2,j'}^{\nu,\tau} \left. G(w^\nu, w^\nu \pm \Delta w_{ph}, \mu, \mu') \right|_{\mu'=\mu_{j'}-\frac{1}{2}}^{\mu'=\mu_{j'}+\frac{1}{2}} \right\} \right.$$

$$\left. - n_{i,j}^{\nu,\tau} s^\nu(w^\nu \mp \Delta w_{ph}) \left. G(w^\nu, w^\nu \mp \Delta w_{ph}, \mu, \mu') \right|_{\mu'=-1}^{\mu'=1} \right), \quad \text{(B.4.3-43)}$$

with $G(w^\nu, \acute{w}^\nu, \mu, \mu')$ after (3.2.3-84), which is stemming from the anisotropic strength of the POP scattering. Since the anisotropic nature is not altered by considering/neglecting the Pauli principle, $G(w^\nu, \acute{w}^\nu, \mu, \mu')$ is identical in both cases. Therefore, also the subsequent integration after μ and w^ν in conjunction with $G(w^\nu, \acute{w}^\nu, \mu, \mu')$ is valid here. Thus, for the anisotropic version of POP scattering, the final discretized collision term (B.4.0-38) reads here

$$\sum_{\acute{r}=1}^{2} \int_{w_l^\nu(\acute{r})}^{w_u^\nu(\acute{r})} \int_{\mu_{j-\frac{1}{2}}}^{\mu_{j+\frac{1}{2}}} (\sqrt{m_\nu^*})^3 s^\nu(w^\nu)\overline{C^{POP}} d\mu\, dw^\nu \approx \frac{\sqrt{m_\nu^*}K^{POP}\pi}{\Delta w \Delta \mu} N_{ph}^\pm$$

$$\sum_{\acute{r}=1}^{2} \left[\left. S^\nu(w^\nu) \right|_{w_l^\nu(\acute{r})}^{w_u^\nu(\acute{r})} \sum_{j'=1}^{N_\mu} \left\{ n_{i'_{in}+\acute{r}-2,j'}^{\nu,\tau} \left[H_{POP}(w_m^\nu(\acute{r}), w_m^\nu(\acute{r}) \pm \Delta w_{ph}) \right|_{\mu_{j'}-\frac{1}{2}}^{\mu_{j'}+\frac{1}{2}} \right]_{\mu_{j-\frac{1}{2}}}^{\mu_{j+\frac{1}{2}}} \right\}$$

$$\left. - n_{i,j}^{\nu,\tau}\, \left. S^\nu(\acute{w}^\nu) \right|_{\acute{w}_l^\nu(\acute{r})}^{\acute{w}_u^\nu(\acute{r})} \left[\left. H_{POP}(w_m^\nu(\acute{r}), w_m^\nu(\acute{r}) \mp \Delta w_{ph}) \right|_{-1}^{1} \right]_{\mu_{j-\frac{1}{2}}}^{\mu_{j+\frac{1}{2}}} \right], \quad \text{(B.4.3-44)}$$

where H_{POP} (integration of G w.r.t. μ) given in appendix B.6 and the *mean value theorem of integration* after (3.2.3-88) for the final integration after w^ν have been employed. Note, due to the inelastic nature of POP scattering, also $\acute{r}$ (counter for two adjacent w^ν intervals) is included in (B.4.3-44). For the meaning of the remaining quantities included in (B.4.3-44), the reader is referred to section 3.2.3.4.

Isotropic approximation

Based on the equations (3.2.3-90)-(3.2.3-92) in section 3.2.3.4, the equivalent of (3.2.3-93) with neglected Pauli principle becomes with (B.4.0-37)

$$(\sqrt{m_\nu^*})^3 s^\nu(w^\nu)\overline{C^{\mathrm{POP,ISO}}} =$$

$$\frac{\sqrt{m_\nu^*}K^{\mathrm{POP}}\pi}{2\Delta w}N_{\mathrm{ph}}^{\pm}\left[\tilde{g}(w^\nu \pm \Delta w_{\mathrm{ph}}, w^\nu)s^\nu(w^\nu)\sum_{j'=1}^{N_\mu}\left\{n_{i'_{\mathrm{in}}+\acute{r}-2,j'}^{\nu,\tau}\right\}\right.$$

$$\left.- 2\frac{n_{i,j}^{\nu,\tau}}{\Delta\mu}\tilde{g}(w^\nu, w^\nu \mp \Delta w_{\mathrm{ph}})s^\nu(w^\nu \mp \Delta w_{\mathrm{ph}})\right] .\text{(B.4.3-45)}$$

Using for the integration after w^ν the *mean value theorem of integration* (3.2.3-88) (here for $\tilde{g}$), the collision term after (B.4.0-38) evaluates to

$$\sum_{\acute{r}=1}^{2}\int_{w_{\mathrm{l}}^\nu(\acute{r})}^{w_{\mathrm{u}}^\nu(\acute{r})}\int_{\mu_{j-\frac{1}{2}}}^{\mu_{j+\frac{1}{2}}}(\sqrt{m_\nu^*})^3 s^\nu(w^\nu)\overline{C^{\mathrm{POP,ISO}}}d\mu\, dw^\nu \approx \frac{\sqrt{m_\nu^*}K^{\mathrm{POP}}\pi}{2\Delta w}N_{\mathrm{ph}}^{\pm}\Delta\mu$$

$$\sum_{\acute{r}=1}^{2}\left[\tilde{g}(w_{\mathrm{m}}^\nu(\acute{r}) \pm \Delta w_{\mathrm{ph}}, w_{\mathrm{m}}^\nu(\acute{r}))\, S^\nu(w^\nu)|_{w_{\mathrm{l}}^\nu(\acute{r})}^{w_{\mathrm{u}}^\nu(\acute{r})}\sum_{j'=1}^{N_\mu}\left\{n_{i'_{\mathrm{in}}+\acute{r}-2,j'}^{\nu,\tau}\right\}\right.$$

$$\left.- 2\frac{n_{i,j}^{\nu,\tau}}{\Delta\mu}\tilde{g}(w_{\mathrm{m}}^\nu(\acute{r}), w_{\mathrm{m}}^\nu(\acute{r}) \mp \Delta w_{\mathrm{ph}})\, S^\nu(w^\nu \mp \Delta w_{\mathrm{ph}})|_{w_{\mathrm{l}}^\nu(\acute{r})}^{w_{\mathrm{u}}^\nu(\acute{r})}\right] ,\qquad \text{(B.4.3-46)}$$

where for (B.4.3-46) the integration scheme for inelastic scattering processes as described in section 3.2.3.2 is employed. Like for the previously considered anisotropic version, also here $\acute{r}$ counts the two adjacent w^ν intervals involved in the last integration after w^ν in (B.4.0-38).

B.5 Impurity scattering adjustment

In order to improve the agreement between the simulated and measured electron mobilities of SI, the strength of IMP-scattering is re-adjusted in [31, 37] by a fit factor, which is assumed to be doping dependent. Based on the graphs placed in [31, 37], analytic functions are developed:

$$f_c(x) = ax^3 + bx^2 + cx + d , \qquad \text{(B.5.0-47)}$$

$$f_e(x) = a \exp\left(-\left|\frac{x-b}{c}\right|^d\right) , \qquad \text{(B.5.0-48)}$$

with $x = \log_{10}\left(\frac{N_A + |N_D|}{1 \cdot 10^{15} \text{cm}^{-3}}\right)$. Based on the functions (B.5.0-47) and (B.5.0-48) the parameters a-d are adjusted to the extracted data taken from [31, 37]. Here, no special effort has been payed for obtaining a-d to generate differentiable/exactly continuous transitions between ranges of validity, since the derivatives of these functions over the whole range are not needed.

	a	b	c	d	$x \in$	
w/o Pauli.,	0.0306	0.1921	0.0654	1.1847	$[0, 2.25]$	f_c
	4.0377	5.5298	4.0032	4.2578	$(2.25, 3.75]$	f_e
e^- maj.	5.0215	-59.1563	233.5593	-304.8873	$(3.75, 5)$	f_c
	93.9425	5.9952	0.4464	0.9302	$[5, 6]$	f_e
w/o Pauli.,	-0.1227	0.4625	-0.3923	2.5179	$[0, 2.25]$	f_c
	0.3919	-3.5731	10.0005	-6.3161	$(2.25, 4.5)$	f_c
e^- min.	4.2947	-56.4194	249.0017	-367.3078	$[4.5, 5.2)$	f_c
	90.3304	6.4507	0.4219	0.9370	$[5.2, 6]$	f_e
w/ Pauli.,	0.0345	0.2021	0.0102	1.2379	$[0, 2.075]$	f_c
	3.7693	3.4729	2.2376	1.7322	$(2.075, 3.405]$	f_e
e^- min./maj.	4.0685	4.4440	14.4396	0.9716	$(3.405, 4.6)$	f_e
	4.0206	4.5885	1.2297	1.7774	$[4.6, 6]$	f_e

Table B.1: Parameters for the functions (B.5.0-47) and (B.5.0-48) for the impurity scattering adjustment. Note, if the Pauli principle is taken into account, no differentiation between electrons as minorities/majorities is made. Latter stems from the lack of data and should be investigated in more detail.

With (B.5.0-47)/(B.5.0-48) and the parameters from table B.1, the calculated fit factors for impurity scattering are compared with the extracted ones in figure B.1.

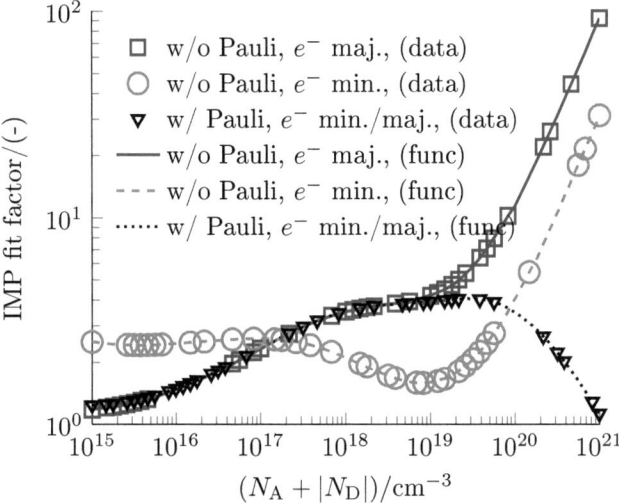

Figure B.1: Calculated and extracted fit factors for IMP-scattering.

B.6 Analytic primitives for POP-scattering

The three primitives for the integration over the μ'-μ-domain are categorized (see figure 3.4 in section 3.2.3.4) in a

1) primitive for the interior except for the corners of the μ'-μ-domain,

2) primitive for the horizontal region and a

3) primitive for a region underneath the main-diagonal of the μ'-μ-domain.

Note, for the equations below, $h_\nu(w^\nu)/h_\nu(\acute{w}^\nu)$ after (3.2.3-81) is used. The primitive for the interior of the μ'-μ-domain reads

$$H^{\text{int}}(w^\nu, \acute{w}^\nu, \mu, \mu') = \int G(w^\nu, \acute{w}^\nu, \mu, \mu')d\mu$$

$$= \frac{1}{2\sqrt{h_\nu(w^\nu)h_\nu(\acute{w}^\nu)}} \int \text{arcsinh}\left(\frac{\hat{a} + \hat{b}\mu}{\hat{c}\sqrt{1-\mu^2}}\right) d\mu$$

$$= \frac{1}{2\sqrt{h_\nu(w^\nu)h_\nu(\acute{w}^\nu)}} \left(\mu \, \text{arcsinh}\left(\frac{\hat{a} + \hat{b}\mu}{\hat{c}\sqrt{1-\mu^2}}\right) + \frac{1}{4}\log\left[\left(\frac{(\hat{b}^2 - \hat{a}^2)^2(1-\mu^2)}{16(B^2 - C^2)}\right)^2\right] \right.$$

$$\left. + \frac{\hat{a}}{2\sqrt{\hat{b}^2 - \hat{c}^2}} \log\left(4\left[\sqrt{A}\left(\sqrt{\hat{b}^2 - \hat{c}^2} - \hat{a}\right) + C\right]^2\right)\right), \tag{B.6.0-49}$$

with

$$\hat{a} = 2\sqrt{h_\nu(w^\nu)h_\nu(\acute{w}^\nu)}\mu', \tag{B.6.0-50}$$

$$\hat{b} = -(h_\nu(w^\nu) + h_\nu(\acute{w}^\nu)), \tag{B.6.0-51}$$

$$\hat{c} = \sqrt{(h_\nu(w^\nu) - h_\nu(\acute{w}^\nu))^2}, \tag{B.6.0-52}$$

$$A = (\hat{a} + \hat{b}\mu)^2 + \hat{c}^2(1-\mu^2), \tag{B.6.0-53}$$

$$B = \hat{b}\sqrt{A} + \hat{a}\hat{b}\mu + \hat{c}^2 + \hat{a}^2, \tag{B.6.0-54}$$

$$C = \hat{a}\sqrt{A} + (\hat{b}^2 - \hat{c}^2)\mu + \hat{a}\hat{b}. \tag{B.6.0-55}$$

Note, (B.6.0-49) is **only** valid for $\mu \in (-1, 1)$ **and** $\mu' \in (-1, 1)$. However, this limitation can be circumvented by considering the limiting behavior of

(B.6.0-49) for $\mu \to \pm 1$ and assuming $\mu' \neq \pm 1$:

$$\lim_{\mu \to \pm 1} H^{\text{int}}(w^\nu, \acute{w}^\nu, \mu, \mu') = H^{\text{bnd}}(w, w', \mu, \mu')$$

$$= \frac{1}{2\sqrt{h_\nu(w^\nu)h_\nu(\acute{w}^\nu)}} \left(\log\left(\frac{\hat{b}^2 - \hat{a}^2}{4\sqrt{(\hat{c}^2 - \hat{b}^2 + \hat{a}^2)^2}} \right) \right.$$

$$\left. + \frac{\hat{a}}{2\sqrt{\hat{b}^2 - \hat{c}^2}} \log\left(4\left[\sqrt{A}\left(\sqrt{\hat{b}^2 - \hat{c}^2} - \hat{a} \right) + C \right]^2 \right) \right),$$

$$\text{(B.6.0-56)}$$

where the abbreviations (B.6.0-50)-(B.6.0-55) have been used again. The result (B.6.0-56) can also be applied for the case $\mu' \to \pm 1$ and $\mu \neq \pm 1$ by simply replacing μ with μ'.

For the horizontal region (see figure 3.4, center) the following equation is obtained:

$$H^{\text{hori}}(w^\nu, \acute{w}^\nu, \mu)$$

$$= \frac{1}{2\sqrt{h_\nu(w^\nu)h_\nu(\acute{w}^\nu)}} \log\left(\frac{h_\nu(w^\nu) + h_\nu(\acute{w}^\nu) + 2\sqrt{h_\nu(w^\nu)h_\nu(\acute{w}^\nu)}}{h_\nu(w^\nu) + h_\nu(\acute{w}^\nu) - 2\sqrt{h_\nu(w^\nu)h_\nu(\acute{w}^\nu)}} \right) (\mu + 1) .$$

$$\text{(B.6.0-57)}$$

Note, also for (B.6.0-57) μ can be replaced by μ' and thus it is integrated over a region parallel to the μ-axis.

The last primitive is responsible for the region bounded by the main diagonal and reads

$$H^{\text{diag}}(w^\nu, \acute{w}^\nu, \mu) =$$

$$\frac{1}{2\sqrt{h_\nu(w^\nu)h_\nu(\acute{w}^\nu)}} \left(\mu \mathfrak{f}(w^\nu, \acute{w}^\nu, \mu) + \log\left(\sqrt{t_0^2}\sqrt{1 - \mu^2}\sqrt{h_\nu(w^\nu)h_\nu(\acute{w}^\nu)} \right) \right.$$

$$\left. + \frac{1}{4} \log\left[\left(\frac{t_1^2 + \sqrt{t_0^2(1 - \mu^2) + t_1^4\mu^2}}{t_1^2 - \sqrt{t_0^2(1 - \mu^2) + t_1^4\mu^2}} \right)^2 \right] \right),$$

$$\text{(B.6.0-58)}$$

with the terms

$$
\mathfrak{f}(w^\nu, \acute{w}^\nu, \mu) \;=\; \operatorname{arcsinh}\left(\frac{2\sqrt{h_\nu(w^\nu)h_\nu(\acute{w}^\nu)} + \mu(h_\nu(w^\nu) + h_\nu(\acute{w}^\nu))}{\sqrt{t_0^2}\sqrt{1 - \mu^2}}\right)
$$

$$
- \operatorname{arcsinh}\left(\frac{\mu\, t_1^2}{\sqrt{t_0^2}\sqrt{1 - \mu^2}}\right) \tag{B.6.0-59}
$$

$$
t_0 \;=\; (h_\nu(w^\nu) - h_\nu(\acute{w}^\nu)), \tag{B.6.0-60}
$$

$$
t_1 \;=\; (\sqrt{h_\nu(w^\nu)} - \sqrt{h_\nu(\acute{w}^\nu)}). \tag{B.6.0-61}
$$

However, due to the denominators of the arguments involved in the arcsinh-functions of $\mathfrak{f}(w^\nu, \acute{w}^\nu, \mu)$ in (B.6.0-58), a division by zero occurs for $\mu = \pm 1$. Therefore, the limiting behavior of (B.6.0-58) is investigated at these poles:

$$
\lim_{\mu \to 1} H^{\mathrm{diag}}(w, w', \mu)
$$

$$
= \frac{1}{2\sqrt{h_\nu(w^\nu)h_\nu(\acute{w}^\nu)}} \log\left(\frac{(\sqrt{h_\nu(w^\nu)} + \sqrt{h_\nu(\acute{w}^\nu)})^2 \sqrt{t_0^2}\,\sqrt[4]{h_\nu(w^\nu)h_\nu(\acute{w}^\nu)}}{\sqrt{t_1^2}}\right),
$$

$$
\tag{B.6.0-62}
$$

$$
\lim_{\mu \to -1} H^{\mathrm{diag}}(w, w', \mu) = \lim_{\mu \to 1} H^{\mathrm{diag}}(w, w', \mu) - \frac{H^{\mathrm{hori}}(w, w', \mu = 1)}{2}
$$

$$
= \frac{1}{2\sqrt{h_\nu(w^\nu)h_\nu(\acute{w}^\nu)}} \log\left(\sqrt{t_1^2}\sqrt{t_0^2}\,\sqrt[4]{h_\nu(w^\nu)h_\nu(\acute{w}^\nu)}\right), \tag{B.6.0-63}
$$

where for (B.6.0-63) the fact is used, that (B.6.0-62) has to give half of the total integral except for a constant offset, which is $\lim_{\mu \to -1} H^{\mathrm{diag}}(w, w', \mu)$.

B.7 Analytic primitives for PZ-scattering

Starting point is equation (3.2.3-126) of the anisotropic PZ-formulation in section 3.2.3.7:

$$
G_{\mathrm{PZ}}(w^\nu, \mu, \mu') = \frac{\mu - \mu'\,(1 + c)}{2\,(1 + \frac{c}{2})\,\sqrt{R}} + \operatorname{arcsinh}\left(\frac{\mu' - \mu\,(1 + c)}{\sqrt{2c\,(1 + \frac{c}{2})}\sqrt{1 - \mu^2}}\right)
$$

$$
\tag{B.7.0-64}
$$

with c after (3.2.3-123) and R after (3.2.3-125). For $H_{\text{PZ}}(w^\nu, \mu, \mu')$ introduced in section 3.2.3.7, equation (3.2.3-128), $G_{\text{PZ}}(w^\nu, \mu, \mu')$ after (B.7.0-64) is integrated over μ. The first term on the right-hand side of (B.7.0-64) can be integrated without any difficulties, in contrast to the second term:

$$\int G_{\text{PZ}}(w^\nu, \mu, \mu')d\mu = \frac{\sqrt{R}}{2\left(1 + \frac{c}{2}\right)} + \int \operatorname{arcsinh}\left(\frac{\mu' - \mu(1+c)}{\sqrt{2c\left(1 + \frac{c}{2}\right)}\sqrt{1 - \mu^2}}\right) d\mu .$$

$$\text{(B.7.0-65)}$$

Like in case of POP-scattering (see section B.6), also here primitives for

1) the interior except for the corners of the μ'-μ-domain,

2) the horizontal region and for

3) a region underneath the main-diagonal of the μ'-μ-domain,

are needed. Regarding the first item, the following primitive is found for the interior of the μ'-μ-domain:

$$H^{\text{int}}(w^\nu, \mu, \mu') = \mu \operatorname{arcsinh}\left(\frac{\mu' - \mu(1+c)}{\sqrt{2c\left(1 + \frac{c}{2}\right)}\sqrt{1 - \mu^2}}\right)$$

$$+ \frac{1}{4}\log\left[\left(\frac{(\hat{b}^2 - \hat{a}^2)^2(1 - \mu^2)}{16(B^2 - C^2)}\right)^2\right] + \frac{\hat{a}}{2}\log\left(4\left[\sqrt{A}(1 - \hat{a}) + C\right]^2\right) ,$$

$$\text{(B.7.0-66)}$$

with

$$\hat{a} = \mu' , \tag{B.7.0-67}$$

$$\hat{b} = -(1 + c) , \tag{B.7.0-68}$$

$$\hat{c} = \sqrt{2c\left(1 + \frac{c}{2}\right)} , \tag{B.7.0-69}$$

$$A = (\hat{a} + \hat{b}\mu)^2 + \hat{c}^2(1 - \mu^2) , \tag{B.7.0-70}$$

$$B = \hat{b}\sqrt{A} + \hat{a}\hat{b}\mu + \hat{c}^2 + \hat{a}^2 , \tag{B.7.0-71}$$

$$C = \hat{a}\sqrt{A} + \mu + \hat{a}\hat{b} . \tag{B.7.0-72}$$

Also here, (B.7.0-66) is **only** valid for $\mu \in (-1, 1)$ **and** $\mu' \in (-1, 1)$. Investigating the limiting behavior of (B.7.0-66) for $\mu \to \pm 1$ and assuming $\mu' \neq \pm 1$ leads to

$$\lim_{\mu \to \pm 1} H^{\text{int}}(w, \mu, \mu') = H^{\text{bnd}}(w, \mu, \mu')$$

$$= \log\left(\frac{\hat{b}^2 - \hat{a}^2}{4\sqrt{(\hat{a}^2 - 1)^2}}\right) + \frac{\hat{a}}{2} \log\left(4\left[\sqrt{A}(1 - \hat{a}) + C\right]^2\right), \quad \text{(B.7.0-73)}$$

where the abbreviations (B.7.0-67)-(B.7.0-72) have been employed again. For the horizontal region, the following function is found:

$$H^{\text{hori}}(w, \mu) = \log\left(\frac{2}{c} + 1\right)(\mu + 1). \quad \text{(B.7.0-74)}$$

Note, the w-dependence of (B.7.0-74) stems from c, which contains $h_\nu(w^\nu)$ after (3.2.3-81). The integral over the region bounded by the main diagonal of the μ'-μ-domain can be obtained by

$$H^{\text{diag}}(w, \mu) = \frac{1}{2} \log\left[\frac{c}{2}\left(1 + \frac{c}{2}\right)(1 - \mu^2)\frac{\sqrt{2c}\sqrt{1 - \mu^2 + \frac{c}{2}} + c}{\sqrt{2c}\sqrt{1 - \mu^2 + \frac{c}{2}} - c}\right]$$
$$+ \mu \, \mathfrak{f}(w, \mu), \quad \text{(B.7.0-75)}$$

with

$$\mathfrak{f}(w, \mu) = \text{arcsinh}\left(\frac{\mu(1 - (1 + c))}{\sqrt{2c\left(1 + \frac{c}{2}\right)}\sqrt{1 - \mu^2}}\right)$$
$$- \text{arcsinh}\left(\frac{-1 - (1 + c)\mu}{\sqrt{2c\left(1 + \frac{c}{2}\right)}\sqrt{1 - \mu^2}}\right). \quad \text{(B.7.0-76)}$$

In order to complete the *diagonal*-component, also here the limiting behavior of H^{diag} for $\mu \to \pm 1$ has to be investigated, since otherwise a division by zero occurs in the arcsinh-terms of (B.7.0-76) and thus in (B.7.0-75). One obtains

$$\lim_{\mu \to -1} H^{\text{diag}}(w, \mu) = \log\left(c\sqrt{1 + \frac{c}{2}}\right), \tag{B.7.0-77}$$

$$\lim_{\mu \to 1} H^{\text{diag}}(w, \mu) = \log\left(2\left(1 + \frac{c}{2}\right)^{\frac{3}{2}}\right), \text{ and} \tag{B.7.0-78}$$

$$\lim_{\mu \to 1} H^{\text{diag}}(w, \mu) - \lim_{\mu \to -1} H^{\text{diag}}(w, \mu) = \log\left(\frac{2}{c} + 1\right). \tag{B.7.0-79}$$

Equation (B.7.0-79) is listed as *proof of concept*, since it has to give half of the integral over the total $\mu/\mu^{'}$-domain, which can easily be verified with (B.7.0-74) for $\mu = 1$.

To conclude this section, the desired function for PZ-scattering is composed by

$$H_{\text{PZ}}(w^{\nu}, \mu, \mu^{'}) = \frac{\mu - \mu^{'}(1 + c)}{2\left(1 + \frac{c}{2}\right)\sqrt{R}} + H^{\times}(w^{\nu}, \mu(, \mu^{'})), \tag{B.7.0-80}$$

where $H^{\times}(w^{\nu}, \mu(, \mu^{'}))$ is one of the three considered primitives (B.7.0-66), (B.7.0-73), (B.7.0-74) or (B.7.0-75)-(B.7.0-78).

APPENDIX C

Supplementary material for chapter 4

C.1 Accumulated electron/hole transit time and electron velocity

For the derivation of (4.2.2-26) in section 4.2.2, a 1D NPN-HBT or NPN-BJT is assumed. Starting point is the accumulated hole transit time:

$$\tau_p(x) \;=\; \frac{q}{g_{\mathrm{m}}} \int_0^x \frac{\partial p(\acute{x})}{\partial V_{\mathrm{BE}}}\bigg|_{V_{\mathrm{CE}}=\mathrm{const.}} d\acute{x} \;, \qquad \text{(C.1.0-1)}$$

with the transconductance $g_{\mathrm{m}} = \frac{\partial J_{\mathrm{C}}}{\partial V_{\mathrm{BE}}}\big|_{V_{\mathrm{CE}}=\mathrm{const.}}$, the elementary charge q and the hole density p. Since the device under investigation has to fulfill the charge neutrality, the following relations hold:

$$\int_0^L p(x) - N_{\mathrm{A}}^-(x)dx \;=\; \int_0^L n(x) - N_{\mathrm{D}}^+(x)dx \;, \qquad \text{(C.1.0-2)}$$

$$\frac{\partial}{\partial V_{\mathrm{BE}}}\bigg|_{V_{\mathrm{CE}}=\mathrm{const.}} \int_0^L p(x)dx \;=\; \frac{\partial}{\partial V_{\mathrm{BE}}}\bigg|_{V_{\mathrm{CE}}=\mathrm{const.}} \int_0^L n(x)dx \;, \qquad \text{(C.1.0-3)}$$

$$\int_0^L \frac{\partial p(x)}{\partial V_{\mathrm{BE}}}\bigg|_{V_{\mathrm{CE}}=\mathrm{const.}} dx \;=\; \int_0^L \frac{\partial n(x)}{\partial V_{\mathrm{BE}}}\bigg|_{V_{\mathrm{CE}}=\mathrm{const.}} dx \;, \qquad \text{(C.1.0-4)}$$

with the electron density n and the total length L of the device. Thus, due to the charge neutrality the quasi-static hole density integrated over the device has to equal the integrated quasi-static electron density. Therefore, the hole transit time $\tau_p(L)$ equals $\tau_n(L)$. With (C.1.0-4) and (C.1.0-1) follows

$$\tau_p(L) = \tau_n(L) = \frac{q}{g_m} \int_0^L \frac{\partial n(\acute{x})}{\partial V_{\mathrm{BE}}}\bigg|_{V_{\mathrm{CE}}=\mathrm{const.}} d\acute{x} \ . \tag{C.1.0-5}$$

In order to include the average electron velocity in (C.1.0-5), it is assumed that the 1D (in x-direction) collector current density is a pure electron current density and thus equals the transfer current density:

$$J_{\mathrm{C}} \quad = \quad J_{\mathrm{T}} = q\, n(\acute{x}) \langle v_n^x(\acute{x}) \rangle \ , \tag{C.1.0-6}$$

where J_{T} is position independent due to compensatory contributions of the position dependent $n(x)$ and $\langle v_n^x(x) \rangle$, since the current continuity has to be fulfilled. Thus, with (C.1.0-6) the quasi-static electron density in (C.1.0-5) becomes

$$\frac{\partial n(\acute{x})}{\partial V_{\mathrm{BE}}}\bigg|_{V_{\mathrm{CE}}=\mathrm{const.}} = \frac{1}{\langle v_n^x(\acute{x}) \rangle} \left(\frac{g_m}{q} - n(\acute{x})\, \frac{\partial \langle v_n^x(\acute{x}) \rangle}{\partial V_{\mathrm{BE}}}\bigg|_{V_{\mathrm{CE}}=\mathrm{const.}} \right) . \tag{C.1.0-7}$$

Inserting (C.1.0-7) into (C.1.0-5) and using $\frac{J_{\mathrm{C}}}{\langle v_n^x(\acute{x}) \rangle} = q\, n(\acute{x})$, according to (C.1.0-6), gives

$$\begin{aligned}
\tau_{n,p}(L) &= \int_0^L \frac{1}{\langle v_n^x(\acute{x}) \rangle} \left(1 - \frac{J_{\mathrm{C}}}{\langle v_n^x(\acute{x}) \rangle g_m} \frac{\partial \langle v_n^x(\acute{x}) \rangle}{\partial V_{\mathrm{BE}}}\bigg|_{V_{\mathrm{CE}}=\mathrm{const.}} \right) d\acute{x} \ , \\
&= \int_0^L \frac{1}{\langle v_n^x(\acute{x}) \rangle} \left(1 - \frac{V_{\mathrm{T}}}{\overline{g_m}} \frac{1}{\langle v_n^x(\acute{x}) \rangle} \frac{\partial \langle v_n^x(\acute{x}) \rangle}{\partial V_{\mathrm{BE}}}\bigg|_{V_{\mathrm{CE}}=\mathrm{const.}} \right) d\acute{x} \ ,
\end{aligned}$$
$$\tag{C.1.0-8}$$

with $\overline{g_m} = \frac{V_{\mathrm{T}} g_m}{J_{\mathrm{C}}}$ and the thermal voltage V_{T}.

Bibliography

[1] C. Auer, F. Schürrer, and C. Ertler. Hot phonon effects on the high-field transport in metallic carbon nanotubes. *Physical Review B*, 74:165409, Oct 2006.

[2] K. Banoo. *Direct solution of the Boltzmann transport equation in nanoscale Si devices*. PhD thesis, Purdue University, 2000.

[3] J. J. Barnes, R. J. Lomax, and G. I. Haddad. Finite-element simulation of GaAs MESFET's with lateral doping profiles and submicron gates. *IEEE Transactions on Electron Devices*, 23(9):1042–1048, Sep 1976.

[4] F. Bloch. Über die Quantenmechanik der Elektronen in Kristallgittern. *Zeitschrift für Physik*, 52(7-8):555–600, 1929.

[5] M. Bollhöfer and Y. Saad. Multilevel Preconditioners Constructed From Inverse-Based ILUs. *SIAM Journal on Scientific Computing*, 27(5):1627–1650, November 2005.

[6] M. Bollhöfer and Y. Saad. ILUPACK - preconditioning software package. Available online at http://ilupack.tu-bs.de/, June 2011.

[7] R. Borges, M. Carmona, B. Costa, and W. S. Don. An improved weighted essentially non-oscillatory scheme for hyperbolic conservation laws. *Journal of Computational Physics*, 227(6):3191 – 3211, 2008.

[8] I. N. Bronstein, K. A. Semendjajew, G. Musiol, and H. Mühlig. *Taschenbuch der Mathematik*. Verlag Harri Deutsch, 2005.

[9] D. Brust. Electronic Spectra of Crystalline Germanium and Silicon. *Physical Review*, 134:A1337–A1353, Jun 1964.

[10] F. M Bufler. *Full-band Monte Carlo simulation of electrons and holes in strained Si and SiGe*. Herbert Utz Verlag, 1998.

[11] F. M. Bufler, P. Graf, B. Meinerzhagen, B. Adeline, M. M. Rieger, H. Kibbel, and G. Fischer. Analysis of electron transport properties in unstrained and strained $Si_{1-x}Ge_x$ alloys. *IEEE Transactions on Semiconductor Technology Modeling and Simulation*, pages 1–37, N 1996.

[12] F. M. Bufler, P. Graf, B. Meinerzhagen, B. Adeline, M.M. Rieger, H. Kibbel, and G. Fischer. Low- and high-field electron-transport parameters for unstrained and strained $Si_{1-x}Ge_x$. *Electron Device Letters, IEEE*, 18(6):264–266, June 1997.

[13] C. Canali, C. Jacoboni, F. Nava, G. Ottaviani, and A. Alberigi-Quaranta. Electron drift velocity in silicon. *Physical Review B*, 12:2265–2284, Sep 1975.

[14] M. L. Cohen and T. K. Bergstresser. Band Structures and Pseudopotential Form Factors for Fourteen Semiconductors of the Diamond and Zinc-blende Structures. *Physical Review*, 141:789–796, Jan 1966.

[15] S. Datta, Shen Shi, K. P. Roenker, M. M. Cahay, and W. E. Stanchina. Simulation and design of InAlAs/InGaAs pnp heterojunction bipolar transistors. *IEEE Transactions on Electron Devices*, 45(8):1634–1643, Aug 1998.

[16] R. Driad, J. Rosenzweig, R. E. Makon, R. Losch, V. Hurm, H. Walcher, and M. Schlechtweg. InP DHBT-Based IC Technology for 100-Gb/s Ethernet. *IEEE Transactions on Electron Devices*, 58(8):2604–2609, Aug 2011.

[17] E. Fatemi and F. Odeh. Upwind Finite Difference Solution of Boltzmann Equation Applied to Electron Transport in Semiconductor Devices. *Journal of Computational Physics*, 108(2):209 – 217, 1993.

[18] W. Fawcett and G. Hill. Temperature dependence of the velocity/field characteristic of electrons in InP. *Electronics Letters*, 11(4):80–81, February 1975.

[19] M. Galler. *Multigroup equations for the description of the particle transport in semiconductors*, volume 70. World Scientific, 2005.

[20] M. Galler. *Multigroup methods for the description of the particle transport in semiconductors*. PhD thesis, Technische Universität Graz, 2005.

[21] M. Galler and F. Schürrer. A Direct multigroup-WENO Solver for the 2D Non-stationary Boltzmann-Poisson System for GaAs Devices: GaAs-MESFET. *Journal of Computational Physics*, 212(2):778–797, March 2006.

[22] P. Garcias-Salva, J. M. Lopez-Gonzalez, and L. Prat. A comparison between Monte Carlo and extended drift-diffusion models for abrupt InP/InGaAs HBTs. *IEEE Transactions on Electron Devices*, 48(6):1045–1053, Jun 2001.

[23] N. Goldsman, L. Henrickson, and J. Frey. A physics-based analytical/numerical solution to the Boltzmann transport equation for use in device simulation. *Solid-State Electronics*, 34(4):389 – 396, 1991.

[24] T. Grasser, H. Kosina, and S. Selberherr. Consistent comparison of drift-diffusion and hydro-dynamic device simulations. In *Simulation of Semiconductor Processes and Devices, 1999. SISPAD '99. 1999 International Conference on*, pages 151–154, 1999.

[25] H. K. Gummel. On the definition of the cutoff frequency fT. *Proceedings of the IEEE*, 57(12):2159–2159, Dec 1969.

[26] P. Harrison. *Quantum Wells, Wires and Dots: Theoretical and Computational Physics of Semiconductor Nanostructures*. Wiley, 2005.

[27] A. Harten, B. Engquist, S. Osher, and S. R Chakravarthy. Uniformly high order accurate essentially non-oscillatory schemes, {III}. *Journal of Computational Physics*, 71(2):231 – 303, 1987.

[28] S. Hasan, M. A. Alam, and M. S. Lundstrom. Simulation of Carbon Nanotube FETs Including Hot-Phonon and Self-Heating Effects. *IEEE Transactions on Electron Devices*, 54(9):2352–2361, Sept 2007.

[29] A. K. Henrick, T. D. Aslam, and J. M. Powers. Mapped weighted essentially non-oscillatory schemes: Achieving optimal order near critical points. *Journal of Computational Physics*, 207(2):542 – 567, 2005.

[30] C. Herring and E. Vogt. Transport and Deformation-Potential Theory for Many-Valley Semiconductors with Anisotropic Scattering. *Physical Review*, 101:944–961, Feb 1956.

[31] S.-M. Hong, A.-T. Pham, and C. Jungemann. *Deterministic Solvers for the Boltzmann Transport Equation*. Springer, 2011.

[32] Institute for Microelectronics, TU Vienna, Institute for Microelectronics, TU Vienna Gußhausstraße 27-29 A-1040 Wien Austria/Europe. *Minimos-NT*, release 2.1 edition, 2004.

[33] C. Jacoboni, C. Canali, G. Ottaviani, and A. Alberigi Quaranta. A review of some charge transport properties of silicon. *Solid-State Electronics*, 20(2):77 – 89, 1977.

[34] C. Jacoboni and L. Reggiani. The Monte Carlo method for the solution of charge transport in semiconductors with applications to covalent materials. *Reviews of Modern Physics*, 55:645–705, Jul 1983.

[35] S. C. Jain, J. M. McGregor, and D. J. Roulston. Band-gap narrowing in novel III-V semiconductors. *Journal of Applied Physics*, 68(7):3747–3749, 1990.

[36] G.-S. Jiang and C.-W. Shu. Efficient Implementation of Weighted ENO Schemes. *Journal of Computational Physics*, 126(1):202 – 228, 1996.

[37] C. Jungemann and B. Meinerzhagen. *Hierarchical Device Simulation - The Monte-Carlo Perspective*. Springer, 2003.

[38] C. Jungemann, A. T. Pham, B. Meinerzhagen, C. Ringhofer, and M. Bollhöfer. Stable discretization of the Boltzmann equation based on spherical harmonics, box integration, and a maximum entropy dissipation principle. *Journal of Applied Physics*, 100(2), 2006.

[39] C. Kittel. *Introduction to Solid State Physics*. Wiley, 1995.

[40] J. Korn, H. Rucker, B. Heinemann, A. Pawlak, G. Wedel, and M. Schroter. Experimental and theoretical study of fT for SiGe HBTs with a scaled vertical doping profile. In *Bipolar/BiCMOS Circuits and Technology Meeting - BCTM, 2015 IEEE*, pages 117–120, Oct 2015.

[41] H. Kosina. A method to reduce small-angle scattering in Monte Carlo device analysis. *IEEE Transactions on Electron Devices*, 46(6):1196–1200, Jun 1999.

[42] R.J. LeVeque. *Finite Volume Methods for Hyperbolic Problems*. Cambridge Texts in Applied Mathematics. Cambridge University Press, 2002.

[43] M. Levinshtein. *Handbook Series on Semiconductor Parameters*. World Scientific.

[44] H. Liu and X. Jiao. WLS-ENO: Weighted-Least-Squares Based Essentially Non-Oscillatory Schemes for Finite Volume Methods on Unstructured Meshes. *ArXiv e-prints*, February 2016.

[45] X.-D. Liu, S. Osher, and T. Chan. Weighted Essentially Non-oscillatory Schemes. *Journal of Computational Physics*, 115(1):200 – 212, 1994.

[46] M. Lundstrom. *Fundamentals of carrier transport*. Cambridge University Press, 2000.

[47] M. Lundstrom. Drift-diffusion and computational electronics - still going strong after 40 years! In *Simulation of Semiconductor Processes and Devices (SISPAD), 2015 International Conference on*, pages 1–3, Sept 2015.

[48] A. Majorana and R. M. Pidatella. A Finite Difference Scheme Solving the Boltzmann-Poisson System for Semiconductor Devices. *Journal of Computational Physics*, 174(2):649 – 668, 2001.

[49] F. Mousty, P. Ostoja, and L. Passari. Relationship between resistivity and phosphorus concentration in silicon. *Journal of Applied Physics*, 45(10):4576–4580, 1974.

[50] B. K. Ridley. Reconciliation of the Conwell-Weisskopf and Brooks-Herring formulae for charged-impurity scattering in semiconductors: Third-body interference. *Journal of Physics C Solid State Physics*, 10:1589–1593, May 1977.

[51] B. K. Ridley. *Quantum Processes in Semiconductors*. Oxford University Press, 1999.

[52] M. M. Rieger and P. Vogl. Electronic-band parameters in strained $Si_{1-x}Ge_x$ alloys on $Si_{1-y}Ge_y$ substrates. *Physical Review B*, 48:14276–14287, Nov 1993.

[53] M. Rohner. *Physical limitations of InP/InGaAs heterojunction-bipolar transistors*. PhD thesis, ETH Zürich, 2002.

[54] D. Schroeder. Three-dimensional nonequilibrium interface conditions for electron transport at band edge discontinuities. *IEEE Transactions on Computer-Aided Design of Integrated Circuits and Systems*, 9(11):1136–1140, Nov 1990.

[55] D. Schroeder. The inflow moments method for the description of electron transport at material interfaces. *Journal of Applied Physics*, 72(3):964–970, 1992.

[56] M. Schröter. Semiconductor Electronics, 2007. Lecture notes, Technische Universität Dresden.

[57] M. Schröter, G. Wedel, B. Heinemann, C. Jungemann, J. Krause, P. Chevalier, and A. Chantre. Physical and Electrical Performance Limits of High-Speed SiGeC HBTs-Part I: Vertical Scaling. *IEEE Transactions on Electron Devices*, 58(11):3687–3696, Nov 2011.

[58] S. Selberherr. *Analysis and Simulation of Semiconductor Devices.* Springer-Verlag, 1984.

[59] J. Serre and A. Ghazali. From band tailing to impurity-band formation and discussion of localization in doped semiconductors: A multiple-scattering approach. *Physical Review B*, 28:4704–4715, Oct 1983.

[60] Y. Shi, G. Niu, J. D. Cressler, and D. L. Harame. On the consistent modeling of band-gap narrowing for accurate device-level simulation of scaled SiGe HBTs. *IEEE Transactions on Electron Devices*, 50(5):1370–1377, May 2003.

[61] SILVACO International, 4701 Patrick Henry Drive, Bldg. 1 Santa Clara, CA 95054. *ATLAS User's Manual*, October 2004.

[62] J. W. Slotboom and H. C. de Graaff. Bandgap narrowing in silicon bipolar transistors. *IEEE Transactions on Electron Devices*, 24(8):1123–1125, Aug 1977.

[63] J. Smit, M. van Sint Annaland, and J. A. M. Kuipers. Grid adaptation with WENO schemes for non-uniform grids to solve convection-dominated partial differential equations. *Chemical Engineering Science*, 60(10):2609 – 2619, 2005.

[64] M. Sotoodeh, A. H. Khalid, and A. A. Rezazadeh. Empirical low-field mobility model for III-V compounds applicable in device simulation codes. *Journal of Applied Physics*, 87(6):2890–2900, 2000.

[65] Synopsys, Inc., 690 E. Middlefield Road Mountain View, CA 94043. *Sentaurus Device User Guide*, version k-2015.06 edition, June 2015.

[66] J. M. Thijssen. *Computational Physis.* Cambridge University Press; 2 edition, 2007.

[67] J. L. Thobel, L. Baudry, A. Cappy, P. Bourel, and R. Fauquembergue. Electron transport properties of strained $In_xGa_{1-x}As$. *Applied Physics Letters*, 56(4):346–348, 1990.

[68] R. E. Thomas. Carrier mobilities in silicon empirically related to doping and field. *Proceedings of the IEEE*, 55(12):2192–2193, Dec 1967.

[69] K. Tomizawa. *Numerical Simulation of Submicron Semiconductor Devices.* Artech House Boston/London, 1993.

[70] S.-E. Ungersböck. Numerische Berechnung der Bandstruktur von Halbleitern. Diploma thesis, Technische Universität Wien, 2002.

[71] R. Wang, H. Feng, and R. J. Spiteri. Observations on the fifth-order WENO method with non-uniform meshes. *Applied Mathematics and Computation*, 196(1):433 – 447, 2008.

[72] G. Wedel. Einfluß der Ladungstransportmodellierung auf das Verhalten moderner SiGe HBTs. Diploma thesis, Technische Universität Dresden, 2008.

[73] G. Wedel and M. Schröter. Hydrodynamic simulations for advanced SiGe HBTs. In *Bipolar/BiCMOS Circuits and Technology Meeting (BCTM), 2010 IEEE*, pages 237–244, Oct 2010.

[74] S. Wei and M. Y. Chou. Phonon dispersions of silicon and germanium from first-principles calculations. *Physical Review B*, 50:2221–2226, Jul 1994.